Contemporary Clinical Neuroscience

More information about this series at http://www.springer.com/series/7678

Gülin Öz
Editor

Magnetic Resonance Spectroscopy of Degenerative Brain Diseases

Springer

Editor
Gülin Öz
Center Magnetic Resonance Research
University of Minnesota Medical School
Minneapolis, Minnesota, USA

Contemporary Clinical Neuroscience
ISBN 978-3-319-33553-7 ISBN 978-3-319-33555-1 (eBook)
DOI 10.1007/978-3-319-33555-1

Library of Congress Control Number: 2016944143

Printed on acid-free paper

This Springer imprint is published by Springer Nature
The registered company is Springer International Publishing AG Switzerland

Foreword

Nuclear magnetic resonance (NMR) was discovered by physicists, exploited by chemists and swiftly commandeered as magnetic resonance imaging (MRI) by medical radiologists for its diagnostic value. Magnetic resonance spectroscopy (MRS), the chemical analytical counterpoise, has thus largely become a stepchild of diagnostic MRI – and is included as a minor after-thought, if at all, in prevailing neurology textbooks. The present volume is an innovation and corrects the prior omission. True to its roots as a non-destructive tool for chemical and biochemical analysis, MRS has its adherents in neurology and the neurosciences. Biochemistry, they reason, must to a greater rather than a smaller extent underlie brain 'function'; MRS may therefore unlock the secrets of brain, cognition and even mind. Neurodegenerative diseases attack all of these normal functions. The present volume assembles many of the 'keys' that have been discovered by a quarter-of a century of dedicated investigation and presents them in a readable, coherent and convenient form, for the coming generations of neurologists, psychiatrists, neuroscientists and biomedical engineers who will one day reap that harvest. Confronted by a virtual epidemic of neurodegenerative disorders and an ageing population, the potential for MRS as a new diagnostic tool has become compelling.

The achievement of editor and author, Dr. Gülin Öz and the 15 other contributors to the new volume "Magnetic Resonance Spectroscopy of Degenerative Brain Disorders" is timely and significant. It makes accessible to the expert and novice alike, the excitement of 25 years of research and development of MRS as metabolic imaging of the living, functioning human brain. Thanks to the success of MRI as a first-line diagnostic tool in neurology and latterly also in psychiatry, these expensive and ungainly machines are universally available and have become familiar and well-tolerated by patients. MRI has migrated from the Clinic to become available to an entire generation of physicians and scientists interested in brain function, dysfunction and disease previously confined to exploring their subjects indirectly, through post-mortem tissue or secondary tests on blood and cerebrospinal fluid. Latterly, and in a variant of reverse-engineering, MRI and MRS 'scanners' based on successful human applications, but suited to the 1000-fold smaller brain of the mouse, have been designed to take advantage of the revolution in genetics and animal modeling of human degenerative brain diseases. This continuous iteration of preclinical with clinical research is another special feature of the present volume. But the real value of MRS, explored through the medium of human degenerative brain disorders, lies in the non-invasive access to neurochemistry that this technology provides: while we should not forget that the major innovation in 'structural MRI', which distinguishes it from computerized X-ray tomography, comes from its power to image neurochemistry of brain water in all its myriad forms, one metabolite N-acetyl aspartate, NAA, dramatically advances the role of MRS in the brain. Largely as a consequence of the industry of many of the Authors assembled in this new volume, NAA can now safely be viewed as a 'neuronal marker'. MRS of NAA, deserves its place as the new 'tendon-hammer' (invented in 1841 and still going strong) for diagnostic neurologists by defining the 'number' of viable neurons and axons in any pre-selected region of the human brain. In this Volume we are informed how NAA predicts (pre-clinical diagnosis), documents progression, and gives early warning of response in most, if not all neurodegenerative diseases.

But this Volume takes us well beyond NAA, demystifying previously unfamiliar MR 'spectra' to reveal an additional dozen brain metabolites, each the potential 'biomarkers' of some feature of pathobiology. The concept of neurodegeneration itself is explained in an Introductory chapter by the Editor, touching on the genetic, biochemical, intracellular and apoptotic origins of diseases as various as Alzheimer's, Huntington's, Parkinson's, motor neuron disease (Lou Gehrig's disease) and a handful of cerebellar diseases epitomized by Friedreich's ataxia. Technical details crucial to credible clinical MRS – and often ignored in more general "MRI" tomes devoted principally to the butterfly collection of disease-specific images – are here stressed to allow explorers to evaluate their reliability as tools in the broader endeavor of the entire field of now broadly recognized as neurospectroscopy. Armed with these skills, the reader is then offered detailed evaluation of neurospectroscopy as a harbinger of degenerative disease and a bell-weather of successful diagnosis by a succession of specialists in each of disorders included in that concept. Although the Editor and some of the authors of individual Chapters call for 'standardization' in MRS technique, for future multi-center clinical trials this may be premature for two reasons; the two Chapters devoted to MRS technique illustrate that a considerable diversity of MRS methods may be necessary if the analytical depths of magnetic resonance are to be plumbed: we should recall that is precisely the path taken by diagnostic MRI where every new technique is incorporated. Next, with the increased recognition and reliance on biomarkers, not least those revealed through MRS itself, the era of massive clinical trials may itself be evolving into personalized medicine. Might not diversity in MRS-methodology become a strength? In any event, the authors of the Chapters which follow have not heeded the 'siren-call' of standardization nor have they been restricted by any Editorial limitation on the MRS techniques they describe – and, as befits a work of scholarship, we are offered everything available under the heading of proton (^{1}H) MRS, provided that it contributes to the central purpose of explaining neurodegenerative disorder. The real pleasure in reading this compelling volume comes with the carefully considered Chapters each by experts in their field as the richness of MRS plays out to explore the commonalities and the differences between the disorders classified here as 'neurodegenerative'.

While they may be 'linked' by loss of NAA there is much more to the story. Alzheimer's disease, with the longest track record, more than 20 years of MRS study, is yielding to careful cross-sectional and longitudinal studies with added-value in the taxonomy of a once-uniform diagnosis. MRS is well positioned to become a meaningful biomarker for this dominant group of dementias, when therapeutic option(s) finally emerge. In successive Chapters on the key disorders of this spectrum – Huntington's, Parkinson's, Motor Neuron Disease and Ataxias the authors skillfully blend in vivo pre-clinical studies in an ever increasing cast of disease models, with copious clinical experience of the real disease permitting an excellent evaluation of the relevance of the model through their MRS patterns. Each Chapter is accompanied by References pruned to eliminate the irrelevant or frankly wrong (all too common in the published MRS literature), with inclusion of crucial lost literature from the 'dark-age' which preceded the Internet. The slimmest Chapter, that on prion disease highlights one of the limitations, even of so robust a tool as clinical MRS – confining a deteriorating patient within a monster tube – today's MR scanner – can be an almost insurmountable challenge! Can we hope this is one that MR engineers will surmount?

Three Chapters not traditionally covered by the adjective 'degenerative' – on multiple sclerosis, HIV and epilepsy – are included. Here too the Editor's decision was wise – each Chapter is authoritatively written and informative with the bonus that two, MS and HIV, unlike the classical degenerative disorders, have a number of efficacious treatments. Putative MRS biomarkers sometimes respond to treatments – as expected – thereby providing an opportunity to test their hypotheses of etiology, evolution and reversal of disease. The final chapter, that on epilepsy also more than earns its place, notably emphasizing that epilepsy syndromes may involve neural re-generation, rather than de-generation. So there might be 'light at the end of the tunnel' of MRS of ever-progressing degenerative brain diseases. Neural re-generation, through the medium of pluripotent stem cells, while not directly mentioned anywhere in this Volume, is believed by many to represent the future. MRS detection and monitoring of stem cell populations has been mooted and seems well-within the bounds of possibility. If so, this book represents an important milestone on the long road ahead in ridding Society of degenerative brain disorders. Because degenerative brain disease remains a poorly understood accumulation of historically distinct disorders, readers who stray beyond their disease of interest to read more widely about MRS in other diseases, challenges and answers, will be richly rewarded by the cross-fertilization of the neighboring Chapters.

Reading this Volume thoroughly is highly recommended.

Brian Ross: Division of Medical Engineering, California Institute of Technology, Pasadena.

Contents

Contributors

Niamh Cawley NMR Research Unit, Queen Square MS Centre, UCL Institute of Neurology, London, UK

Olga Ciccarelli NMR Research Unit, Queen Square MS Centre, UCL Institute of Neurology, London, UK

NIHR, UCL/UCLH Biomedical Research Centre, London, UK

Brent Clark Department of Laboratory Medicine and Pathology, University of Minnesota, Minneapolis, MN, USA

Janet M. Dubinsky Department of Neuroscience, University of Minnesota, Minneapolis, MN, USA

Ulrike Dydak School of Health Sciences, Purdue University, West Lafayette, IN, USA

Department of Radiology and Imaging Sciences, Indiana University School of Medicine, Indianapolis, IN, USA

David A. Edmonsdon Department of Radiology and Imaging Sciences, Indiana University School of Medicine, Indianapolis, IN, USA

Damien Galanaud Department of Neuroradiology & Institut du Cerveau et de la Moëlle, Pitié Salpêtrière Hospital, Paris, France

Varan Govind Department of Radiology, University of Miami School of Medicine, Miami, FL, USA

Jonathan Graff-Radford Department of Neurology, Mayo Clinic, Rochester, MN, USA

Pierre-Gilles Henry Center for Magnetic Resonance Research, Department of Radiology, University of Minnesota, Minneapolis, MN, USA

Kejal Kantarci Department of Neurology, Mayo Clinic, Rochester, MN, USA

Fanny Mochel Institut du Cerveau et de la Moelle épinière, ICM, Inserm U 1127, CNRS UMR 7225, Sorbonne Universités, UPMC Univ Paris 06 UMR S 1127, Paris, France

Département de Génétique, Salpêtrière University Hospital, Paris, France

Groupe de Recherche Clinique Neurométabolique, Université Pierre and Marie Curie, Paris, France

Gülin Öz Center for Magnetic Resonance Research, Department of Radiology, University of Minnesota, Minneapolis, MN, USA

Jullie W. Pan Department of Neurology and Radiology, University of Pittsburgh, Kauffman, PA, USA

Bart W.J. Philips Radboud University Medical Center, Radiology and Nuclear Medicine, Nijmegen, Netherlands

Eva-Maria Ratai Neuroradiology Division, Department of Radiology and A.A. Martinos Center for Biomedical Imaging, Massachusetts General Hospital and Harvard Medical School, Boston, MA, USA

Tom W.J. Scheenen Radboud University Medical Center, Radiology and Nuclear Medicine, Nijmegen, Netherlands

Ivan Tkáč Center for Magnetic Resonance Research, Department of Radiology, University of Minnesota, Minneapolis, MN, USA

S. Elizabeth Zauber Department of Neurology, Indiana University School of Medicine, Indianapolis, IN, USA

Reviewers

Robert Bartha Centre for Functional and Metabolic Mapping, Robarts Research Institute, Department of Medical Biophysics, University of Western Ontario, London, ON, Canada

Thomas Ernst Department of Medicine, John A. Burns School of Medicine, University of Hawai'i, Honolulu, HI, USA

Sanjay Kalra Department of Medicine (Neurology), Neuroscience and Mental Health Institute (NMHI), University of Alberta, Edmonton, AB, Canada

Blair Leavitt Department of Medical Genetics, and Division of Neurology, Department of Medicine, Centre for Molecular Medicine and Therapeutics, University of British Columbia, Vancouver, BC, Canada

Mario Mascalchi Quantitative and Functional Neuroradiology Research Program, "Mario Serio" Department of Experimental and Clinical Biomedical Sciences, University of Florence, Florence, Italy

Susanne G. Mueller Department of Radiology and Biomedical Imaging, School of Medicine, University of California San Francisco, San Francisco, CA, USA

Petra J.W. Pouwels Department of Physics and Medical Technology, VU University Medical Center, Amsterdam, The Netherlands

About the Editor

Gülin Öz, PhD, is Associate Professor of Radiology at the Center for Magnetic Resonance Research (CMRR), University of Minnesota. Following BS degrees in Physics and Chemistry and a PhD in Biochemistry, she continued with postdoctoral training at the CMRR where she later joined the faculty. She spearheaded the utilization of neurochemical profiles obtained by high field MRS to assess cerebral changes in patients with neurological diseases and their response to treatment. She also co-led a MRS consensus effort together with Dr. Risto Kauppinen to facilitate the use and standardization of robust MRS methodology in the clinical setting. She uses high field, multi-nuclear magnetic resonance spectroscopy (MRS) to explore neurochemical and metabolic alterations in diseases that affect the brain. Her research is primarily focused on applications of advanced MRS methods in neurodegenerative diseases and diabetes.

Chapter 1
Imaging Neurodegeneration: What Can Magnetic Resonance Spectroscopy Contribute?

Gülin Öz

Abstract With increased prevalence of neurodegenerative diseases with age and an aging society, neuroimaging for diagnosis, prognosis, and therapy monitoring in these diseases has become more important than ever. There is particularly a great need for robust biomarkers and surrogate markers of cerebral pathology that can facilitate development of effective treatments in these conditions. Many radionuclide and MRI modalities are currently used in clinical research, with some already accepted among diagnostic criteria for neurodegenerative diseases. Others are being evaluated for their potential to monitor the pathogenic events during neurodegeneration at multiple levels from the global network level down to the subcellular and molecular levels. This chapter places magnetic resonance spectroscopy (MRS) within the context of other imaging modalities for evaluating neurodegeneration and summarizes its unique role in simultaneously assessing multiple relevant pathophysiological events, including neuronal loss/dysfunction, gliosis, demyelination, impaired energetics, increased membrane turnover, demyelination, synaptic dysfunction, and oxidative stress. Finally, the steps that still need to be taken to facilitate wider utility of advanced MRS methodology are outlined.

Keywords Structural MRI • Diffusion MRI • Resting state fMRI • Positron emission tomography • Magnetization transfer MRI • Manganese-enhanced MRI • Susceptibility-weighted imaging • *N*-Acetylaspartate • Choline • Creatine • *Myo*-inositol • Lactate • Glutamate • Glutamine • GABA • Glutathione • Ascorbate

G. Öz, Ph.D. (✉)
Department of Radiology, Center for Magnetic Resonance Research, University of Minnesota, 2021 6th St. S.E., Minneapolis, MN 55455, USA
e-mail: gulin@cmrr.umn.edu

G. Öz (ed.), *Magnetic Resonance Spectroscopy of Degenerative Brain Diseases*, Contemporary Clinical Neuroscience, DOI 10.1007/978-3-319-33555-1_1

Imaging Cerebral Pathology in Degenerative Brain Diseases

Neurological diseases affect as many as one billion people worldwide. Of these, neurodegenerative diseases cause progressive neuronal degeneration and death and are a major cause of disability and human suffering. With increased prevalence of these diseases with age and an aging society in developed countries (between 2000 and 2030, the number of individuals 65 years of age and older will double [1]), neuroimaging for diagnosis, prognosis, and therapy monitoring in these diseases has become more important than ever.

This volume on *MRS of Degenerative Brain Diseases* focuses on chronic, progressive neurodegenerative diseases, including both preclinical (animal model) and clinical applications. Thus, acute insults that cause neuronal damage and degeneration (as in stroke and traumatic brain injury) are not covered in the volume. Chronic neurodegenerative diseases, such as Alzheimer's disease (AD), Parkinson's disease (PD), Huntington's disease (HD), amyotrophic lateral sclerosis (ALS), and spinocerebellar ataxia (SCA), represent a complex family of neurological disorders in which vulnerable neuron populations are progressively lost. Current treatments for these conditions are supportive as disease-modifying treatments have been elusive. A major impediment in therapy development has been the lack of robust biomarkers and surrogate markers of cerebral pathology [2–5]. Management currently relies on structural MRI and clinical measures, which are also the primary outcome measures in clinical trials. Assessing whether therapies impact the progression of neurodegeneration is particularly challenging because the slow progression and phenotypic variability in these diseases necessitate long clinical trials with large sample sizes. Clinical outcome measures typically used in these trials have many limitations: they do not distinguish between disease-modifying and purely symptomatic drug effects, they are of no use in the earliest, preclinical stage (when neuroprotective agents are likely to be most effective), and they usually have poor test-retest reliability [5]. Therefore, while an essential component of any treatment trial, these measures need to be supplemented with noninvasive neuroimaging to directly assess treatment effects on the brain.

Commensurate with the complexity of these diseases, there are a number of imaging modalities that can be utilized to monitor various aspects of the pathophysiology of neurodegeneration. Mechanistic investigations into the process of neurodegeneration uncovered that while the genetic and environmental triggers may be different for different neurodegenerative conditions, multiple pathogenic pathways are common to many of them. These include accumulation of aberrant or misfolded proteins, oxidative stress, mitochondrial injury, excitotoxicity, synaptic failure, altered metal homeostasis, axonal and dendritic transport failure, neurovascular deficits, and neuroinflammation [6–9]. These pathogenic processes are active for years and result in neuronal dysfunction, even prior to onset of overt clinical symptoms [10], and lead to the ultimate demise of neurons when the cells can no longer cope with the loss of homeostatic balance.

Table 1.1 lists examples of commonly used modalities to monitor these pathogenic events at multiple levels from the global network level down to the subcellular (axonal/dendritic/synaptic) and molecular levels. Only proton (^{1}H) MR modalities that can be performed with standard equipment are listed in the table. However, it should be noted that there are non-proton MR methods such as ^{31}P, ^{13}C, and ^{17}O MRS that can provide even more detailed mechanistic information on metabolic aspects of neurodegeneration [11–13]. Also note that some of these modalities have been substantially more validated and are used in modern diagnostic criteria [14], while others require further validation. Thus, rates of global and regional brain atrophy measured with *structural MRI* have been validated in longitudinal multi-site studies and now are being accepted as therapeutic outcome measures in some neurodegenerative diseases [2, 5, 15]. Structural MRI is advantageous because MR scanners are widely available in the clinical setting and volumetric measurements are based on relatively standardized and simple data acquisition with protocols that are part of routine clinical MRI.

While gross morphometric changes detectable by structural MRI have found wide utility and undoubtedly contributed to understanding of the temporal progression of neurodegeneration, microstructural damage in the tissue occurs earlier in the neurodegenerative sequence of events and is accessible by *diffusion MRI* [16]. Recent years marked a surge in development of methods to increase the reliability of tractography and to enhance understanding of the contributions to the diffusion MR signal in order to extract tangible quantitative properties of the tissue, such as axon diameter and density. In addition, functional network connectivity in the brain can now be relatively easily assessed using *resting state functional MRI* (rsfMRI), which measures spontaneous fluctuations in the blood-oxygen-level-dependent (BOLD) signal that are correlated across anatomically distinct brain regions [17]. Importantly, increasing evidence supports the involvement of structural and functional network connectivity deficits in progressive neurodegeneration [18, 19], and hence these methods are expected to find wide utility in neurodegenerative disease trials.

Among non-MR modalities, *positron emission tomography* (*PET*) has been used by many to monitor multiple pathophysiological events during neurodegeneration because it can detect early metabolic and cellular abnormalities, such as toxic protein accumulation [20], defects in glucose uptake [21], synaptic dysfunction [22], and, more recently, neuroinflammation [23] (Table 1.1). Importantly, new NIH and Alzheimer Association guidelines include amyloid ligand PET in the diagnostic criteria for AD [24]. The drawbacks to PET scans are that they are significantly more expensive and less readily available than MR and involve radiation exposure, which makes repeat scanning problematic.

With yet other methods available to assess other pathophysiologic aspects of neurodegeneration, such as demyelination using *magnetization transfer MRI* [25], axonal transport deficiencies using *manganese-enhanced MRI* (MEMRI) [26], and iron content using various T_2 and susceptibility-based MRI methods [27], it is clear that multimodal investigations have the best chance to comprehensively evaluate progressive neurodegeneration and if interventions slow down or reverse the

Table 1.1 Neuroimaging modalities that can be utilized to assess pathophysiological events during neurodegeneration

Pathophysiological event	Imaging modality
Whole brain/network level	
Regional atrophy	Structural MRI
Reduced structural connectivity	Diffusion MRI
Reduced functional connectivity	Resting state fMRI
Cellular level	
Neuronal loss	*MRS* (*NAA*, *Glu*)
Reactive astrocytosis	*MRS* (*Ins*, *Gln*)
Microglial activation/inflammation	PET (TSPO), *MRS* (*Ins*, *Gln*), DCE-MRI (BBB breakdown)
Neurovascular dysfunction	Perfusion MRI (ASL), fMRI
Subcellular level	
Axonal/dendritic transport failure	MEMRI
Demyelination	MT-MRI, diffusion MRI, *MRS* (*Cho*)
Axonal loss	Diffusion MRI, *MRS* (*NAA*, *Glu*)
Synaptic dysfunction	PET (dopamine), *MRS* (*Glu*, *Gln*, *GABA*)
Molecular level	
Toxic protein accumulation	PET (Aβ)
Impaired energetics/mitochondrial injury	PET (FDG), *MRS* (*Lac*, *NAA*, *Cr*)
Oxidative stress	*MRS* (*GSH*, *Asc*)
Altered metal homeostasis	Relaxometry, SWI (Fe)

The list is not meant to be comprehensive and only includes examples of commonly used modalities and measured compounds, e.g., amyloid β accumulation with PET
NAA *N*-acetylaspartate, *Ins* *myo*-inositol, *Gln* glutamine, *TSPO* translocator protein, *DCE-MRI* dynamic contrast-enhanced MRI, *BBB* blood-brain barrier, *ASL* arterial spin labeling, *MEMRI* manganese-enhanced MRI, *MT-MRI* magnetization transfer MRI, *Cho* choline, *Glu* glutamate, *GABA* γ-aminobutyric acid, *FDG* fluoro-deoxyglucose, *Lac* lactate, *Cr* creatine, *GSH* glutathione, *Asc* ascorbate, *SWI* susceptibility-weighted imaging, *Fe* iron

progression of pathology. PD presents a good example for a need for multimodal imaging. Namely, radionuclide imaging detects and monitors dopamine dysfunction with high specificity and sensitivity in PD, but alternative methods are needed to monitor non-dopaminergic aspects of the disease [22], as well as to monitor the nigral pathology [28], which is the hallmark of PD. This need has spurred interest in MR-based biomarkers for PD despite great success with PET/SPECT imaging of dopamine function.

Added Value of MRS in Clinical Research and Care in Neurodegenerative Diseases

When regional morphometric changes in conventional T_1 and T_2 MRI are typically observed in degenerative brain conditions and are easily accessible via routine clinical exams, the need for more specialized techniques such as *magnetic resonance spectroscopy* (*MRS*) can be questioned. The macrostructural changes however are the end point of a cascade of events that lead to neuronal death and atrophy, and it is widely accepted that the underlying cellular and biochemical changes start years before symptoms and irreversible structural damage detectable by conventional imaging [10]. Therefore, biomarkers and surrogate markers that are sensitive to these early pathological changes and that can gauge drug effectiveness objectively and quickly are still needed and can have a high impact on development of neuroprotective therapies. In addition, there are cases where standard CT or MRI evaluation does *not* demonstrate gross structural nervous system changes even in symptomatic patients, such as in PD [29] and ALS [30]. Methods that uncover chemical changes in the tissue, such as MRS, are ideally suited in these cases to assess pathological changes in the absence of gross morphological alterations.

MRS has been repeatedly shown to be sensitive to events preceding neuronal loss [31–33] and can be included in a standard brain MRI exam along with structural MRI. It has been a powerful complementary tool to conventional MRI for diagnosis and monitoring of disease progression and response to therapy because it can detect changes in cell density, cell type, or biochemical composition [34]. Namely, ^{1}H MRS enables quantification of endogenous neurochemicals, including potential markers of neuronal health (*N*-acetylaspartate (NAA), glutamate), glial proliferation (*myo*-inositol, glutamine), demyelination/increased membrane turnover (choline), deficits in energy metabolism (lactate, NAA, creatine), neurotransmitter abnormalities (glutamate, glutamine, GABA), and oxidative stress (glutathione and ascorbate) [35], and thereby can contribute to monitoring multiple pathogenic events in neurodegeneration alongside other neuroimaging modalities (Table 1.1). Furthermore, a number of clinical and preclinical applications have demonstrated that MRS can assess neuronal dysfunction and loss, as well as accompanying cellular processes, and thereby provide robust and noninvasive biomarkers of neurodegeneration [36–46], as detailed in this volume. Importantly, neurochemical alterations measured noninvasively by MRS are independent of [47] and precede atrophy [31–33] and therefore provide additional information over structural MRI, especially early in the neurodegenerative disease course.

Consistently, a recent international MRS consensus effort concluded that MRS is expected to contribute to patient management in neurodegenerative diseases [34]. In addition, consensus was reached that MRS is “clinic-ready” for diagnostic, prognostic, and treatment assessment of brain tumors, various neonatal and pediatric disorders, demyelinating disorders, and infectious brain lesions. The breakdown for the utility of MRS into the clinic-ready vs. promising disease categories primarily stemmed from the differences in effect sizes observed in these diseases.

Namely, clinic-ready applications involve large effect sizes such that biochemical changes in brain lesions are detectable reliably in individual patients. On the other hand, neurochemical changes are more subtle in the "promising" category of diseases that include neurodegenerative diseases, making robust and highly reproducible MRS acquisition and analysis protocols critical for clinical utility.

Perspectives: What Still Needs to Be Done

While standard MR hardware and software are highly optimized and automated for imaging protocols, MRS protocols that are currently provided by clinical scanner manufacturers are not state of the art. Namely, the data quality that can be obtained with the vendor-provided protocols is sufficient to quantify three to five metabolites (NAA, creatine, choline, and, depending on acquisition parameters, *myo*-inositol and lactate); however, the data quality typically does not allow quantification of metabolites such as glutamate, glutamine, GABA, and glutathione, metabolites of high interest for neurodegenerative conditions. Due to inadequacies of automated shimming protocols, many expert researchers utilize manual shimming on clinical hardware for best shimming results, which is not feasible or efficient in the clinical setting. In addition, vendor-provided localization sequences have several deficiencies such as large chemical shift displacement at 3 T and higher field strengths, poor water suppression performance, and generation of unwanted coherences, which prevent acquisition of consistently high-quality data that can reliably demonstrate subtle neurochemical alterations in disease. For example, in a systematic comparison of the standard vendor protocol (PRESS + advanced 3D shimming) vs. an in-house developed MRS protocol (sLASER + FASTMAP) with randomized acquisitions in the same scanning session at 3 T, we found that almost half of the spectra obtained with the conventional protocol (13/28) did not fit our quality criteria (water linewidth $\leq$ 10 Hz, [48]), whereas only 1/28 spectra obtained with the sLASER + FASTMAP protocol did not fit these criteria in the elderly population [49]. Note that this linewidth criterion was set to avoid bias in concentrations of some weakly represented metabolites [48]. These considerations underline the advantages of utilizing advanced MRS methodology at 3 T when investigating neurodegenerative diseases. Namely, high-quality MRS data with excellent within- and between-site reproducibility of 10–15 neurochemicals can be obtained with standard clinical 3 T hardware as long as optimized acquisition and analysis techniques are used, including robust protocols for B_0 and B_1 adjustment [48]. Figure 1.1 shows examples of 3 T spectra acquired with an advanced MRS protocol from brain regions that are affected by the progressive neurodegenerative diseases that are covered in this volume.

Another critical need for wider utility of advanced MRS methodology is *automation* of MRS acquisition and analysis protocols to make them "MR technologist friendly." For example, one of the biggest challenges of single voxel MRS (SVS) in the clinical environment is the operator dependence of voxel selection. Methods

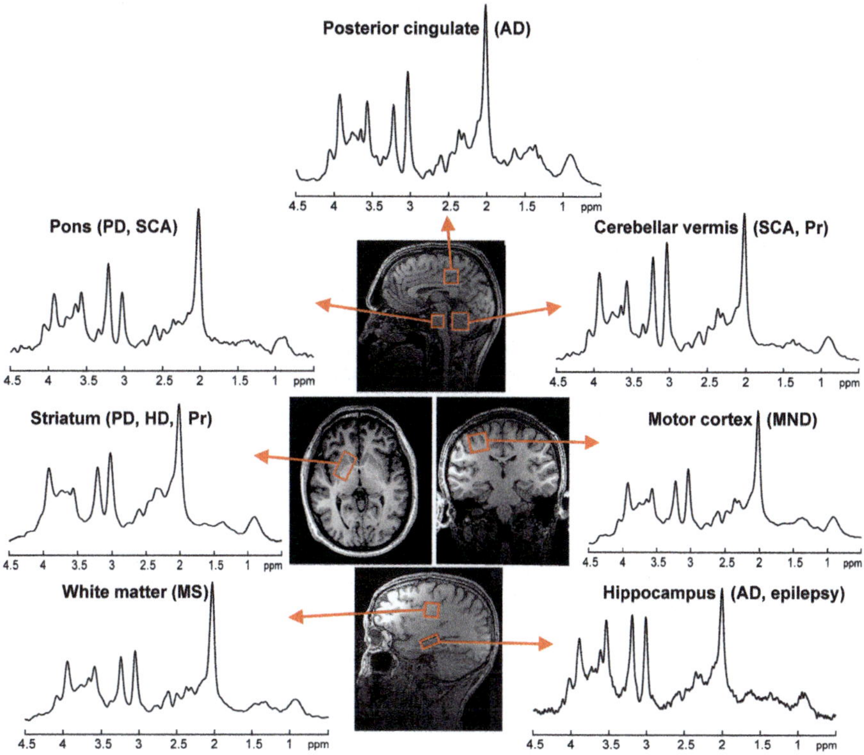

Fig. 1.1 Proton MR spectra (semi-LASER [57], TR/TE = 5000/28 ms) obtained at 3 T from brain regions that are of interest for various degenerative brain diseases as listed in *parentheses*. The spectra were processed and weighted identically (Gaussian apodization $\sigma = 0.13$ s) prior to Fourier transformation. Voxel positions are shown on T_1-weighted images. *AD* Alzheimer's disease, *PD* Parkinson's disease, *HD* Huntington's disease, *SCA* spinocerebellar ataxia, *MND* motor neuron disease, *MS* multiple sclerosis, *Pr* prion disease. Spectra and images courtesy of Drs. Dinesh Deelchand, James Joers, Pierre-Gilles Henry, Fanny Mochel, and Petr Bednařík

have been developed to automatically place the volume of interest (VOI) in longitudinal scans to match the VOI at the baseline scan of the same subject [50, 51]; however, atlas-based automatic voxel positioning to improve VOI consistency both between- and within-subjects has not yet been implemented for SVS. On the analysis end, spectral fitting using a metabolite basis set needs to be incorporated into the workflow on the scanner as data analysis is currently a major hurdle in the clinical utility of MRS [34].

Next, it will be important to *standardize* such automated, advanced MRS acquisition and analysis protocols across vendors and validate their multi-site reproducibility [48, 52]. In fact, lack of standardization has been an impediment for all biomarker discovery efforts [53] and its importance recognized for all neuroimaging modalities [54]. Clearly such standardization efforts would be most effectively pursued in collaboration with MR scanner manufacturers, and it is encouraging that representatives from major manufacturers are involved in the MRS consensus effort [34], demonstrating their interest in improving the vendor-provided MRS packages in future products.

Finally, following standardization of acquisition and analysis methodology in multi-site settings [48, 52], the added value of MRS needs to be evaluated systematically for different neurodegenerative diseases and clinical questions [55] to provide sufficient data for future evidence-based medicine (EBM) assessments [56].

Acknowledgments The preparation of this chapter was in part supported by the National Institute of Neurological Disorders and Stroke (NINDS) grant R01 NS070815. The Center for MR Research is supported by the National Institute of Biomedical Imaging and Bioengineering (NIBIB) grant P41 EB015894 and the Institutional Center Cores for Advanced Neuroimaging award P30 NS076408. The author acknowledges valuable feedback from Drs. Christophe Lenglet, Pierre-Gilles Henry, and David A. Okar and thanks Drs. Dinesh Deelchand, James Joers, Pierre-Gilles Henry, Fanny Mochel, and Petr Bednařík for providing images and spectra for the figure.

References

1. Vincent GK, Velkoff VA (2010) The next four decades: the older population in the United States: 2010 to 2050. Current Population Reports. US Census Bureau
2. Andre R, Scahill RI, Haider S, Tabrizi SJ (2014) Biomarker development for Huntington's disease. Drug Discov Today 19(7):972–979
3. Dorsey ER, Holloway RG, Ravina BM (2006) Biomarkers in Parkinson's disease. Expert Rev Neurother 6(6):823–831
4. Turner MR, Grosskreutz J, Kassubek J, Abrahams S, Agosta F, Benatar M, Filippi M, Goldstein LH, van den Heuvel M, Kalra S, Lule D, Mohammadi B (2011) Towards a neuroimaging biomarker for amyotrophic lateral sclerosis. Lancet Neurol 10(5):400–403
5. Mueller SG, Schuff N, Weiner MW (2006) Evaluation of treatment effects in Alzheimer's and other neurodegenerative diseases by MRI and MRS. NMR Biomed 19(6):655–668
6. Bossy-Wetzel E, Schwarzenbacher R, Lipton SA (2004) Molecular pathways to neurodegeneration. Nat Med 10(Suppl):S2–S9
7. Jellinger KA (2009) Recent advances in our understanding of neurodegeneration. J Neural Transm 116(9):1111–1162
8. Ramanan VK, Saykin AJ (2013) Pathways to neurodegeneration: mechanistic insights from GWAS in Alzheimer's disease, Parkinson's disease, and related disorders. Am J Neurodegener Dis 2(3):145–175
9. Zlokovic BV (2011) Neurovascular pathways to neurodegeneration in Alzheimer's disease and other disorders. Nat Rev Neurosci 12(12):723–738
10. DeKosky ST, Marek K (2003) Looking backward to move forward: early detection of neurodegenerative disorders. Science 302(5646):830–834
11. Zhu XH, Du F, Zhang N, Zhang Y, Lei H, Zhang X, Qiao H, Ugurbil K, Chen W (2009) Advanced in vivo heteronuclear MRS approaches for studying brain bioenergetics driven by mitochondria. Methods Mol Biol 489:317–357
12. Blüml S, Moreno A, Hwang JH, Ross BD (2001) 1-^{13}C glucose magnetic resonance spectroscopy of pediatric and adult brain disorders. NMR Biomed 14(1):19–32
13. Gruetter R, Adriany G, Choi IY, Henry PG, Lei H, Öz G (2003) Localized in vivo ^{13}C NMR spectroscopy of the brain. NMR Biomed 16(6–7):313–338

14. Jack CR Jr, Holtzman DM (2013) Biomarker modeling of Alzheimer's disease. Neuron 80 (6):1347–1358
15. Kantarci K, Jack CR Jr (2003) Neuroimaging in Alzheimer disease: an evidence-based review. Neuroimaging Clin N Am 13(2):197–209
16. Goveas J, O'Dwyer L, Mascalchi M, Cosottini M, Diciotti S, De Santis S, Passamonti L, Tessa C, Toschi N, Giannelli M (2015) Diffusion-MRI in neurodegenerative disorders. Magn Reson Imaging 33(7):853–876
17. Barkhof F, Haller S, Rombouts SA (2014) Resting-state functional MR imaging: a new window to the brain. Radiology 272(1):29–49
18. Pievani M, Filippini N, van den Heuvel MP, Cappa SF, Frisoni GB (2014) Brain connectivity in neurodegenerative diseases—from phenotype to proteinopathy. Nat Rev Neurol 10 (11):620–633
19. Iturria-Medina Y, Evans AC (2015) On the central role of brain connectivity in neurodegenerative disease progression. Front Aging Neurosci 7:90
20. Jack CR Jr, Barrio JR, Kepe V (2013) Cerebral amyloid PET imaging in Alzheimer's disease. Acta Neuropathol 126(5):643–657
21. Perani D (2014) FDG-PET and amyloid-PET imaging: the diverging paths. Curr Opin Neurol 27(4):405–413
22. Stoessl AJ, Lehericy S, Strafella AP (2014) Imaging insights into basal ganglia function, Parkinson's disease, and dystonia. Lancet 384(9942):532–544
23. Zimmer ER, Leuzy A, Benedet AL, Breitner J, Gauthier S, Rosa-Neto P (2014) Tracking neuroinflammation in Alzheimer's disease: the role of positron emission tomography imaging. J Neuroinflammation 11:120
24. Jack CR Jr (2012) Alzheimer disease: new concepts on its neurobiology and the clinical role imaging will play. Radiology 263(2):344–361
25. Filippi M, Rocca MA (2007) Magnetization transfer magnetic resonance imaging of the brain, spinal cord, and optic nerve. Neurotherapeutics 4(3):401–413
26. Silva AC, Bock NA (2008) Manganese-enhanced MRI: an exceptional tool in translational neuroimaging. Schizophr Bull 34(4):595–604
27. Dusek P, Dezortova M, Wuerfel J (2013) Imaging of iron. Int Rev Neurobiol 110:195–239
28. Lehericy S, Sharman MA, Dos Santos CL, Paquin R, Gallea C (2012) Magnetic resonance imaging of the substantia nigra in Parkinson's disease. Mov Disord 27(7):822–830
29. Brooks DJ (2000) Morphological and functional imaging studies on the diagnosis and progression of Parkinson's disease. J Neurol 247(Suppl 2):II11–II18
30. Foerster BR, Welsh RC, Feldman EL (2013) 25 years of neuroimaging in amyotrophic lateral sclerosis. Nat Rev Neurol 9(9):513–524
31. Godbolt AK, Waldman AD, MacManus DG, Schott JM, Frost C, Cipolotti L, Fox NC, Rossor MN (2006) MRS shows abnormalities before symptoms in familial Alzheimer disease. Neurology 66(5):718–722
32. Öz G, Nelson CD, Koski DM, Henry PG, Marjanska M, Deelchand DK, Shanley R, Eberly LE, Orr HT, Clark HB (2010) Noninvasive detection of presymptomatic and progressive neurodegeneration in a mouse model of spinocerebellar ataxia type 1. J Neurosci 30 (10):3831–3838
33. Kantarci K, Boeve BF, Wszolek ZK, Rademakers R, Whitwell JL, Baker MC, Senjem ML, Samikoglu AR, Knopman DS, Petersen RC, Jack CR Jr (2010) MRS in presymptomatic MAPT mutation carriers: a potential biomarker for tau-mediated pathology. Neurology 75 (9):771–778
34. Öz G, Alger JR, Barker PB, Bartha R, Bizzi A, Boesch C, Bolan PJ, Brindle KM, Cudalbu C, Dincer A, Dydak U, Emir UE, Frahm J, González RG, Gruber S, Gruetter R, Gupta RK, Heerschap A, Henning A, Hetherington HP, Howe FA, Hüppi PS, Hurd RE, Kantarci K, Klomp DW, Kreis R, Kruiskamp MJ, Leach MO, Lin AP, Luijten PR, Marjańska M, Maudsley AA, Meyerhoff DJ, Mountford CE, Nelson SJ, Pamir MN, Pan JW, Peet AC, Poptani H, Posse S, Pouwels PJ, Ratai EM, Ross BD, Scheenen TW, Schuster C, Smith IC, Soher BJ,

Tkáč I, Vigneron DB, Kauppinen RA, The MRS Consensus Group (2014) Clinical proton MR spectroscopy in central nervous system disorders. Radiology 270(3):658–679
35. Duarte JM, Lei H, Mlynárik V, Gruetter R (2012) The neurochemical profile quantified by in vivo ^{1}H NMR spectroscopy. Neuroimage 61(2):342–362
36. Jenkins BG, Kraft E (1999) Magnetic resonance spectroscopy in toxic encephalopathy and neurodegeneration. Curr Opin Neurol 12(6):753–760
37. Viau M, Marchand L, Bard C, Boulanger Y (2005) ^{1}H magnetic resonance spectroscopy of autosomal ataxias. Brain Res 1049(2):191–202
38. Schuff N, Meyerhoff DJ, Mueller S, Chao L, Sacrey DT, Laxer K, Weiner MW (2006) N-acetylaspartate as a marker of neuronal injury in neurodegenerative disease. Adv Exp Med Biol 576:241–262, discussion 361–363
39. Marjańska M, Curran GL, Wengenack TM, Henry PG, Bliss RL, Poduslo JF, Jack CR Jr, Ugurbil K, Garwood M (2005) Monitoring disease progression in transgenic mouse models of Alzheimer's disease with proton magnetic resonance spectroscopy. Proc Natl Acad Sci U S A 102(33):11906–11910
40. Tkáč I, Dubinsky JM, Keene CD, Gruetter R, Low WC (2007) Neurochemical changes in Huntington R6/2 mouse striatum detected by in vivo ^{1}H NMR spectroscopy. J Neurochem 100 (5):1397–1406
41. Mohamed MA, Barker PB, Skolasky RL, Selnes OA, Moxley RT, Pomper MG, Sacktor NC (2010) Brain metabolism and cognitive impairment in HIV infection: a 3-T magnetic resonance spectroscopy study. Magn Reson Imaging 28(9):1251–1257
42. Öz G, Hutter D, Tkáč I, Clark HB, Gross MD, Jiang H, Eberly LE, Bushara KO, Gomez CM (2010) Neurochemical alterations in spinocerebellar ataxia type 1 and their correlations with clinical status. Mov Disord 25(9):1253–1261
43. Öz G, Iltis I, Hutter D, Thomas W, Bushara KO, Gomez CM (2011) Distinct neurochemical profiles of spinocerebellar ataxias 1, 2, 6, and cerebellar multiple system atrophy. Cerebellum 10(2):208–217
44. Öz G, Tkáč I, Charnas LR, Choi IY, Bjoraker KJ, Shapiro EG, Gruetter R (2005) Assessment of adrenoleukodystrophy lesions by high field MRS in non-sedated pediatric patients. Neurology 64(3):434–441
45. Ratai E, Kok T, Wiggins C, Wiggins G, Grant E, Gagoski B, O'Neill G, Adalsteinsson E, Eichler F (2008) Seven-Tesla proton magnetic resonance spectroscopic imaging in adult X-linked adrenoleukodystrophy. Arch Neurol 65(11):1488–1494
46. Kantarci K, Petersen RC, Boeve BF, Knopman DS, Tang-Wai DF, O'Brien PC, Weigand SD, Edland SD, Smith GE, Ivnik RJ, Ferman TJ, Tangalos EG, Jack CR Jr (2004) ^{1}H MR spectroscopy in common dementias. Neurology 63(8):1393–1398
47. Zimmerman ME, Pan JW, Hetherington HP, Katz MJ, Verghese J, Buschke H, Derby CA, Lipton RB (2008) Hippocampal neurochemistry, neuromorphometry, and verbal memory in nondemented older adults. Neurology 70(18):1594–1600
48. Deelchand DK, Adanyeguh IM, Emir UE, Nguyen TM, Valabregue R, Henry PG, Mochel F, Öz G (2015) Two-site reproducibility of cerebellar and brainstem neurochemical profiles with short-echo, single voxel MRS at 3 T. Magn Reson Med 73(5):1718–1725
49. Deelchand DK, Kantarci K, Eberly LE, Öz G (2015) Towards translation of advanced MRS methodology to clinical setting. In: Proc Intl Soc Mag Reson Med, Toronto, Canada. p 4660
50. Hancu I, Blezek DJ, Dumoulin MC (2005) Automatic repositioning of single voxels in longitudinal ^{1}H MRS studies. NMR Biomed 18(6):352–361
51. Dou W, Speck O, Benner T, Kaufmann J, Li M, Walter M (2015) Automatic voxel positioning for MRS at 7 T. MAGMA 28:259–270
52. van de Bank BL, Emir UE, Boer VO, van Asten JJ, Maas MC, Wijnen JP, Kan HE, Öz G, Klomp DW, Scheenen TW (2015) Multi-center reproducibility of neurochemical profiles in the human brain at 7 T. NMR Biomed 28(3):306–316
53. Poste G (2011) Bring on the biomarkers. Nature 469(7329):156–157

54. European Society of Radiology (ESR) (2010) White paper on imaging biomarkers. Insights Imaging 1(2):42–45
55. Lin A, Ross BD, Harris K, Wong W (2005) Efficacy of proton magnetic resonance spectroscopy in neurological diagnosis and neurotherapeutic decision making. NeuroRx 2(2):197–214
56. Lin AP, Tran TT, Ross BD (2006) Impact of evidence-based medicine on magnetic resonance spectroscopy. NMR Biomed 19(4):476–483
57. Öz G, Tkáč I (2011) Short-echo, single-shot, full-intensity proton magnetic resonance spectroscopy for neurochemical profiling at 4 T: validation in the cerebellum and brainstem. Magn Reson Med 65(4):901–910

Chapter 2
Methodology of MRS in Animal Models: Technical Challenges and Solutions

Ivan Tkáč

Abstract In vivo ^{1}H MR spectroscopy is a unique technique, which is capable of providing neurochemical information from a selected volume of tissue noninvasively. However, the richness and reliability of neurochemical information gained by MRS depends heavily on the data acquisition and processing techniques utilized. What makes the use of MRS in neuroscience and medical research even more challenging is the fact that the most advanced MRS techniques developed in the last 15 years are not routinely provided by MR scanner vendors. This chapter provides an overview of the MRS methodology for studying animal models of human neurodegenerative diseases. The chapter's subsections focus on MRS data acquisition, processing, and metabolite quantification. The data acquisition section outlines some basic hardware requirements, B_0 shimming, water suppression, and localization techniques. The data processing section describes methods applied on acquired MRS data before metabolite quantification, such as frequency, phase, and eddy current correction. The quantification section focuses specifically on LCModel analysis. Finally, some examples, demonstrating the potentials of high-field ^{1}H MRS for neurochemical profiling in mice, are presented.

Keywords Localization • PRESS • STEAM • LASER • SPECIAL • B_0 shimming • FASTMAP • Chemical shift displacement error • Water suppression • Eddy currents • Quantification

Introduction

Animal models of human neurodegenerative diseases are fundamental for preclinical studies. In addition, enormous progress has been made in the last decade in developing new mouse models of neurodegenerative diseases by novel genetic engineering tools. An examination of these models requires advanced experimental techniques that provide accurate information about underlying

I. Tkáč, Ph.D. (✉)
Center for Magnetic Resonance Research, University of Minnesota, 2021 Sixth Street SE, Minneapolis, MN 55455, USA
e-mail: ivan@cmrr.umn.edu

G. Öz (ed.), *Magnetic Resonance Spectroscopy of Degenerative Brain Diseases*, Contemporary Clinical Neuroscience, DOI 10.1007/978-3-319-33555-1_2

neuropathological processes and that can be used for monitoring disease progression and evaluating the effectiveness of novel therapeutic approaches to assist drug discovery and development. In vivo ^{1}H MR spectroscopy is a unique method because it is capable of providing neurochemical information from a selected volume of brain tissue noninvasively [1]. This feature makes the technique ideal for longitudinal animal studies, which substantially reduces the number of used animals and avoids unnecessary animal euthanasia. Figure 2.1 shows the in vivo ^{1}H MR spectrum of the mouse hippocampus and demonstrates the range and type of neurochemical information achievable at 9.4 T. A reliable quantification up to 20 brain metabolites is currently feasible from the rat or mouse brain, and the metabolite detection threshold is about 0.5 μmol/g. The ^{1}H MR spectrum in Fig. 2.1 clearly illustrates that a substantial spectral overlap is present at magnetic field strengths as high as 9.4 T. Consequently, the full scale of the neurochemical information can be obtained only if carefully optimized advanced spectroscopy acquisition and processing techniques are used. A suboptimal pulse sequence design or a modest misadjustment of acquisition parameters can substantially deteriorate the resulting spectral quality, which makes reliable quantification of metabolites impossible. As the most advanced MRS techniques and processing tools are not routinely provided by MR scanner vendors, reaching a high spectral quality and reliability of metabolite quantification can be challenging for MRS users.

The primary goal of this chapter is to review the basic principles of MRS in animal models, to improve the understanding of the challenges that MRS users might be facing, and to help potential new MRS users to choose the most appropriate approaches in order to maximize the neurochemical information that can be obtained by this technique. The chapter specifically provides an overview of the ^{1}H MRS methodology for studying animal models of human neurodegenerative diseases. The chapter's subsections focus on MRS data acquisition, processing, and metabolite quantification. At the end of the chapter, some examples demonstrating the potentials of high-field ^{1}H MRS for neurochemical profiling in mice are presented.

MRS Data Acquisition

Hardware Requirements

The hardware of modern small animal MRI scanners, in general, meets the requirements for advanced MRS. But spectroscopy has high demands especially on the performance of the second-order shim system and the design of the RF coils. Sensitivity is one of the limiting factors for MRS, because the concentration of MR detectable metabolites is by 4 orders of magnitude lower than the concentration

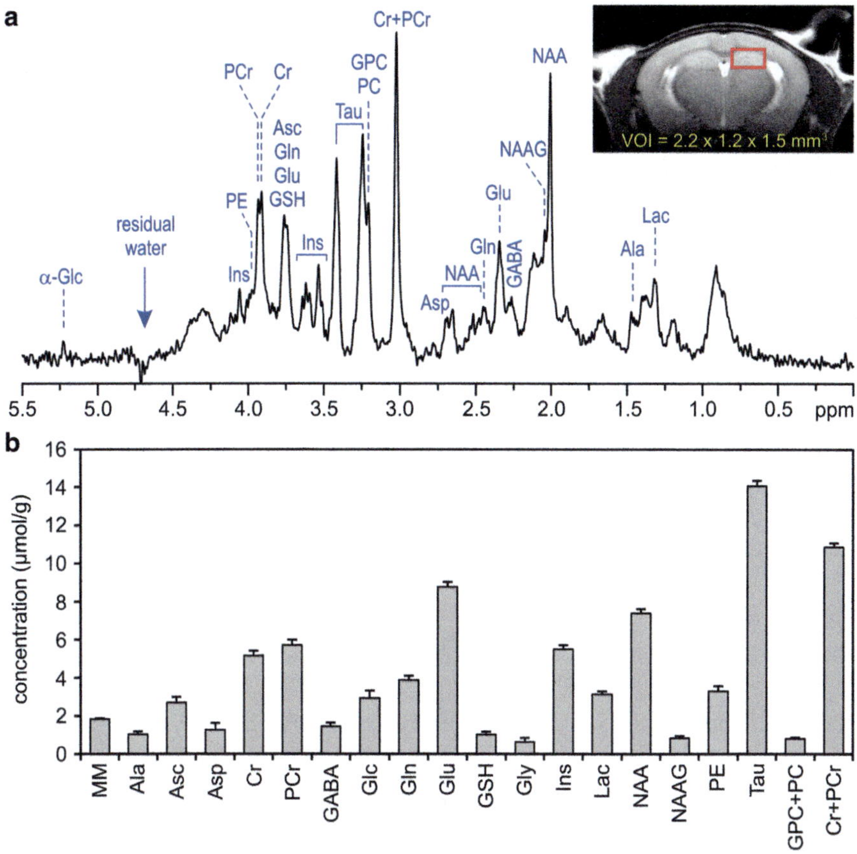

Fig. 2.1 In vivo ^{1}H MRS of the mouse brain at 9.4 T. (**a**) A representative ^{1}H MR spectrum acquired from the hippocampus (STEAM with OVS and VAPOR water suppression, TE = 2 ms, TR = 5 s, NT = 320). (**b**) The neurochemical profile of the hippocampus quantified from the spectrum shown in (**a**), LCModel, error bars indicate the estimated error of this fit (Cramér-Rao lower bounds). Abbreviations of metabolites: *MM* macromolecules, *Ala* alanine, *Asc* ascorbate, *Asp* aspartate, *Cr* creatine, *PCr* phosphocreatine, *GABA* γ-aminobutyric acid, *Glc* glucose, *Gln* glutamine, *Glu* glutamate, *GSH* glutathione, *Gly* glycine, *Ins* myo-inositol, *Lac* lactate, *NAA* *N*-acetylaspartate, *NAAG* *N*-acetylaspartylglutamate, *PE* phosphoethanolamine, *Tau* taurine, *GPC* glycerophosphocholine, *PC* phosphocholine

of water, which provides the source signal for MRI. As both the MR sensitivity and the chemical shift dispersion (in Hz) scale with the strength of the static magnetic field (B_0), ultrahigh-field MR systems ($\geq$7 T) are highly beneficial for small animal MRS. Currently 9.4 T MR scanners appear to be the best compromise between cost and performance for advanced ^{1}H MRS of rodents.

RF Coils

Appropriate RF coils can substantially boost the sensitivity and increase the signal-to-noise ratio (SNR) of acquired data. In order to increase the sensitivity of the RF coil, the coil has to be as close as possible to the examined volume of interest (VOI) in the brain. Therefore, surface RF coils might be preferred over volume coils for brain MRS of rodents, because it is much easier to place a small surface RF coil directly on top of the head in close proximity of the VOI. An additional advantage of small surface RF coils is their spatially limited sensitivity, which significantly helps to minimize the unwanted signals arising from remote areas outside of the VOI. However, this spatially limited sensitivity may cause difficulties in acquiring MRS data from deep brain structures. One solution is to use a volume RF coil for transmission, which can generate a spatially homogeneous B_1 field, and a separate small surface RF coil only for reception. Another alternative is to use a quadrature transmit/receive surface RF coil with two geometrically decoupled single-turn loops of appropriate diameter (10–15 mm), which might be the best compromise for mouse brain MRS, providing high sensitivity and reasonable spatial homogeneity for transmit as well as for receive RF fields.

Shim System

The B_0 field homogeneity is absolutely essential for MRS, because it determines the spectral resolution. B_0 inhomogeneity over the selected VOI increases the spectral linewidth that consequently decreases the SNR and increases the overlap of metabolite resonances, which together result in compromised metabolite quantification. B_0 inhomogeneities primarily originate from susceptibility differences between air and tissue. These inhomogeneities are scaled with the strength of the B_0 field and become highly nonlinear at ultrahigh magnetic fields. Animal MR scanners are typically equipped with the hardware (shim coils and drivers) capable of generating magnetic fields of appropriate strengths and symmetries to compensate these inhomogeneities. The process of adjusting the B_0 field homogeneity is known as B_0 shimming or simply shimming. The spatial magnetic field inhomogeneities are, in general, very complex, but for small VOIs they can be sufficiently well approximated by first- and second-order shim terms.

The strength of linear shims (X, Y, Z) is not a limiting factor, because powerful gradient coils and gradient amplifiers are widely used in MR scanners for the first-order shim corrections. Therefore, the strength of the second-order shim system is critical for successful MRS at high fields. The smaller the animal head, the bigger are the field deformations, which implies that a substantially more powerful shim system is required for brain ^{1}H MRS in mice relative to rats. The maximum strengths of the second-order shim system recommended for the 9.4 T small animal MR scanner are at least 2000 Hz/cm^2 (50 μT/cm^2) for XZ, YZ, and Z2 shim coils and 1000 Hz/cm^2 (25 μT/cm^2) for XY and X2Y2 shim coils [2].

B_0 Shimming

As it was mentioned in the previous section, B_0 shimming is a key factor for MRS and reliable metabolite quantification because the separation of some metabolite resonances (signals) is only a few Hz. Successful B_0 shimming requires not only a powerful second-order shim system but also an efficient B_0 mapping technique for automatic adjustment of currents in shim coils. Methods developed for B_0 field mapping can be grouped into two categories: methods based on 3D B_0 mapping [3] and B_0 mapping along projections [4]. In both cases, B_0 mapping is based on phase evolution of the transverse magnetization in an inhomogeneous B_0 field. Small VOIs selected for ^{1}H MRS require high spatial resolution of the B_0 mapping method, which favors projection techniques, such as FASTMAP [4, 5] over 3D mapping techniques.

FASTMAP is a fully automatic B_0 shimming method. It maps the field along thin bars through the center of the shimmed volume. Typically, three to four iterations are necessary for very fine shimming and the entire process takes less than 2 min. The FASTMAP technique allows adjustment of B_0 homogeneity in volumes as small as 4 μL (Figs. 2.1 and 2.2a). The natural signal linewidth (assuming perfect B_0 homogeneity) is determined by the T_2 relaxation and the microscopic heterogeneity of the tissue. The linewidths of unsuppressed water signals ranging from 10 to 12 Hz are achievable at 9.4 T for most mouse brain regions. However, in some brain areas, such as the cerebellum, the spectral linewidth is broader despite perfect shimming due to intrinsic properties of the tissue [6, 7]. In general, the natural signal linewidth is smaller in neonates relative to adults due to higher water content and less myelin. Consequently, water signal linewidths as low as 8–10 Hz are achievable at 9.4 T in mouse pups [8].

Poor B_0 shimming not only increases the spectral linewidths (Fig. 2.2e) but may also significantly affect the signal lineshape. Distorted and asymmetric signal lineshapes make metabolite quantification difficult and negatively affect the reliability of assessed metabolite levels.

Localization Techniques

The basic difference between in vivo ^{1}H MRS and the traditional high-resolution ^{1}H NMR spectroscopy widely used in organic chemistry and biochemistry is the capability of MRS to acquire data from a well-defined volume of sample. Spatial localization is fundamental for in vivo applications, because MR spectra acquired from the whole animal would not provide meaningful information. Spatially selective excitation is typically based on a combination of three orthogonal slice-selective pulses (band-selective RF pulses simultaneously applied with field gradients), and the signal of interest is acquired from the intersection of these slices. In addition, localization techniques must include water suppression because the detection of brain

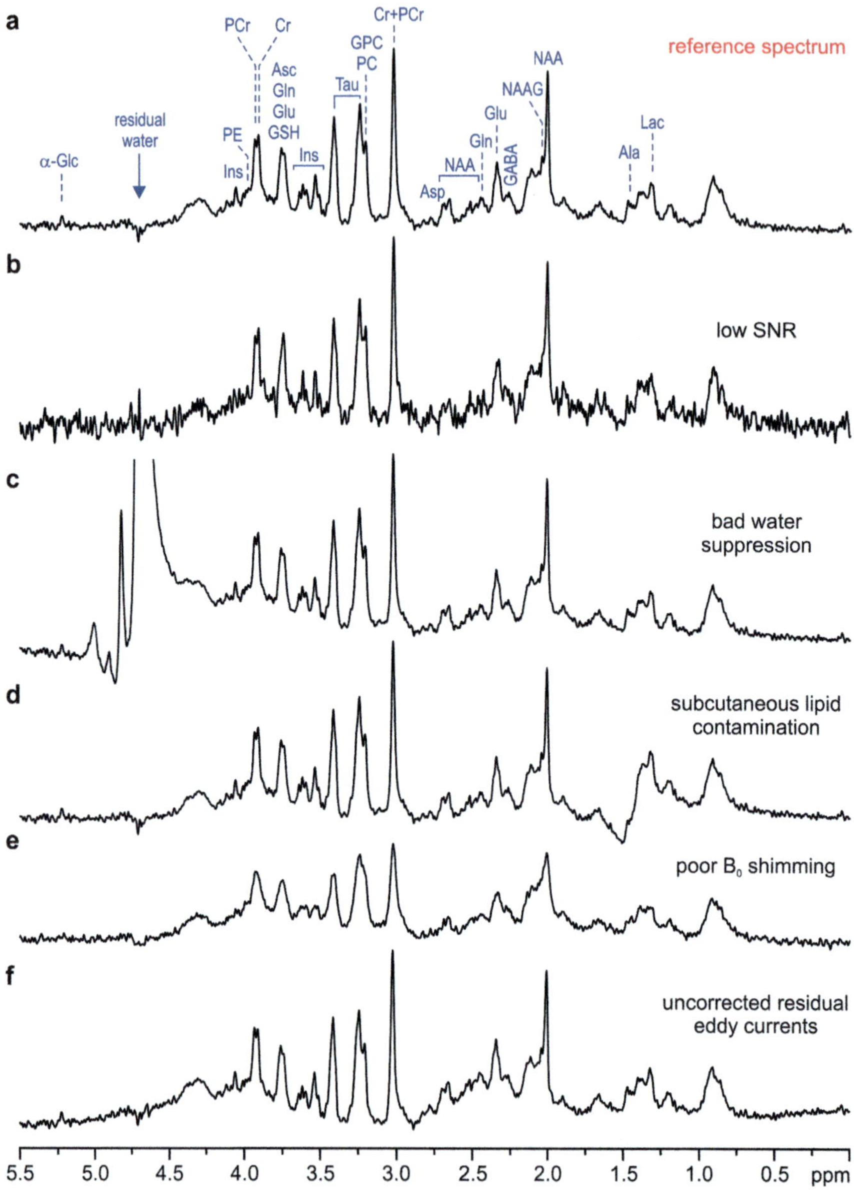

Fig. 2.2 Most important factors affecting spectral quality. (**a**) Reference spectrum acquired from the mouse hippocampus using STEAM localization sequence with OVS and VAPOR water suppression at 9.4 T (TE = 2 ms, TR = 5 s), (**b**) spectrum with low signal-to-noise ratio, (**c**) spectrum affected by misadjusted water suppression, (**d**) spectrum acquired without OVS, (**e**) poorly adjusted B_0 field homogeneity, (**f**) spectrum affected by uncorrected residual eddy currents

metabolites (in mM concentration range) is extremely difficult in the presence of a large water signal (brain water content ~70–80 %), 4–5 orders of magnitude stronger than the signals of metabolites. The design and properties of pulse sequences most often used for single-voxel ^{1}H MRS studies of small animals will be discussed in more detail in the following paragraphs. This chapter does not include the methodology of MR spectroscopic imaging (MRSI) [9, 10] due to the lack applications of this technique in animal models of neurodegeneration.

Pulse Sequences for ^{1}H MRS

The two most frequently used localization pulse sequences for in vivo ^{1}H MRS are PRESS (point-resolved spectroscopy) [11] and STEAM (stimulated echo acquisition mode) [12]. The PRESS sequence is based on the combination of three orthogonal slice-selection pulses with a double spin echo (90°-180°-180°). On the other hand, the STEAM sequence is based on the combination of slice-selective pulses with a stimulated echo (90°-90°-90°). Each technique has its pros and cons. The PRESS sequence provides twice as much signal as the STEAM from the same VOI, but the VOI is not as well defined as in STEAM (slice-selection profile of 180° is typically worse than that of 90° pulses). In addition, PRESS is more susceptible for unwanted coherences, and the chemical shift displacement error (see the next paragraph) might be a big problem for high-field ^{1}H MRS when the peak transmit RF power is limited. The STEAM sequence provides only 50 % of available signal from the VOI (because of the stimulated echo), but the volume selection is superior to PRESS. This sequence is also less susceptible for unwanted coherences that originate from outside of the VOI and enables better water suppression due to an additional water suppression RF pulse during the TM period (between second and third 90° RF pulses). Finally, STEAM allows shorter echo times (TE) than PRESS, which has the advantages of suppressing J-modulation in metabolite spectra with coupled spin systems and detecting signals of metabolites with short T_2 relaxation more reliably. The localization performance of both sequences can be improved by outer volume suppression (OVS) applied prior to the VOI selection (Fig. 2.2a, d), which has been well demonstrated with the ultrashort echo-time STEAM sequence developed for ^{1}H MRS of rodents at 9.4 T [13].

The LASER (localization by adiabatic selective refocusing) pulse sequence [14] provides full intensity from selected VOI similar to the PRESS sequence, but the volume selection, achieved by a pair of adiabatic RF pulses for selection of each slice, is far superior relative to PRESS. In addition, the LASER sequence is fully adiabatic, which means that the VOI is properly selected independent of the applied RF power (above the required adiabaticity threshold). This property of the sequence is extremely useful for MRS studies using surface RF coils with highly heterogeneous transmit B_1 profiles. This advantage in volume selection comes at the cost of increased minimum echo time that might require T_2 correction for absolute metabolite quantification, although TEs as short as 15 ms can be easily achieved with this sequence using standard 9.4 T hardware [7, 15].

More recently, the SPECIAL (spin echo full intensity acquired localization) pulse sequence was introduced [16], which is based on a single spin echo combined with a slice inversion applied on alternative scans (180°-90°-180°), similar to 1D ISIS (image-selected in vivo spectroscopy) localization technique [17]. This approach allows acquiring full intensity signal with very short TE. However, these advantages are partially hampered by a potential for spectral contamination resulting from incomplete subtraction of unwanted coherences since SPECIAL is not a single-shot MRS technique.

Chemical Shift Displacement Error

Chemical shift displacement error (CSDE) is a general problem of all single-voxel MRS localization techniques that use slice-selective pulses (simultaneously applied band-selective RF pulse and the field gradient) for VOI selection. CSDE results from a spatial displacement of selected slices for off-resonance signals. It means that metabolite resonances with different chemical shifts originate from volumes that are spatially displaced from the nominal VOI (Fig. 2.3). This spatial displacement is proportional to the ratio of the spectrum frequency range (in Hz) to the RF pulse bandwidth. Most of the signals of interest in ^{1}H MRS are grouped within a 3 ppm range, but this spectral range corresponds to a frequency range that is linearly proportional to applied magnetic field B_0. Therefore, CSDE becomes a serious problem at high magnetic fields. For example, the 3 ppm range corresponds to 600 Hz at 4.7 T, but it increases to 1200 Hz at 9.4 T. The commonly used 90° sinc RF pulse of 1 ms has a 3 kHz bandwidth, which results in 20 % VOI displacement for resonances separated by 3 ppm just along one axis at 4.7 T. However, the bandwidth of 180° sinc RF pulse generated by the same peak transmit power is only 2.25 kHz, resulting in 54 % VOI displacement at 9.4 T. This shows the CSDE challenge for the PRESS sequence at high magnetic fields. This drawback in volume selection can be minimized by increasing the bandwidth of RF pulses by

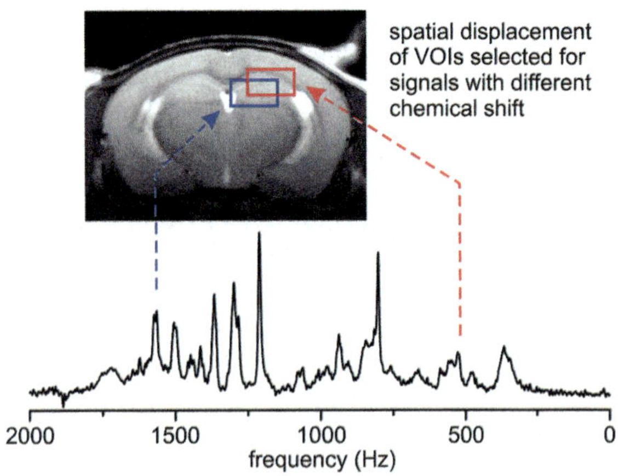

Fig. 2.3 Illustration of the chemical shift displacement error. The volumes from which the metabolite signals originate are displaced from the nominal position of the VOI depending on the frequency offset from the nominal frequency used for the VOI selection

decreasing their duration, as in ultrashort echo-time STEAM [13], or by substituting amplitude-modulated RF pulses by pairs of broadband frequency-modulated adiabatic RF pulses, as it is used in the LASER sequence [14]. These approaches limit the CSDE at 9.4 T to an acceptable 10 % per each slice selection. Although a major CSDE is not recognizable in a spectrum, it may lead to a significant bias and misinterpretation of acquired MRS data.

Water Suppression

Highly effective water suppression is extremely important for reliable quantification of metabolites because the residual water signal, which is rather often out of phase, may cause serious baseline distortions (Fig. 2.2c). Even though the residual water signal can be removed from the spectra using the Hankel singular value decomposition (HSVD) approach [18], this approach cannot eliminate its sidebands (typically caused by gradient coil vibration) overlapping with metabolite resonances. Therefore, in order to minimize unwanted coherences in ^{1}H MRS data, highly efficient water suppression is always preferred over post-processing water signal removal techniques. The RF pulses applied in the water suppression pulse train have to be frequency selective, suppressing resonances only in a narrow spectral range ($\pm$0.4 ppm) centered on the water resonance at 4.7 ppm. The widely used WET (water suppression enhanced through T_1 effects) technique [19] has been recently more often replaced by the VAPOR (variable pulse power and optimized relaxation delays) water suppression method [13]. This technique, which consists of seven frequency-selective RF pulses, is relatively insensitive to B_1 and T_1 variation and can routinely suppress the residual water signal down to the noise level (Fig. 2.2) [2, 8, 20].

Localization Performance

The localization performance of the sequence can be characterized by two factors: first, how accurately the received FID is localized to the selected VOI and, secondly, to which extent the received FID is contaminated by unwanted coherences from outside of the VOI. The accuracy of localization, outer edges of the VOI, and the CSDE depend on the excitation profiles and bandwidths of the RF pulses utilized in the localization sequence. The small sidelobes on both sides of the central excitation profile can excite spins outside of the VOI. If the sidelobes excite extracerebral subcutaneous tissue, unwanted signal from lipids (around 1.5 ppm and typically out of phase) will contaminate the ^{1}H MRS spectrum of the brain (Fig. 2.2d). These imperfections in the localization performance can be minimized by appropriate OVS (Fig. 2.2a, d). The OVS scheme must utilize broadband RF pulses (the frequency-modulated hyperbolic secant pulses are the best choice) to keep the CSDE of OVS in a similar range as the CSDE of the VOI.

The second challenge that accompanies the volume selection originates from the fact that the series of RF pulses also creates other types of coherences (FIDs, spin and stimulated echoes) that have to be eliminated by crusher gradients (gradient dephasing) and phase cycling of RF pulses and receiver. Unwanted coherence suppression by gradient dephasing is preferable because the elimination of strong unwanted coherences by subtraction does not work reliably under in vivo conditions. The dephased magnetization is apparently lost. However, this apparently lost magnetization can be rephased back during signal detection by inhomogeneous B_0 fields outside of the VOI and can generate unwanted echoes.

Because of an inherently low detection sensitivity of MRS and strong overlap of metabolite resonances in an overcrowded spectral range (1–4 ppm), the spectral quality is critical for reliable metabolite quantification. Unwanted signals, deteriorating the spectral quality, always originate from outside of the VOI. A sufficient SNR (Fig. 2.2a, b) is very important for accurate and precise metabolite quantification. But the quality of the spectra, which depends on B_0 shimming, the efficiency of water suppression, and the elimination of unwanted coherences, is even more important than the SNR (Fig. 2.2). Only well-designed and optimized pulse sequences can provide high-quality spectra, which are critical for meaningful analysis and neurochemical profiling. There are many examples in literature for high-quality ^{1}H MR spectra measured by STEAM [2, 8, 20–22], LASER [6, 7, 15, 23], and SPECIAL [24–26] pulse sequences.

Eddy Currents

Eddy currents are induced in the conducting structures of the magnet by fast magnetic field gradient switching and result in undesirable time-varying magnetic field gradients and magnetic field shifts. These gradients and field shifts cause line broadening and distortions of lineshape. Despite compensation for eddy current effects by hardware, some residual eddy currents always persist and affect spectral linewidth and lineshape (Fig. 2.2f) and thereby may cause major problems in accuracy of metabolite quantification.

Animal Handling and Data Acquisition Strategies

Successful in vivo ^{1}H MRS requires keeping optimal and stable physiological conditions during the entire experiment. To achieve that, a well-designed animal holder, which firmly holds the rat or mouse head in stable position and at the same time allows smooth respiration, is required. Continuous monitoring of respiration and body temperature is essential. The respiration rate for spontaneously breathing mice should be kept in a range of 80–100 per minute by adjusting the isoflurane level in anesthetic gas mixture ($N_2O/O_2 \sim 1{:}1$). The deoxyhemoglobin is a paramagnetic molecule and is a major source of microscopic B_0 heterogeneity of the tissue, which affects the signal linewidth. Without optimal blood oxygen saturation, high spectral resolution is not possible despite perfect B_0 shimming.

Animal respiration and small body motions (not the head) result in frequency fluctuations, which cause line broadening if these fluctuations are not corrected. Even very small head movements, caused, e.g., by gasping, result in signal phase fluctuations, which decrease the signal intensity of the averaged spectrum. Therefore, a single-scan averaging mode (FID of the each scan is saved separately in memory) is preferred in order to allow post-processing phase and frequency corrections before data summation. If the SNR of a single-scan spectrum is too low (VOI is too small) for performing any correction, then data acquisition in small blocks is recommended. For example, instead of acquiring 160×1 scan, 20×8 scans can be acquired. This approach enables frequency correction at least between these 20 FIDs (spectra) before summation. In addition, this MRS data collection approach allows eliminating corrupted blocks of data from summation to avoid deterioration of the final spectral quality due to intermittent instability.

MRS Data Processing and Quantification

MRS Data Pre-processing

Frequency and Phase Correction

As it was mentioned before, respiration and small body movements can cause frequency and phase fluctuations in the acquired data. In addition, maintaining a long-term stability of the B_0 field within ± 1 Hz is difficult. Therefore, a single-scan averaging approach is highly recommended because it allows correction of any frequency or phase change on each single scan (FID). Preferentially, these corrections have to be applied on time-domain data (FIDs). If the SNR of single-scan data is not sufficient, then the single-scan data should be first summed in small blocks (e.g., eight scans per block) and then the frequency correction can be applied on these summed data. This approach cannot correct any phase or frequency instability that occurs within these short time periods, but can efficiently remove a long-term frequency drift. Uncompensated small frequency fluctuations result in broader lines, but signal integrals are preserved. However, uncompensated phase fluctuations reduce the signal integrals and may lead to an underestimation of metabolite concentrations. Therefore, blocks with obviously decreased intensity must be eliminated from the final summation. If we already know that the SNR of single-scan data will not be sufficient to perform these corrections, saving data acquired in relatively short time periods (e.g., eight scans) is a reasonable approach.

Correction for Residual Eddy Currents

Lineshape distortions caused by residual eddy currents can be easily removed from metabolite spectra using characteristics of the lineshape distortion taken from the unsuppressed water signal [27]. This unsuppressed water signal must be acquired exactly with the same set of parameters, including all gradients, as those used for metabolite spectra acquisition, except for RF pulses used for water suppression. This eddy current correction can be applied on individual scans or simply on the final summed FID.

Metabolite Quantification

Analysis of short echo-time in vivo ^{1}H MR spectra of the brain is rather complex and requires sophisticated fitting methods. Despite increased chemical shift dispersion at ultrahigh magnetic fields and optimal B_0 shimming, spectra of individual metabolites are highly overlapped. Therefore, a simple integration method that has been routinely used for long TE spectra with three or four peaks is not applicable for short TE spectra. A big advantage for meaningful quantification of ^{1}H MRS spectra is the knowledge about metabolites that make a dominant contribution to in vivo spectrum of the brain. The homeostasis of the brain is extremely well maintained, therefore except in cases of major metabolic defects, the set of metabolites contributing to brain spectra is well defined. In addition, the pH and temperature of the brain are typically maintained at a very narrow range, which means that ^{1}H MR spectra of metabolites can be easily measured under identical conditions. The database of these metabolite spectra forms the prior knowledge for all spectral quantification techniques. In order to get meaningful neurochemical information, fitting algorithms decompose the in vivo spectrum into appropriately broadened metabolite spectra from the database. It is a rather simplified model, but it works reasonably well.

Quantification Methods

The two most commonly used spectral fitting software packages are jMRUI [28, 29], working in the time domain, and LCModel [30], working in the frequency domain. If the same prior knowledge is used, both methods should provide very similar quantification results [31]. Quantification errors are estimated by Cramér-Rao lower bounds (CRLB). It should be emphasized that CRLB are estimated on the basis of the assumption that the model (spectral basis set) is correct and complete. Obviously, this is not possible and reasonable simplifying assumptions have to be made. The metabolite spectra included in the database can be experimentally measured or simulated based on published information about metabolite chemical shifts and *J*-couplings [32]. The basic principle of LCModel analysis is demonstrated in Fig. 2.4. The in vivo ^{1}H MR spectrum (also shown in Fig. 2.1a) is decomposed into a linear combination of metabolite spectra from the prior knowledge database (basis set). The fitting routine is scaled in such a way that the coefficients of this linear combination are the estimated metabolite concentrations.

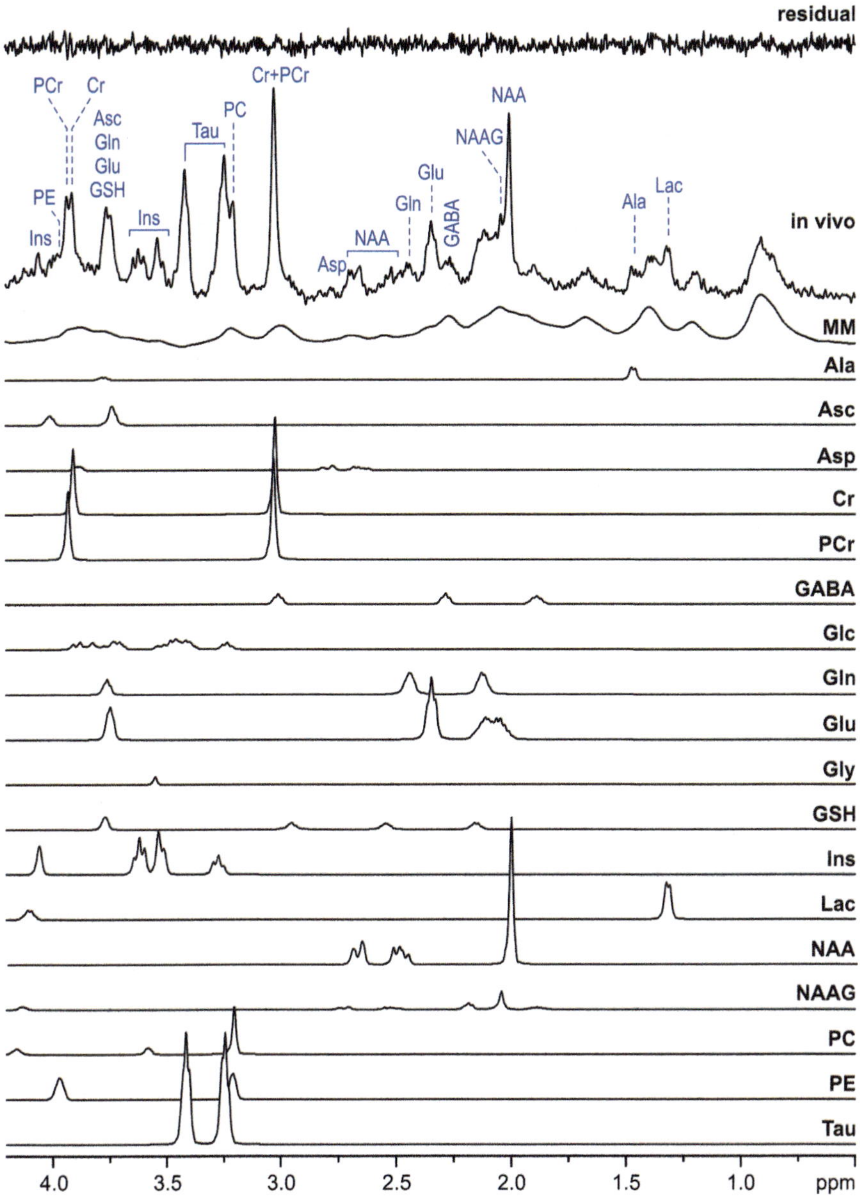

Fig. 2.4 LCModel analysis of the spectrum acquired from the mouse hippocampus at 9.4 T (shown on Fig. 2.1). LCModel analysis was performed using a simulated basis set that includes experimentally measured spectrum of fast-relaxing macromolecules

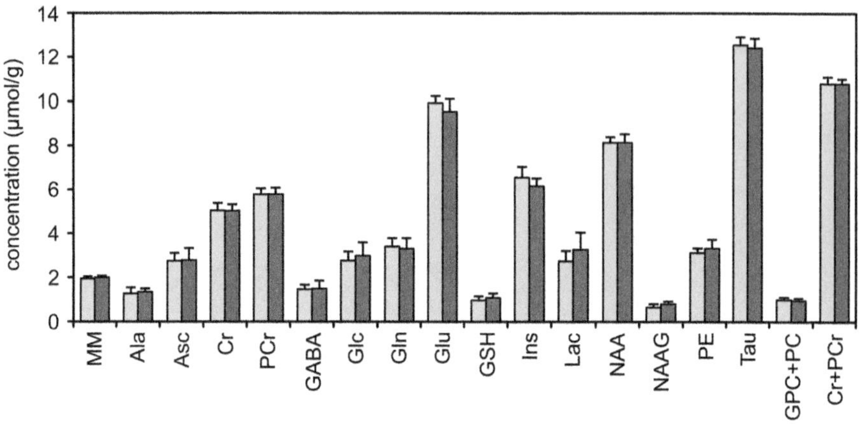

Fig. 2.5 Reproducibility of neurochemical profiling in the mouse brain at 9.4 T. The neurochemical profiles of the hippocampus measured from two different groups of male C57BL/6 mice ($N = 10$ in each group) are shown in *light* and *dark gray*. Error bars indicate the SD

The result of the LCModel analysis is shown as a bar diagram in Fig. 2.1. The reproducibility of the neurochemical profiling in mice is demonstrated in Fig. 2.5, where the metabolite concentrations quantified from the hippocampus of two different groups of C57BL/6 mice are compared.

Macromolecule Background

Short TE ^{1}H MR spectra have significant signal contribution from fast-relaxing macromolecules that are easily recognizable between 0.5 and 1.8 ppm (Figs. 2.1 and 2.4). A dominant contribution to these macromolecule signals in the healthy brain comes from mobile proteins in cytosol. These broad signals must be taken into account in metabolite quantification and should not be fitted as a baseline. Including a measured macromolecule spectrum in the basis set is a very robust approach, which improves the quantification of weakly represented metabolites, such as GABA [33, 34]. The macromolecule spectrum can be experimentally measured using an inversion-recovery experiment [34]. The small residual metabolite signals can be suppressed using diffusion weighting [35], or they can be removed from the spectra in post-processing [36].

Referencing

Metabolite quantification requires appropriate referencing to obtain values in concentration units. The signal of total creatine (Cr + PCr) has been widely used as an internal reference, assuming a concentration of 8 μmol/g. However, using total creatine as an internal reference is far from being optimal because its content varies

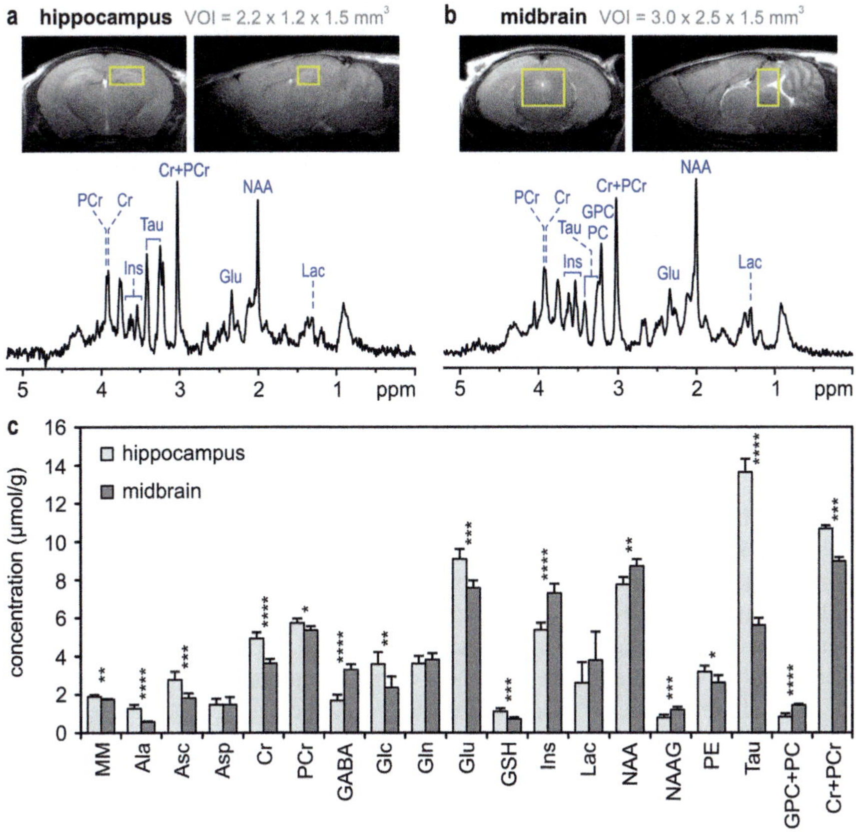

Fig. 2.6 Comparison of the neurochemical profiles acquired from the hippocampus and midbrain of C57BL/6 mice ($N = 6$) at 9.4 T. Error bars indicate the SD, $*p < 0.05$, $**p < 0.01$, $***p < 0.001$, $****p < 0.0001$

between brain regions (Fig. 2.6) and changes significantly during brain development [24, 25, 37] and as a consequence of neurodegenerative processes [20, 21]. Using the unsuppressed water signal as a reference is a very useful approximation and has worked extremely well in many in vivo ^{1}H MRS applications [1, 6–8, 15, 20, 21, 23, 26]. An alternative referencing method called ERETIC (Electric Reference To access In vivo Concentrations) has been described [38].

Relaxation

It is extremely difficult to precisely measure T_1 and T_2 relaxation times of all brain metabolites in vivo. In addition, relaxation times are not properties of molecules, but each proton in a molecule has its own T_1 and T_2 values. The relaxation parameters of metabolites have been recently assessed based on the simplified assumption that all CH_2 and CH protons within the same molecule have approximately the same relaxation properties [39–41]. When MRS data are acquired using shorter repetition time (TR) or longer TE, then corrections for relaxation effects must be applied for "absolute" metabolite quantification. But these corrections are unnecessary when an ultrashort TE localization sequence and a long TR are used [8, 20–22], making this a robust approach for reliable neurochemical profiling.

Neurochemical Profiling

The primary goal of neurochemical profiling is to extend the range of quantifiable metabolites and to improve the precision and the accuracy of their quantification. Reliable quantification of more than 15 brain metabolites is feasible from the rodent brain when advanced high-field MR instrumentation, carefully optimized pulse sequences and acquisition strategies, and sophisticated data processing methods are used [1, 6–8, 15, 20–22, 26, 33]. The metabolite distribution in the rat or mouse brain is rather heterogeneous, which is illustrated in Fig. 2.6 where the neurochemical profiles of the hippocampus and the midbrain are compared. This figure also demonstrates the power of ^{1}H MRS to detect differences in tissue composition.

In conclusion, reliable, noninvasive quantification of an extended range of brain metabolites in rodents is feasible. To achieve this goal, high-quality in vivo ^{1}H MR spectra and sophisticated processing tools with well-optimized prior knowledge are needed and are now being used successfully by an increasing number of laboratories worldwide.

References

1. Duarte JM, Lei H, Mlynárik V, Gruetter R (2012) The neurochemical profile quantified by in vivo ^{1}H NMR spectroscopy. Neuroimage 61(2):342–362. doi:10.1016/j.neuroimage.2011.12.038
2. Tkáč I, Henry PG, Andersen P, Keene CD, Low WC, Gruetter R (2004) Highly resolved in vivo ^{1}H NMR spectroscopy of the mouse brain at 9.4 T. Magn Reson Med 52(3):478–484
3. Miyasaka N, Takahashi K, Hetherington HP (2006) Fully automated shim mapping method for spectroscopic imaging of the mouse brain at 9.4 T. Magn Reson Med 55(1):198–202. doi:10.1002/mrm.20731
4. Gruetter R (1993) Automatic, localized in vivo adjustment of all first- and second-order shim coils. Magn Reson Med 29(6):804–811
5. Gruetter R, Tkáč I (2000) Field mapping without reference scan using asymmetric echo-planar techniques. Magn Reson Med 43(2):319–323
6. Öz G, Kittelson E, Demirgoz D, Rainwater O, Eberly LE, Orr HT, Clark HB (2015) Assessing recovery from neurodegeneration in spinocerebellar ataxia 1: comparison of in vivo magnetic resonance spectroscopy with motor testing, gene expression and histology. Neurobiol Dis 74:158–166. doi:10.1016/j.nbd.2014.11.011
7. Öz G, Vollmers ML, Nelson CD, Shanley R, Eberly LE, Orr HT, Clark HB (2011) In vivo monitoring of recovery from neurodegeneration in conditional transgenic SCA1 mice. Exp Neurol 232(2):290–298. doi:10.1016/j.expneurol.2011.09.021

8. Wallin DJ, Tkáč I, Stucker S, Ennis KM, Sola-Visner M, Rao R, Georgieff MK (2015) Phlebotomy-induced anemia alters hippocampal neurochemistry in neonatal mice. Pediatr Res 77(6):765–771. doi:10.1038/pr.2015.41
9. Alf MF, Lei H, Berthet C, Hirt L, Gruetter R, Mlynárik V (2012) High-resolution spatial mapping of changes in the neurochemical profile after focal ischemia in mice. NMR Biomed 25(2):247–254. doi:10.1002/nbm.1740
10. Mlynárik V, Kohler I, Gambarota G, Vaslin A, Clarke PG, Gruetter R (2008) Quantitative proton spectroscopic imaging of the neurochemical profile in rat brain with microliter resolution at ultra-short echo times. Magn Reson Med 59(1):52–58. doi:10.1002/mrm.21447
11. Bottomley PA (1987) Spatial localization in NMR-spectroscopy in vivo. Ann N Y Acad Sci 508:333–348. doi:10.1111/j.1749-6632.1987.tb32915.x
12. Frahm J, Merboldt KD, Hanicke W (1987) Localized proton spectroscopy using stimulated echoes. J Magn Reson 72(3):502–508. doi:10.1016/0022-2364(87)90154-5
13. Tkáč I, Starcuk Z, Choi IY, Gruetter R (1999) In vivo ^{1}H NMR spectroscopy of rat brain at 1 ms echo time. Magn Reson Med 41(4):649–656
14. Garwood M, DelaBarre L (2001) The return of the frequency sweep: designing adiabatic pulses for contemporary NMR. J Magn Reson 153(2):155–177. doi:10.1006/jmre.2001.2340
15. Öz G, Nelson CD, Koski DM, Henry PG, Marjanska M, Deelchand DK, Shanley R, Eberly LE, Orr HT, Clark HB (2010) Noninvasive detection of presymptomatic and progressive neurodegeneration in a mouse model of spinocerebellar ataxia type 1. J Neurosci 30 (10):3831–3838. doi:10.1523/JNEUROSCI.5612-09.2010
16. Mlynárik V, Gambarota G, Frenkel H, Gruetter R (2006) Localized short-echo-time proton MR spectroscopy with full signal-intensity acquisition. Magn Reson Med 56(5):965–970. doi:10.1002/mrm.21043
17. Ordidge RJ, Connelly A, Lohman JAB (1986) Image-selected in vivo spectroscopy (ISIS)—a new technique for spatially selective NMR-spectroscopy. J Magn Reson 66(2):283–294. doi:10.1016/0022-2364(86)90031-4
18. Vanhamme L, Fierro RD, Van Huffel S, de Beer R (1998) Fast removal of residual water in proton spectra. J Magn Reson 132(2):197–203
19. Ogg RJ, Kingsley PB, Taylor JS (1994) WET, a T1- and B1-insensitive water-suppression method for in vivo localized 1H NMR spectroscopy. J Magn Reson B 104(1):1–10
20. Zacharoff L, Tkáč I, Song Q, Tang C, Bolan PJ, Mangia S, Henry PG, Li T, Dubinsky JM (2012) Cortical metabolites as biomarkers in the R6/2 model of Huntington's disease. J Cereb Blood Flow Metab 32(3):502–514. doi:10.1038/jcbfm.2011.157
21. Tkáč I, Dubinsky JM, Keene CD, Gruetter R, Low WC (2007) Neurochemical changes in Huntington R6/2 mouse striatum detected by in vivo 1H NMR spectroscopy. J Neurochem 100 (5):1397–1406
22. Tkáč I, Henry PG, Zacharoff L, Wedel M, Gong W, Deelchand DK, Li T, Dubinsky JM (2012) Homeostatic adaptations in brain energy metabolism in mouse models of Huntington disease. J Cereb Blood Flow Metab 32(11):1977–1988. doi:10.1038/jcbfm.2012.104
23. Emir UE, Brent Clark H, Vollmers ML, Eberly LE, Öz G (2013) Non-invasive detection of neurochemical changes prior to overt pathology in a mouse model of spinocerebellar ataxia type 1. J Neurochem. doi:10.1111/jnc.12435
24. das Neves Duarte JM, Kulak A, Gholam-Razaee MM, Cuenod M, Gruetter R, Do KQ (2012) N-acetylcysteine normalizes neurochemical changes in the glutathione-deficient schizophrenia mouse model during development. Biol Psychiatry 71(11):1006–1014. doi:10.1016/j.biopsych.2011.07.035
25. Kulak A, Duarte JM, Do KQ, Gruetter R (2010) Neurochemical profile of the developing mouse cortex determined by in vivo ^{1}H NMR spectroscopy at 14.1 T and the effect of recurrent anaesthesia. J Neurochem 115(6):1466–1477. doi:10.1111/j.1471-4159.2010.07051.x
26. Mlynárik V, Cacquevel M, Sun-Reimer L, Janssens S, Cudalbu C, Lei H, Schneider BL, Aebischer P, Gruetter R (2012) Proton and phosphorus magnetic resonance spectroscopy of a

mouse model of Alzheimer's disease. J Alzheimers Dis 31(Suppl 3):S87–S99. doi:10.3233/JAD-2012-112072
27. Klose U (1990) In vivo proton spectroscopy in presence of eddy currents. Magn Reson Med 14(1):26–30
28. Naressi A, Couturier C, Castang I, de Beer R, Graveron-Demilly D (2001) Java-based graphical user interface for MRUI, a software package for quantitation of in vivo/medical magnetic resonance spectroscopy signals. Comput Biol Med 31(4):269–286
29. Ratiney H, Sdika M, Coenradie Y, Cavassila S, van Ormondt D, Graveron-Demilly D (2005) Time-domain semi-parametric estimation based on a metabolite basis set. NMR Biomed 18(1):1–13. doi:10.1002/nbm.895
30. Provencher SW (1993) Estimation of metabolite concentrations from localized in vivo proton NMR spectra. Magn Reson Med 30(6):672–679
31. Cudalbu C, Bucur A, Graveron-Demilly D, Beuf O, Cavassila S (2007) Comparison of two strategies of background-accommodation: influence on the metabolite concentration estimation from in vivo Magnetic Resonance Spectroscopy data. Conf Proc IEEE Eng Med Biol Soc 2007:2077–2080. doi:10.1109/IEMBS.2007.4352730
32. Govindaraju V, Young K, Maudsley AA (2000) Proton NMR chemical shifts and coupling constants for brain metabolites. NMR Biomed 13(3):129–153
33. Mlynárik V, Cudalbu C, Xin L, Gruetter R (2008) ^{1}H NMR spectroscopy of rat brain in vivo at 14.1 Tesla: improvements in quantification of the neurochemical profile. J Magn Reson 194(2):163–168. doi:10.1016/j.jmr.2008.06.019
34. Pfeuffer J, Tkáč I, Provencher SW, Gruetter R (1999) Toward an in vivo neurochemical profile: quantification of 18 metabolites in short-echo-time ^{1}H NMR spectra of the rat brain. J Magn Reson 141(1):104–120
35. Kunz N, Cudalbu C, Mlynárik V, Hüppi PS, Sizonenko SV, Gruetter R (2010) Diffusion-weighted spectroscopy: a novel approach to determine macromolecule resonances in short-echo time ^{1}H-MRS. Magn Reson Med 64(4):939–946. doi:10.1002/mrm.22490
36. Cudalbu C, Mlynárik V, Gruetter R (2012) Handling macromolecule signals in the quantification of the neurochemical profile. J Alzheimers Dis 31(Suppl 3):S101–S115. doi:10.3233/JAD-2012-120100
37. Tkáč I, Rao R, Georgieff MK, Gruetter R (2003) Developmental and regional changes in the neurochemical profile of the rat brain determined by in vivo 1H NMR spectroscopy. Magn Reson Med 50(1):24–32
38. Heinzer-Schweizer S, De Zanche N, Pavan M, Mens G, Sturzenegger U, Henning A, Boesiger P (2010) In-vivo assessment of tissue metabolite levels using 1H MRS and the Electric REference To access In vivo Concentrations (ERETIC) method. NMR Biomed 23(4):406–413. doi:10.1002/nbm.1476
39. Cudalbu C, Mlynárik V, Xin L, Gruetter R (2009) Comparison of T1 relaxation times of the neurochemical profile in rat brain at 9.4 tesla and 14.1 tesla. Magn Reson Med 62(4):862–867. doi:10.1002/mrm.22022
40. de Graaf RA, Brown PB, McIntyre S, Nixon TW, Behar KL, Rothman DL (2006) High magnetic field water and metabolite proton T1 and T2 relaxation in rat brain in vivo. Magn Reson Med 56(2):386–394. doi:10.1002/mrm.20946
41. Xin L, Gambarota G, Mlynárik V, Gruetter R (2008) Proton T2 relaxation time of J-coupled cerebral metabolites in rat brain at 9.4 T. NMR Biomed 21(4):396–401. doi:10.1002/nbm.1205

Chapter 3
Methodology of Clinical MRS: Technical Challenges and Solutions

Bart W.J. Philips and Tom W. Scheenen

Abstract For certain applications in the brain, proton MRS has evolved from a research tool into a clinically viable application, useful for studying diseases and clinically relevant disease parameters. It is however technically challenging to obtain high-quality MR spectra and precisely measure subtle neurochemical changes in degenerative brain diseases. In this chapter these technical challenges will be discussed and potential solutions will be outlined. After introducing the sensitivity challenge of MRS, different methods for localization of a volume of interest are summarized. Subsequently, solutions to suppress the large water signal and unwanted residual signals from outside the volume of interest are discussed together with shimming, culminating in a description of the different calibration steps covering all these issues and an overview of methods for estimating metabolite concentrations from MRS data.

Keywords Single-voxel spectroscopy • MR spectroscopic imaging • Localization • Pulse sequences • PRESS • STEAM • ISIS • LASER • Semi-LASER • SPECIAL • Chemical shift displacement error • Spectral editing • Quantification

Background on Clinical MRS

Magnetic resonance spectroscopy (MRS) can be used to study metabolites in the human body. As long as these metabolites are available in relatively high enough concentrations and feature one or more MR visible nuclei, they can be measured with MRS. Many studies have shown large differences in metabolite levels compared to healthy brain in acute stroke, chronic multiple sclerosis, metabolic disorders, and brain tumors (reviewed in [1]). A recent example that has received considerable interest for imaging brain tumors is the ability to measure 2-hydroxy-glutarate (2-HG), which is normally not measurable in the human tissue.

B.W.J. Philips, M.D. • T.W. Scheenen, Ph.D. (✉)
Department of Radiology and Nuclear Medicine, Radboud University Medical Center, Geert Grooteplein 10, Nijmegen, GA 6525, The Netherlands
e-mail: Bart.Philips@radboudumc.nl; Tom.Scheenen@radboudumc.nl

G. Öz (ed.), *Magnetic Resonance Spectroscopy of Degenerative Brain Diseases*, Contemporary Clinical Neuroscience, DOI 10.1007/978-3-319-33555-1_3

In certain mutations in gliomas, isocitrate dehydrogenase (IDH) is affected, such that 2-HG builds up to measurable quantities in the tumor. 2-HG is therefore a marker of IDH dysfunction, which is associated with a better prognosis in gliomas. MRS can therefore be used in the clinic to measure biomarkers of disease [2]. While brain tumors have been the primary indication for MRS in the clinical setting, consensus was recently reached that MRS is expected to contribute to patient management in neurodegenerative diseases as well [1].

As such, proton MRS has evolved from a research tool into a clinically viable application, useful for studying a number of central nervous system diseases and clinically relevant disease parameters. It is however technically challenging to obtain high-quality MR spectra and precisely measure subtle neurochemical changes in neurodegenerative diseases. In this chapter these technical challenges and potential solutions will be discussed.

Technical Challenges and Solutions

Sensitivity

In magnetic resonance imaging, we are interested in the MR signal of protons from water and lipids. As our brain consists of about 75 % water, this signal is generally abundantly available. In spectroscopy the signals of interest are, in most cases, not the signal of water, but of molecules that are present in our bodies at concentrations much lower than that of water. *N*-acetyl-aspartate (NAA, a marker for healthy neuronal tissue), for example, has a concentration in the order of 10 mmol/L in the brain tissue, whereas water protons are available at a concentration of around 80 mol/L in white matter and around 90 mol/L in gray matter (factor 8×10^3 difference). This results in a substantially lower signal-to-noise ratio (SNR) in spectroscopy when compared to conventional MRI. In combination with the already low intrinsic sensitivity of magnetic resonance, SNR is in most cases the limiting factor in spectroscopy and is the main reason for the often long measurement times.

Using basic statistical physics, a relation can be derived for the magnetization of a sample with an arbitrary number of spins (protons) and temperature T in a static magnetic field B_0. This results in an expression for the sensitivity of magnetic resonance:

$$\text{sensitivity} \approx \frac{\hbar \gamma B_0}{2 k_b T}$$

with $\hbar$, the Planck constant, k_b the Boltzmann constant, and γ the gyromagnetic constant of protons. At room temperature in a magnetic field of 3 Tesla (T), this gives us a sensitivity of about 10^{-5} for hydrogen (which has the highest intrinsic sensitivity of all nuclei). This means that only 1 in 100,000 hydrogen nuclei (or protons) contributes to the magnetic resonance signal. For this reason the sensitivity of magnetic resonance is considered to be low and there are not many ways to influence this. The temperature can obviously not be influenced in a living subject, but the magnetic field strength of an MR system can be increased. Alternatively one could change the nucleus of interest, but as the hydrogen nucleus has the highest gyromagnetic ratio, this will always result in a lower sensitivity of the measurement, which complicates spectroscopy of nuclei other than hydrogen.

Because of the low sensitivity of magnetic resonance and the low concentrations of the molecules of interest in spectroscopy, spectroscopists are often forced to repeat the same measurement many times to increase the SNR to acceptable levels. With increasing repetitions, also more noise is acquired. Noise is stochastic and therefore scales with the square root of measurement time. A measurement with twice as many repetitions therefore gives a SNR that is a factor of $\sqrt{2}$ higher. Increasing SNR with additional repetitions by linearly increasing total measurement time is therefore not very efficient but often necessary. If one, for example, needs to increase the SNR of an experiment by a factor of 4, this would increase the total measurement time by a factor of 16!

Another important factor influencing SNR—apart from hardware considerations in radiofrequency coil setup and signal receive chain—is the voxel size. This results in roughly the following relation for SNR:

$$\text{SNR} \propto \gamma B_0 V \sqrt{t_\text{m}}$$

with V the voxel size and t_m is the measurement time. SNR therefore scales roughly linearly with magnetic field strength, which was one of the main reasons to increase this in modern clinical MR systems over the years. The dependence of SNR on voxel size can be understood by considering that the amount of spins that contribute to the NMR signal is proportional to the voxel volume.

Another reason to increase the main magnetic field, which is mainly of interest to spectroscopists, is the increase in chemical shift between different spectroscopic signals at higher field strengths. The distance between peaks of interest in Hertz increases linearly with the magnetic field. If two metabolite peaks display spectral overlap and cannot be distinguished, the peak separation can be increased by increasing the magnetic field strength. This enables more accurate detection of metabolites and in some cases detection of metabolites that would otherwise not be detectable. An example of interest in neuroscience is the separation of signals from neurotransmitters glutamine and glutamate. These signals are not distinguishable at 1.5 T, whereas they are at 7 T (Fig. 3.1).

Localization

When performing spectroscopy in a human subject, we are interested in the spectrum of a certain part of the body or the brain; hence, we need proper localization of the MR signal. Correct localization is not only essential to obtain

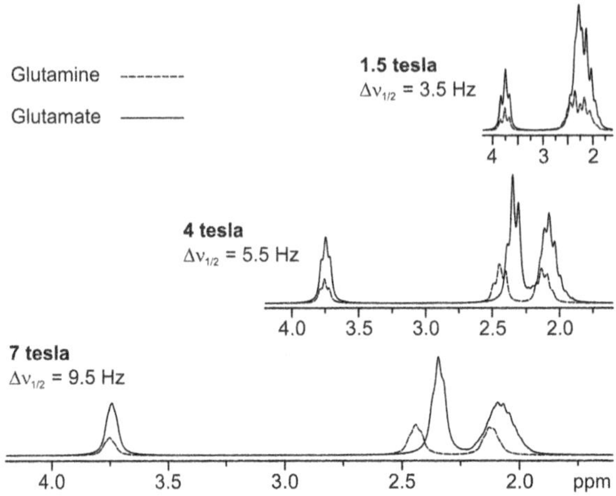

Fig. 3.1 The spectrum of glutamate and glutamine on 1.5, 4, and 7 T. The *x*-axis is scaled with actual frequency. Notice how the glutamine and glutamate peaks around 2.4 ppm are not distinguishable on 1.5 T, whereas a clear distinction can be made on 7 T. Notice also the separation between the peaks at 2.4 and 2.1 ppm of both glutamate and glutamine on 7 T that appear as one peak on 1.5 T. The reason for this is that in the case of 1.5 T, the chemical shift difference between glutamine and glutamate for the peak around 2.4 ppm is small compared to the *J*-coupling effect. Chemical shift scales with the magnetic field strength, whereas *J*-coupling does not. Therefore, at higher field strength, the chemical shift difference between the glutamine and glutamate peaks increases, whereas the width of the peaks does not (the width of the peaks is caused by *J*-coupling in this case and by magnetic field inhomogeneities). Reprinted with permission from [30]

signal from an area of interest, it is also necessary to eliminate signal from outside that area of interest, especially as these unwanted signals originate from water or lipids. For example, when measuring lactate signal in the brain at the chemical shift of 1.3 ppm, it is essential to properly localize the voxel or volume of interest and not receive any signal from subcutaneous lipid tissue surrounding the skull. These lipids have large resonances in the chemical shift range of 0.5–2.0 ppm and can therefore overlap and obscure the lactate signal, as the lipid signals are several magnitudes larger than that of lactate. In spectroscopy we can distinguish two methods for signal localization: single-voxel spectroscopy (SVS) and MR spectroscopic imaging (MRSI, also called chemical shift imaging or CSI).

Single-Voxel Spectroscopy (SVS)

In SVS an MR sequence is used that selects one single volume from which the spectrum is measured. There are several techniques to perform this; three will be discussed here: PRESS, STEAM, and ISIS.

PRESS

Point-resolved spectroscopy (PRESS) [3] localization uses three orthogonal slice-selective radiofrequency (RF) pulses to select a 3D volume. This is depicted in Fig. 3.2, which also shows the PRESS pulse sequence. The sequence starts with a slice-selective 90° pulse that puts the magnetization in the transverse plane. This

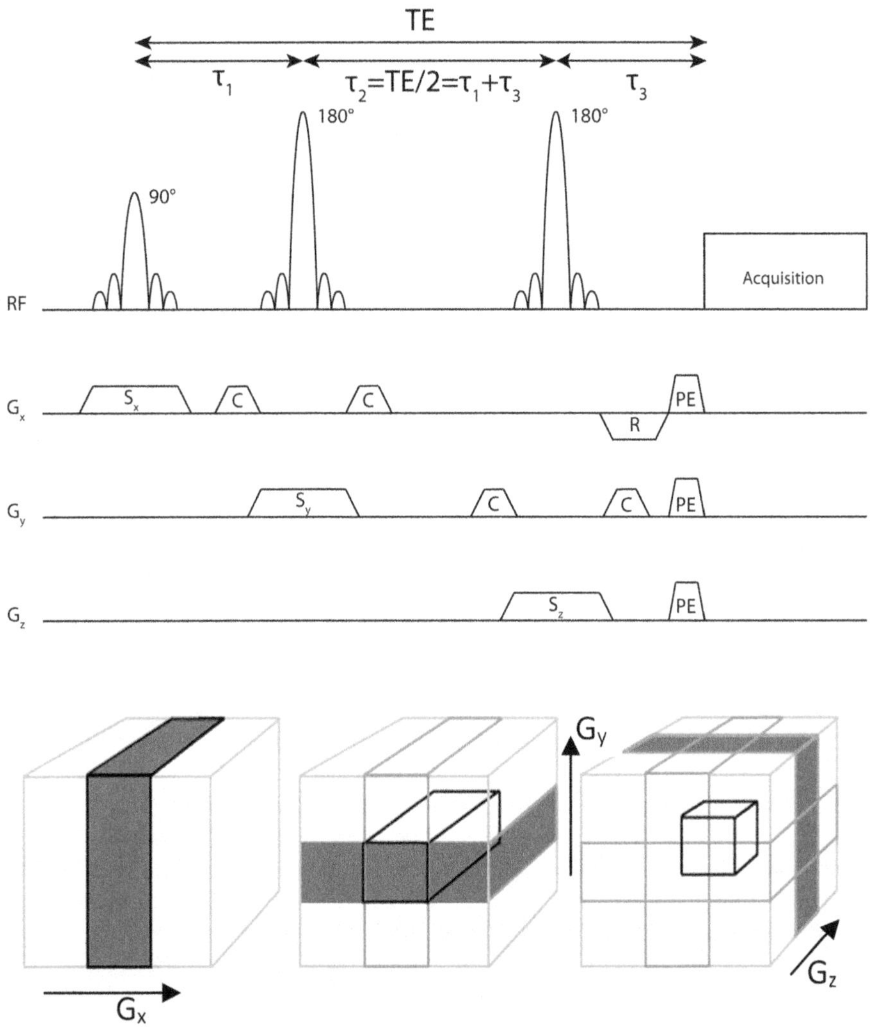

Fig. 3.2 PRESS volume selection with MRSI. Three orthogonal slice-selective RF pulses (one 90° excitation pulse followed by two 180° refocusing pulses) select one volume of interest (*box within bold black lines*). S_x, S_y, and S_z are the slice selection gradients, C are the crusher gradients surrounding the 180° pulses, and PE are the phase encoding gradients that allow spatial localization for MRSI. In SVS mode these PE gradients are not present. The spin echo occurs at time TE, with $\tau_1 + \tau_3 = \tau_2 = TE/2$

pulse selects a slice in one direction. Subsequently, two slice-selective 180° refocusing pulses follow, which invert the spins. These pulses are slice selective in the two other orthogonal directions and, as they invert the magnetization, an echo forms at time $TE = 2 \cdot \tau_2$. This can be understood in the following way. After excitation with the 90° pulse, the magnetization starts dephasing by T2* decay for a time period τ_1. The first 180° pulse inverts the magnetization such that the magnetization starts rephasing in the opposite direction to dephasing that occurred in time period τ_1. This occurs for a time period τ_2. After the second 180° pulse, which again inverts the magnetization, the magnetization rephases in the same direction as in time period τ_1. This means that when $\tau_1 + \tau_3 = \tau_2$, the dephasing of the magnetization by T2* is effectively zero (apart from the T2 decay), and thus a spin echo forms at echo time $TE = 2 \cdot \tau_2$.

By using three orthogonal slice-selective RF pulses, only the magnetization that is located at the intersection of these three orthogonal slices—the volume of interest (VOI)—will form an echo at the abovementioned echo time. Magnetization located outside this intersection will either not form an echo at all or will form an echo at a different time. This way only signals located within this intersection will be measured.

PRESS localization is robust and easy to perform, but has some limitations. One of the main problems is the chemical shift displacement error (CSDE). Signals with different chemical shifts (different positions in the MR spectrum) experience different slice selections: the combination of the bandwidth of the RF pulse, the chemical shift of the metabolite of interest, and the strength of the selective gradient define the exact slice selection for every individual signal. When using large body coils for signal excitation and refocusing, the bandwidths of these conventional RF pulses become quite small. As the chemical shift (in Hertz) increases with magnetic field, the CSDE at 3 T can become very large, if unaccounted for [4]. This causes a mismatch in the selected volume between different metabolites. Consider the following example:

When selecting a VOI of 50 mm with refocusing RF pulses with a bandwidth of 1000 Hz, the effective RF bandwidth per distance is 20 Hz/mm. Suppose we are interested in signals of water (at 4.7 ppm) and lactate (at 1.3 ppm). These signals are separated by 3.4 ppm × 128 MHz = 435 Hz at 3 T. At an RF bandwidth of 20 Hz/mm, this results in a shift in slice selection of these signals of 435/20 = 21.8 mm. Both signals are refocused in a 50 mm slice, but the centers of the two slices are almost 22 mm apart, which is a CSDE of 44 % for these particular signals. With CSDE occurring in both refocusing slices, the position of the volume of interest for the two signals is very different. If the carrier frequency of the slice selection is chosen in between these two signals, the CSDE remains, but both volume selections deviate equally (both 22 % in opposite directions) from the intended volume selection. In our example, a proper choice of carrier frequency for slice-selective pulses would be at an offset from water of half the spectral range of interest (1.75 ppm or 217 Hz offset of water). In Fig. 3.3 the CSDE of a PRESS sequence at 3 T is visualized.

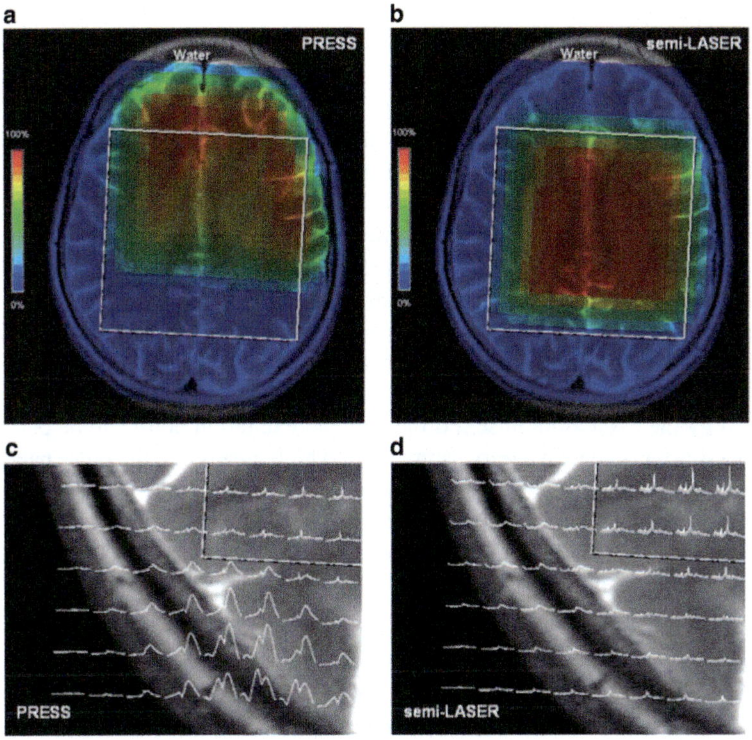

Fig. 3.3 Conventional PRESS compared to the semi-LASER pulse sequence in a healthy volunteer at 3 T. The spectral map of the water signal (at 4.7 ppm) is shown for the PRESS (**a**) and semi-LASER (**b**) approach with a carrier frequency of 2.2 ppm. Notice how the actual selected volume is clearly more displaced for volume selection with a PRESS sequence compared to the semi-LASER sequence. The *white box* indicates the intended volume of interest (VOI), either selected with the PRESS pulses or with the semi-LASER volume selection. The spectral maps show the spectra with range −1.0 to 4.0 ppm of the water-suppressed PRESS (**c**) and semi-LASER (**d**) sequence. Notice residual lipid signals in voxels on the left posterior side of the brain, just outside the VOI that is apparent in the PRESS sequence compared to the semi-LASER sequence. Reprinted with permission from [4]

The reason why PRESS is sensitive to the CSDE is because of its use of conventional refocusing RF pulses, which have an inverse relation between RF pulse time and pulse bandwidth. Large RF transmit coil 180° refocusing pulses generally have quite a long pulse duration, corresponding to a low bandwidth, which makes them more prone to the CSDE.

One of the methods to solve this issue is to refocus magnetization with adiabatic RF pulses, which can have very high bandwidths compared to conventional 180° pulses. This is called the localization by adiabatic selective refocusing (LASER) [5, 6] sequence. The benefit of these types of pulses is that their flip angle is much less sensitive to variations in RF transmit fields (B_1) once the adiabatic condition has been reached and can therefore provide a more homogenous excitation over the volume of interest. These pulses are however very power demanding and need to be played out in pairs to get the same refocusing effect as one conventional 180° refocusing pulse. Moreover, for replacing a conventional slice-selective 90° excitation pulse, three pulses are needed (nonselective excitation and one pair of adiabatic refocusing pulses for one slice selection). This adds up to a total of seven RF pulses within the LASER sequence, instead of the three pulses for PRESS, increasing the minimal echo time in a LASER sequence.

In a so-called semi-LASER [4, 7] sequence, the three first pulses of the full LASER are replaced by one conventional 90° excitation pulse. Shorter echo times can be reached again, and for the refocusing pulses, the advantages of small CSDE and insensitivity to B_1 inhomogeneities remain. It reduces RF power deposition, but it somewhat increases the sequence sensitivity to the CSDE in the direction that is selected by the excitation pulse. Fortunately, an excitation pulse can be performed within a shorter time than a refocusing pulse, which means that the pulse bandwidth can still be kept relatively high (and is therefore not that prone to the CSDE). See Fig. 3.3 for a comparison between the CSDE in the PRESS and semi-LASER sequences.

STEAM

Another widely used method for localization is called stimulated echo acquisition mode (STEAM) [8–10], which is quite similar to PRESS. It also uses three orthogonal slice-selective RF pulses, but all these pulses have a flip angle of 90° (Fig. 3.4). Instead of using the standard spin echo that is caused by a 90° and a 180° RF pulse, it uses the stimulated echo that is caused by three 90° RF pulses. The stimulated echo is then only present in the volume at the intersection of the three slice-selective RF pulses. The actual echo occurs at a time TE/2 after the last RF pulse, in which TE/2 is equal to the time between the first and second RF pulses.

The major advantage of using this sequence is that the magnetization of interest is locked into the longitudinal plane between the second and the third RF pulses. It therefore does not have any T2 decay during this time, which enables measurements at shorter echo times and is especially beneficial for signals with short T2 relaxation times. For signals with a longer T2 time, this will not result in a SNR increase as the stimulated echo only contains half the signal of the spin echo that is used in the PRESS sequence.

One of the other advantages of the STEAM method is that it is less prone to the CSDE than the PRESS method, because 90° RF pulses generally have larger bandwidths than conventional 180° pulses.

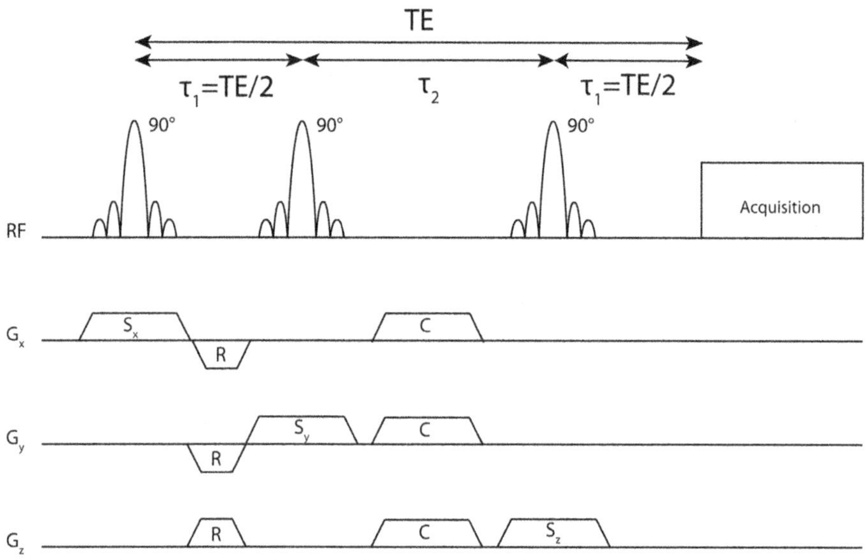

Fig. 3.4 SVS STEAM sequence. Three orthogonal slice-selective 90° RF pulses select one volume. S_x, S_y, and S_z are the slice selection gradients, which all need a refocusing gradient R to compensate signal dephasing by slice selection gradients with a 90° pulse. These refocusing gradients can be placed in between the first and second slice-selective pulses (as shown here) or after the third 90° pulse, before formation of the stimulated echo. Gradients C are the crusher gradients during the mixing time. The stimulated echo occurs at time $TE = 2\tau_1$ with a mixing time, TM, of τ_2. During TM the magnetization is locked into the longitudinal plane and no T2 relaxation occurs. Instead, the magnetization during TM is subject to T1 relaxation. Additional crusher gradient pairs could be placed between the first and second 90° pulse and after the third 90° pulse. When used in MRSI mode, phase encoding gradients can be positioned just before signal acquisition (as is shown in the case for PRESS-MRSI, Fig. *3.2*)

ISIS

Image-selected in vivo spectroscopy (ISIS) [11] is somewhat different from PRESS and STEAM in that it does not directly select a volume. The sequence consists of three orthogonal slice-selective refocusing pulses, which are turned on or off in alternate scans, followed by a 90° excitation pulse after which acquisition starts immediately. With this method signal from the complete volume within the RF coil is acquired, not just the volume of interest. However, the magnetization located within the slices that are selected by the refocusing pulses is inverted, causing signals from that slice to be 180° out of phase. In this sequence one acquisition is performed for all possible combinations of turning the three refocusing pulses on and off. By adjusting the receiver phase to the phase that is expected at the volume of interest, the signal from the volume of interest is always positive, whereas signals from other parts of the body are alternatively positive or negative. In a clever addition of all eight possible acquisitions, this results in the addition of all signals from the volume of interest, whereas signals from outside this region cancel out. In effect one then obtains signal from only the volume of interest.

The advantage of this sequence is that acquisition starts right after the 90° excitation pulse, obtaining the free induction decay (FID) instead of an echo. There is no T2 relaxation from any echo time, which makes this sequence very popular for acquiring signals with the shortest T2.

A major drawback of the sequence is its sensitivity to motion artifacts during the cycle of eight acquisitions: any signal that is not subtracted or added well can lead to large contaminations, especially if the VOI is small compared to the rest of the body. In addition, long repetition times are needed, because the longitudinal magnetization needs to be completely recovered for every repetition.

A hybrid version of ISIS and PRESS is the spin echo full intensity acquired localized (SPECIAL) [12] sequence. In this sequence two orthogonal slices are selected using slice-selective 90° and 180° pulses. The third dimension is selected by a slice-selective 180° pulse, which is applied in alternate scans prior to the 90° excitation pulse and performs the same function as one of the refocusing pulses in the ISIS sequence. As only one direction is selected using a prior refocusing pulse, only two repetitions are needed for volume selection. It is therefore less prone to subtraction artifacts compared to the full ISIS sequence, but it does rely on one refocusing pulse producing a spin echo, although this can be at a very short echo time.

Spectral Editing

The spectral resonances of some metabolites overlap with much larger peaks originating from other metabolites, which make them difficult to visualize and quantify. If these signals are *J*-coupled, elegant ways exist to make use of this coupling from one proton group to another by selectively refocusing the coupling between the resonance groups with additional frequency-selective RF pulses within the normal localization scheme (Fig. 3.5). One example of interest is the measurement of gamma-aminobutyric acid (GABA) in the brain. The concentration of GABA is low in cortical regions (~1 mmol/L), and its spectral pattern largely overlaps with (phospho-)creatine, glutamate, glutamine, and macromolecule resonances. In that particular case, spectral editing can be used to exploit the *J*-coupling of the 3.0 ppm GABA resonance with the GABA resonance at 1.9 ppm to visualize the GABA multiplet at the 3.0 ppm chemical shift position without overlap with other resonances.

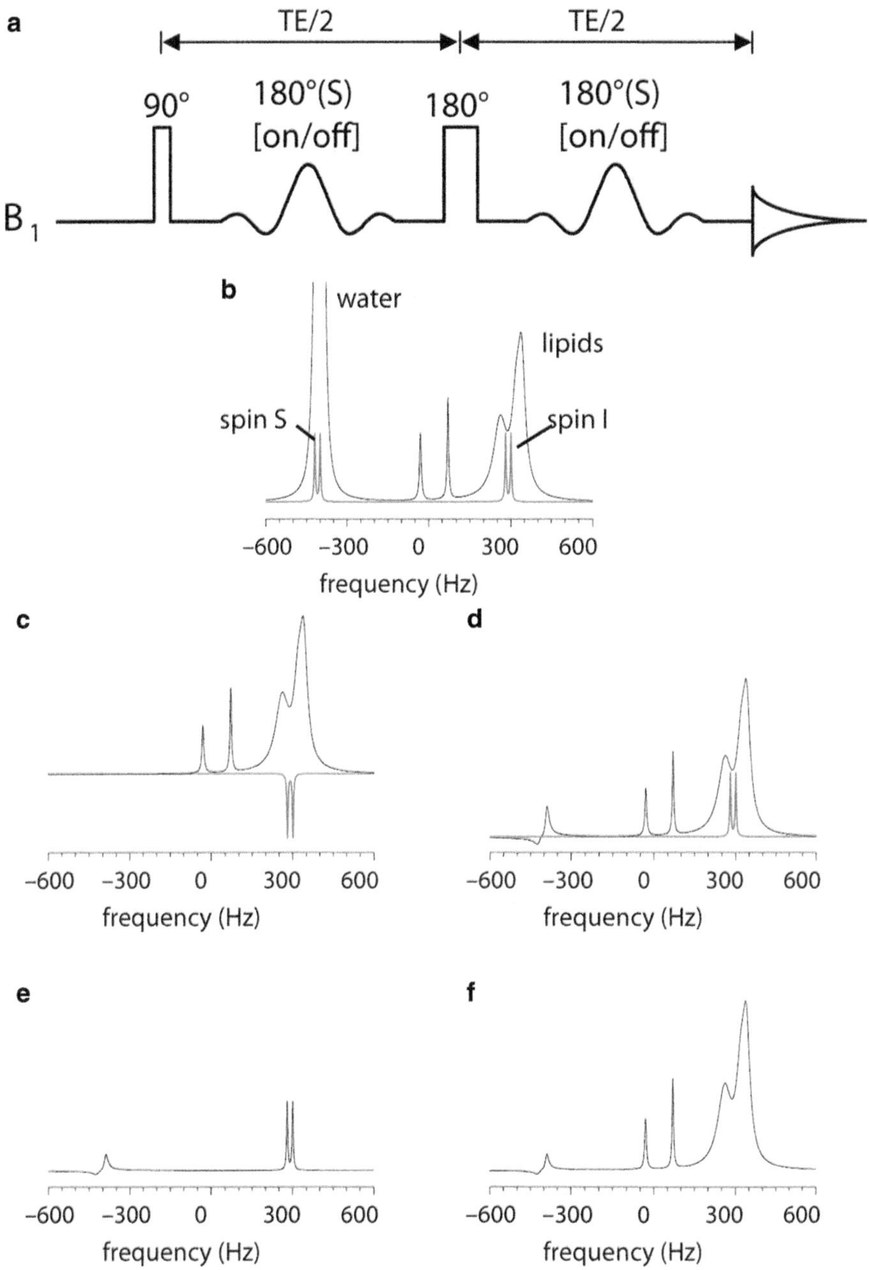

Fig. 3.5 Spectral editing. Consider a metabolite with two resonances that are *J*-coupled, spin S and spin I (see **b**). Both signals overlap with large signals and it is therefore very difficult to visualize and quantify them. (**a**) Depicts a *J*-editing sequence. The 180° (S) pulse selectively inverts the spin S and does nothing with spin I. This RF pulse is alternating between on and off in subsequent acquisitions. When turned off the *J*-coupling is not influenced. The TE is chosen such

MR Spectroscopic Imaging (MRSI)

After localization of a larger VOI with one of the techniques above, the signal within the VOI can be further localized in an imaging approach: MRSI. As traditionally the FID of the signals is used to monitor chemical shift dispersion of different signals, every point in *k*-space (raw data space in imaging acquisition) needs to be acquired with a separate acquisition and is encoded with phase encode gradients in two or three dimensions. This results in a grid, in which every voxel shows the spectrum at that part of the grid. The total measurement time Tacq in a traditional MRSI experiment then becomes:

$$\text{Tacq} = N_a \times N_x \times N_y \times N_z \times \text{TR},$$

in which TR is the repetition time, N_a is the number of averages or repetitions, and $N_{x,y,z}$ is the matrix size in the *x*-, *y*-, and *z*-direction. For even moderately small matrix sizes, this results in long measurement times. For example, with a matrix of 16 by 16 by 16 and TR of 1 s, the measurement would already take more than an hour for 1 average. MRSI matrices are therefore kept small in most cases, which also keeps the spatial resolution of MRSI generally quite low. Several acquisition schemes have been developed to speed up the measurement.

Weighted Elliptical Sampling

By using elliptical sampling, only the *k*-space points within a sphere of radius k_{max} from the center are actually measured. This reduces the measurement acquisition time by about a factor of 3, but it also reduces the effective resolution by about the same factor. This acquisition scheme facilitates weighted averaging of the different phase encoding points, by averaging points that are located close to the center of *k*-space more than the points located far from the center. Because the center of *k*-space contains the highest SNR, it is weighted stronger compared to the edges of *k*-space, increasing the SNR. The weighting function is based upon a certain spatial filter, such as a Hamming filter, to provide optimal SNR [13].

Fig. 3.5 (continued) that TE = 1/*J* = 1/7 Hz = 144 ms and the signal of spin I is completely inverted, which results in spectrum (**c**). If the 180° (S) pulse is turned on, it inverts the magnetization of spin S. This also inverts the *J*-coupling between spin I and S. The time between the two 180° (S) pulses is chosen to be TE/2, which means that the *J*-coupling is completely refocused at the echo time, which results in the spectrum (**d**). Note that the lipid signal is not influenced by the 180° (S) pulse and is therefore equal to the lipid signal in (**c**). If spectrum (**c**) is subtracted from (**d**), this results in spectrum (**e**), in which the lipid signal cancels out and the signal of spin I adds. Now, the signal of interest can be quantified. (**f**) Depicts the addition of spectrum (**c**) and (**d**). Derived from two figures from the book *in vivo NMR spectroscopy*, Wiley, by Robin de Graaf

Acceleration in MRSI Acquisition Schemes

Another method to decrease the long acquisition times associated with MRSI is to accelerate the acquisition scheme. The traditional acquisition scheme described above only samples the spectral dimension during readout. This is inefficient as the spectral dimension can often be sampled with a spectral bandwidth of 1000–2000 Hz, sampling only one or two sample points every millisecond, whereas modern MRI systems can easily obtain 100 samples within 1 ms. Several other acquisition schemes have therefore been developed to sample the spectral dimension and one or two spatial dimensions at the same time, such as echo-planar spectroscopic imaging (EPSI) [14] or spiral MRSI [15].

Volume Selection in MRSI

In MRSI the volume of interest is often substantially greater than the VOI in SVS, as we are interested in the spatial distribution of signals in this case. As signals are spatially localized, one would expect that volume selection is not necessary when performing spectroscopic imaging. This is, however, not at all the case. Matrix sizes have to be kept small, to obtain reasonable measurement times, causing the spatial response function to stretch over the full field of view (see below). Signals from the tissue surrounding the part of the body of interest can then contaminate nearby as well as distant voxels. Especially water and lipid signals can contaminate the spectra of interest. This is, for example, the case in MRSI of the brain. When obtaining the spectra over the whole brain, especially the edges of the brain are located close to lipids. If these are not suppressed, their signal can contaminate voxels in the brain, and since the lipid signals are much stronger than the metabolite signals, they can totally obscure the signals of interest. Accurate localization is therefore also a critical factor in MRSI. The techniques that are used to perform this are the same as the ones that are used to perform SVS.

SNR in MRSI

In MRSI we divide the volume of interest in several voxels in the order of the matrix size. This results in voxels that are generally substantially smaller than the VOI in SVS. As SNR scales linearly with voxel size, SNR becomes even more of an issue in MRSI than in SVS. The long acquisition time of MRSI is therefore also necessary to obtain sufficient SNR. MRSI is always a trade-off between SNR, measurement time, and voxel size. As described earlier, if SNR is low, it takes a lot of additional acquisition time to increase it (scales with the square root of the number of repetitions), so the better choice is to increase voxel size. Accelerated acquisition schemes are also only useful in situations with sufficient SNR.

Spatial Response Function

When viewing spectroscopic imaging data, one is often presented with a spectral grid in which a voxel can be selected for its spectral information. The voxel size depicted in this grid can be quite misleading as the actual voxel size is often much larger. There are two reasons for this. First, the MR system always interpolates the acquisition matrix to the nearest power of 2: 8, 16, 32,... as this results in a more efficient reconstruction with the fast Fourier transformation. The grid therefore does not represent the actual matrix size. The second reason is the use of filters in *k*-space to accommodate efficient weighted elliptical *k*-space sampling and its effect on the width of the spatial response function (SRF). This function determines the smallest detail that an imaging method can resolve and is therefore an accurate measure for spatial resolution. Measures such as the full width at half maximum (FWHM) or integral of the normalized SRF can be used to determine the spatial resolution of the MRSI measurement. In MRSI the SRF of uniformly sampled *k*-space is sinc shaped, but is strongly influenced by filtering (widening the FWHM, decreasing the spatial resolution, but also decreasing contaminating side lobes). This SRF determines the distance from which signals enter the voxels. If the SRF of an experiment, for example, has a FWHM of 10 mm, this means that signals located within a sphere of 10 mm from the center of the voxel are present within that voxel.

Several factors can greatly influence the SRF, such as *k*-space apodization filters and acquisition schemes. Using a 100 % Hamming filter, for example, can increase the actual voxel size by a factor of 3. Therefore, if we would measure a $10 \times 10 \times 10$ matrix that is interpolated to $16 \times 16 \times 16$ and we use a 100 % Hamming filter, the voxels of the spectral grid that the scanner displays are about $(16/10)^{3 \times 3} = 12$ times smaller than the actual voxels. It is therefore important to keep in mind how the different settings of the experiment influence the spatial resolution (Fig. 3.6).

Considerations for SVS and MRSI Sequences

The volume selection of SVS/MRSI depends on the technical capabilities of the MR system and the MR sequence. In many cases artifacts or unwanted magnetization (coherence pathways) may enter the spectra. Using smart sequence designs, these can be reduced as much as possible:

- As discussed before, with the PRESS sequence, we often suffer from the CSDE due to the reciprocal relation between pulse bandwidth and pulse duration. It is therefore advantageous to choose the smallest pulse duration possible within the limits of the RF amplifier and allowed RF power deposition.
- An additional issue with RF pulses is their exact calibration. If refocusing pulses are not exactly 180°, they will excite unwanted magnetization into the transverse plane. To counteract this and the residual magnetization from imperfect slice profiles, the so-called crusher gradients are placed around the

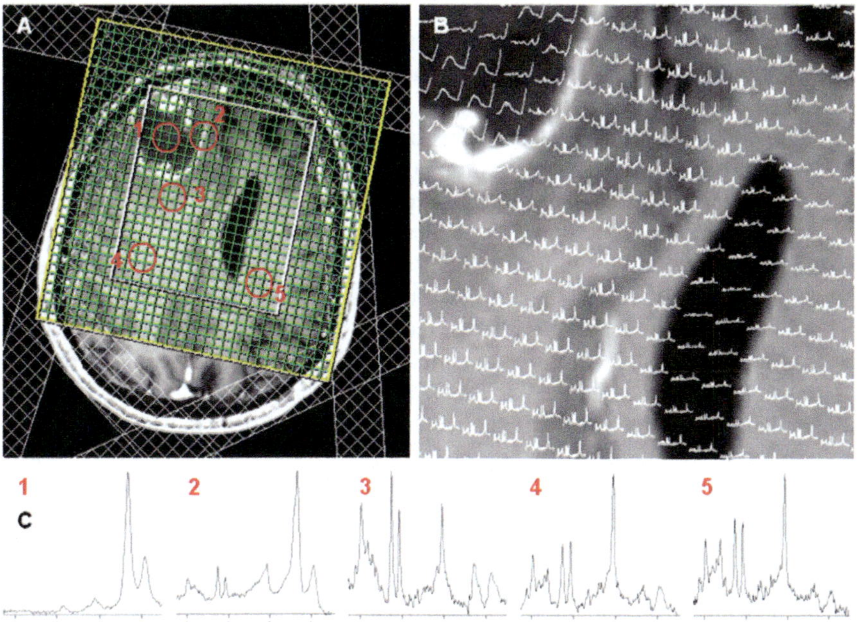

Fig. 3.6 2D MRSI of the brain of a patient with glioblastoma at 3 T. In (**a**) a reconstructed contrast-enhanced axial T1-weighted image of the brain after contrast administration is shown with the location of the outer volume saturation (OVS) slabs over the subcutaneous lipid layer, the volume selection with the semi-LASER pulse sequence in *white*, and the zero-filled acquisition matrix and the FOV in *green*. In (**b**) the range from 1.5 to 4.3 ppm is shown in the individual spectra in the spectral maps overlaid on the T1-weighted image. In (**c**) the spectra from in and around the tumor are shown together with two spectra from the boundaries of the VOI: the true sizes of the corresponding voxel numbers are indicated with *red circles* in (**a**). Measurement parameters: FOV 140 × 140 × 10 mm, matrix size 18 × 18 filled to 32 × 32, acquisition-weighted *k*-space averages 3, TR 2 s, total measurement time 10 min 14 s. Reprinted with permission from [4]

180° pulses. These gradients are of the same magnitude and duration. The magnetization of interest will be dephased by the first gradient before the 180° pulse. After the refocusing pulse, the spins will have inverted, such that the second spoiling gradient after the pulse will exactly rephase the magnetization of interest. Unwanted magnetization that is excited by an incorrect 180° pulse will only be dephased by the gradient pulse after the 180° pulse and therefore spoiled. Only magnetization in the transverse plane is affected by a gradient pulse. This technique is called coherence selection and is used to select only certain coherence pathways.

- During selection of a slice with a symmetrical 90° excitation pulse with its center of gravity in the middle of the pulse and corresponding slice selection gradient, the magnetization is effectively dephased by half the slice selection gradient, which needs rephasing. Slice selections with symmetrical 180° pulses are self-refocusing and do not need this additional rephrasing gradient. Dephasing during excitation can be undone with an additional gradient pulse half the size of the excitation gradient to rephase the initial dephasing. The most effective position in the sequence to perform this gradient pulse is just before the start of the acquisition, after all RF pulses. This ensures that only magnetization that has been excited by the initial excitation pulse is left unspoiled. Magnetization that was not excited into the transverse plane by the slice-selective pulse will undergo additional spoiling by this slice refocusing gradient pulse.

- The phase encoding gradients in MRSI are also best positioned just before acquisition of the MR signal. This ensures that any unwanted residual signal—if present—undergoes phase encoding and will be localized to its spatial origin. If this spatial origin is outside the VOI (e.g., residual lipids), then the unwanted signals will interfere less with signals inside the VOI. If on the other hand phase encoding gradients are positioned early in the RF pulse train, unwanted coherences can spread across all voxels of the MRSI matrix. Localizing unwanted signals to their spatial origins also makes it easier to identify their causes and to apply additional localized signal suppression (see below).
- To optimally attenuate unwanted magnetization, it is best to use different spoiling strengths for the crusher gradients around the different 180° pulses. This ensures optimal coherence selection and prevents recombination of gradients into zero net gradient power. By using prime factors in the spoilers, one can make sure that unwanted magnetization is mostly spoiled.

Water Suppression

As outlined earlier, the concentration of molecules of interest in the brain is in the order of several mmol/L, which is a factor of ~8×10^3 less than the concentration of protons in water itself. In order to reveal the ^{1}H signals from these molecules dissolved in water, the signal of water protons needs to be suppressed. Usually this is done with one or more frequency-selective pulses that excite the water signal and immediately destroy its phase coherence by strong crushing gradients, after which the excitation of the signals of interest can take place.

The simplest way of water suppression is to apply one chemical shift selective (CHESS [16]) RF pulse on the water resonance and spoil the signal with dephasing magnetic field gradient pulses. However, differences or deviations in RF pulse amplitude (B_1) and T1 relaxation times of the water signal make it nearly impossible to fully suppress the water signal with one pulse. Using four frequency-selective excitation pulses with optimized inter-pulse timing and numerically optimized flip angles, derived from a Bloch equation analysis of the longitudinal magnetization over a broad T1 and B_1 range, an improved "water suppression enhanced through T1 effects" (WET [17]) technique was developed. With increasing magnetic field strength, but especially with the use of surface coils for

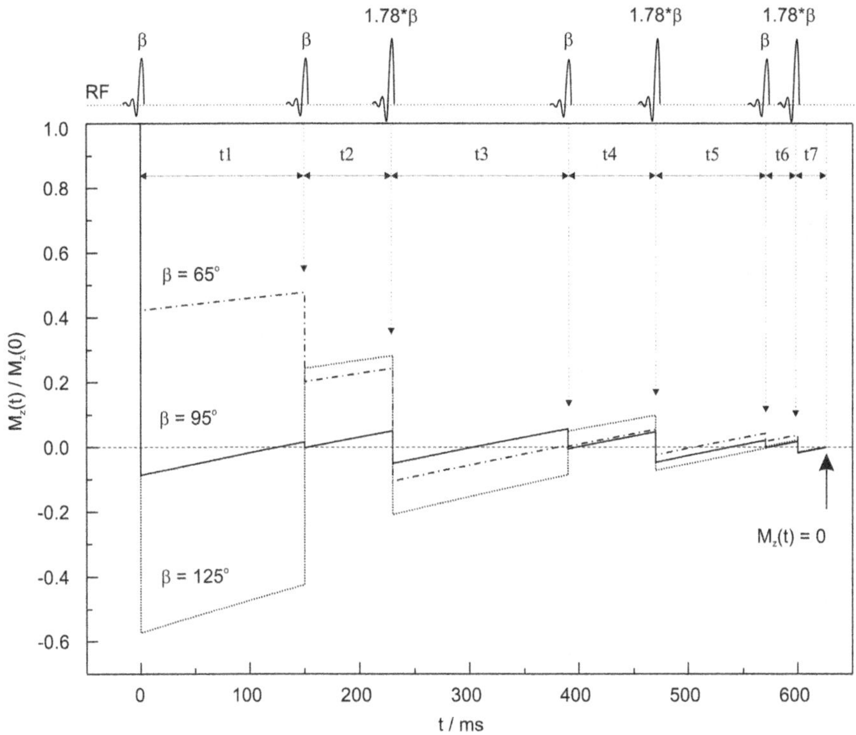

Fig. 3.7 The optimized flip angles and timing of VAPOR. For three different B_1 flip angles (β), the time dependence of the water magnetization $M_z(t)/M_z(0)$ is calculated and equals zero after t7, the moment for excitation of the signals of interest. Optimized time delays t1 = 150 ms, t2 = 80 ms, t3 = 160 ms, t4 = 80 ms, t5 = 100 ms, t6 = 30 ms, t7 = 26 ms. Reprinted with permission from [18]

transmitting the RF signal in, e.g., small animals, the demand rose for a water suppression scheme that was insensitive to an even larger range of B_1 inhomogeneities, which led to another optimized pulse scheme with more pulses: "variable pulse power and optimized relaxation delays" (VAPOR, Fig. 3.7 [18]). This VAPOR water suppression scheme leaves less than 2 % (down to 0 %) of the original water magnetization within a nominal flip angle range from 65 to 125° and a T1 range of water from 1 to 2 s. The crusher gradients between the frequency-selective pulses need to be chosen carefully to not generate unwanted stimulated and spin echoes of the water signal. The bandwidth of the frequency-selective pulses should be small (corresponding to relatively long duration of these RF pulses of, e.g., 20 ms), such that the water suppression scheme does not affect signals in the spectral area of interest. Since it has been around for many years already, the VAPOR scheme has proven to be a reliable and robust water suppression scheme, adopted in many spectroscopy pulse sequences, and remains a preferred method of choice.

Outer Volume Suppression

A relatively simple way to suppress signals at spatial locations is the use of saturation slices (or slabs): a slice-selective excitation of all signals within a slab, followed by a signal dephasing crusher gradient, can suppress the signals from that slice down to a few percent of the original signal. This is a very common way in MR imaging to spatially suppress unwanted signals (e.g., from breathing-related moving tissue causing artifacts throughout an image). This is possible for multiple slices, and if performed outside the desired VOI for spectroscopy, this is known as outer volume suppression (OVS). Positioning of multiple saturation slabs can be done manually or automatically in relation to the position of the VOI.

The exact amount of signal suppression with a saturation slab depends on the accuracy of the flip angle of the pulse, the quality of the slice profile of the slab, and the combination of the T1 of signals in the slab with the time from saturation to excitation of the volume of interest itself. If the intended flip angle of the RF pulse, exciting the slice, is inaccurate, residual magnetization along the z-axis remains in the slice, which can produce residual signal. The slice profile of the RF pulse depends on the shape of the RF pulse in the time domain. Depending on the time allowed for one or more saturation slabs, this duration can be longer, which after optimization can lead to sharp slice profiles. Since there will always be a time between excitation of the signal in the saturation slab and, after dephasing of that signal by crusher gradients, excitation of the volume of interest, T1 relaxation occurs within the saturation slab. This restores magnetization along the z-direction in the slab at a speed depending on the (different) T1 relaxation times of the signals in the slab. Altogether, since slice profiles are not perfect, RF pulse amplitudes have inaccuracies, and T1s of signals in the slab can differ; the amount of signal suppression from an OVS slab is in the order of 95 %. For MR imaging purposes, this can very well be enough, since the signal of interest (water and lipids) in the VOI remains at 100 % intensity. For spectroscopic purposes, suppressing water and/or lipid signals down to ~5 % of the original signal intensity still leaves much more water and lipid signal compared to the spectroscopic signals of interest with their low concentrations. OVS in spectroscopy is therefore only part of multiple solutions to eliminate unwanted signals.

The selection of a volume of interest with a signal localization scheme as described above (Sect. "Localization") is performed with excitation and refocusing pulses that are not perfect. A combination of imperfect slice-selective pulses can cause substantial refocusing of residual signals of water and lipids from outside the VOI, which is why the localization scheme is combined with outer volume suppression pulses. OVS slabs are often positioned manually on the lipid tissue with short T1 around the skull (Fig. 3.6). Positioning of the slabs can also be automated to fit around the volume or voxel of interest, to suppress nearby water signal, or to overcome difficulties with large chemical shift artifacts of low-bandwidth refocusing pulses (fully excited VOI or OVERPRESS [19]). In more advanced schemes, the flip angle of the slab is optimized in combination with the assumed T1 of signals in the slab and the time from slab excitation to VOI excitation. Next to this optimization, the pulse shape of OVS can be optimized to have a sharp slice profile [20], for positioning close to the voxel or VOI without affecting the signal from the voxel itself. Examples exist from full optimization of multiple OVS slabs regarding flip angle, timing, and crusher gradient size and orientation to fully suppress lipid signals from the skull in a whole brain spectroscopic imaging approach [21].

Magnetic Field Homogeneity

The linewidth of spectroscopic signals from protons in molecules is of essential importance for an accurate estimate of the concentration. This spectral linewidth is inversely related to the amount of time the corresponding signal is present in the signal acquisition domain: the longer this signal is present, the narrower the corresponding linewidth in the spectrum. In in vivo spectroscopy, the characteristic decay time of a signal during a free induction decay (the T2*) is proportional to the main magnetic field homogeneity of the corresponding voxel. The magnetic field homogeneity of a certain location depends on the surrounding tissue (lipid, water, tissue interfaces, or presence of air) and can be homogenized by a procedure called shimming. With shimming, gradients in the main magnetic field across the volume or voxel of interest are counteracted by gradients controlled with shim coils. An MRI system is equipped with additional gradient shim coils around its isocenter to control linear (first-order, in three directions) or quadrature (second-order, in five directions) magnetic field gradients. B_0 inhomogeneities in the VOI first need to be measured with a fast mapping method, either with multiple line projections of the B_0 field through the voxel (originally introduced in 1993 [22] and often refined) or with a full 3D B_0 field map [23, 24], which nowadays is often the MR system's standard in a fully automated way [25].

Shimming small voxels or large volumes of interest can require different strategies. Small voxels in SVS are possibly served best by B_0 mapping with line projections in several iterations. Especially in difficult areas of the brain near air-tissue interfaces, higher-order (>2nd order) shimming can be beneficial, although this requires additional hardware (shim coils) with demanding shim current strengths. As the magnetic field homogeneity of a voxel in the brain also varies due to distant motion of the subject caused by breathing, the concept of real-time shimming was introduced to compensate for this [26]. Another recently proposed approach is to monitor the magnetic field continuously by separate small field probes with feedback to control the shim coils [27]. Larger volumes of interest used in MRSI need the current standard of full 3D field mapping to calculate the optimal shim settings. Even though real-time dynamic shim approaches and field probes can also be of aid in MRSI, it will always remain a challenge to achieve similar linewidths in individual voxels of an MRSI examination compared to what is achievable in shimming a single voxel of the same size at one particular location.

Calibrations

Many of the technical considerations in this chapter need subject-specific calibration just before the actual spectroscopic acquisition. Once localization images of the brain of a patient in the MR system have been acquired, usually (magnetization-prepared) T1- and/or T2-weighted images provide an overview of the anatomy of the brain. On this anatomical image, the location of a single voxel for SVS or volume of interest for MRSI can be planned. Most scanners allow the identification of a calibration volume (this can be of the same size as the voxel or volume of interest), on which shimming can be performed with either of the strategies described above (Sect. "Magnetic Field Homogeneity"). Once the linewidth of the water signal of the calibration volume is as narrow as possible (either automated, perhaps iterated, or manually readjusted), the exact frequency of this water signal needs to be established. Subsequently, the RF transmit power (B_1) needs readjustment for the VOI. Since different patients have different head and body sizes, they "load" the MR coil of the system differently, which is why the RF transmitter of the system needs to be adjusted for every individual. Although this is done automatically as the first step when a patient has moved into the MR system based on water signal, an accurate estimate of the amount of RF power needed for the desired flip angle of the spectroscopy pulse sequence at the calibration volume requires a readjustment of the RF transmitter.

Once the RF transmitter readjustment is performed, its value can be used to set the RF amplitudes of the OVS pulses and of the water suppression pulses. Alternatively, since the quality of shimming has influence on the water suppression, the latter can be adjusted in an independent step to minimize the residual water signal in the calibration volume. In case of SVS, a water reference scan can be acquired with a small number of repetitions to establish the lineshape and a reference amplitude for the amount of water in the voxel. The spectral integral of this water reference signal can be used to normalize the metabolite signals of interest derived from the metabolite scan. It can also serve as the basis to add the spectroscopic signals from different receive elements of the used RF coil in a phase-coherent and weighted way. The metabolite scan itself is performed with multiple repetitions in a phase cycle, with an offset in the carrier frequency of the slice-selective excitation and refocusing pulses to minimize CSDE for the spectral range of interest. Addition of signals from different RF coil elements is based on the water reference scan. In MRSI, acquisition of a water reference scan takes too much time; therefore, no information regarding lineshape and integral of the water signal is available, and quantification and coil signal addition needs to be based on (a combination of) signals in the metabolite spectra.

MRS Quantification

As the signal intensity of spectral peaks is proportional to the number of spins that contribute to the magnetization, MR spectroscopy can be used to quantify concentrations of metabolites. In in vivo MRS, this results in a concentration in µmol/g of the tissue. By assuming a certain concentration for one of the MR signals, absolute quantification of resonances from other metabolites is possible. Because the concentration of water in the brain tissue is generally known and is relatively constant, this signal is often used to quantify other metabolites. The procedure for absolute quantification of metabolites by MRS is the following:

- Measure a water-suppressed spectrum of the voxel of interest to acquire the metabolite signals.
- Measure a water spectrum from the same voxel with little T1 and T2 weighting and no CSDE (frequency of slice-selective pulses centered on water).
- Process both spectra and determine the relative metabolite intensities to the water intensity.
- Correct for T1 and T2 relaxation, scalar coupling, and number of protons in the chemical moiety.
- Determine absolute concentration of metabolites by relating it to the assumed water concentration.

Processing and Fitting of Spectra

First the spectral data need to be assessed for quality. Bad quality spectra (distorted baseline, low SNR, bad lineshapes, large residual water or lipid signals) can lead to inaccurate quantification and therefore need to be excluded from the analysis. Then the spectra need to be post-processed and fitted by software such as LCModel [28] or jMRUI [29]. These software packages use a basis set of spectral shape model files, i.e., prior knowledge in resonance position and shapes, to fit the best combination of model shapes to the acquired data. The best fit provides the relative contributions of the different resonances of interest to the spectrum. However, prior to fitting the software packages perform several corrections to the spectrum. First, they perform zero- and first-order phase corrections. Due to experimental imperfections, additions of data from different receive elements of multichannel RF coils, and delays in the start of acquisition, the spectra often contain zero- and first-order phase errors. Second, due to the local chemical environment of the molecules in the voxel or VOI and due to imperfections in the model files or prior knowledge, the spectral peaks are often not exactly on the expected chemical shift position, requiring small chemical shift corrections. Third, spectral lineshapes in in vivo MRS most often do not adhere to exact theoretical lineshapes, due to magnetic field inhomogeneities or eddy currents, which need to be corrected for. A fourth correction that is important in short TE spectroscopy (below ~40 ms) is dealing with the macromolecular baseline. Broad macromolecular signals disturb a flat baseline and can be removed by either using a polynomial fitting procedure or by measuring the actual experimental baseline in a number of volunteers and including the baseline as a model signal in the basis set. The latter method is often more robust. At longer TEs (>80 ms), the macromolecular baseline is almost absent.

Corrections for Absolute Quantification

When the metabolite signal intensities are determined using fitting software, additional corrections are necessary to obtain the absolute concentrations of the metabolites. These consist of T1 and T2 corrections and an assumption of the water content of the tissue at hand. These factors substantially influence the accuracy of the measurement and it is important to provide accurate estimates of T1 and T2 times and tissue water content. Many examples in literature often use TEs longer than 30 ms and up to 144 ms (for measuring lactate). In these cases the difference in T2 times of the metabolites of interest and water can have a large influence on the estimated concentrations. It is important to note the errors in the estimation of the concentrations in the final reports of metabolite concentrations. Fitting software such as LCModel provide the Cramér-Rao lower bounds, which are based on the Fisher information matrix, as estimates of the quantification errors.

A number of corrections are necessary to obtain absolute concentrations for all metabolites. The accuracy of these estimates is largely dependent on the correction factors and performance of the fitting software, and therefore these estimates need to be interpreted with care. The final metabolite concentrations are given in μmol/g of the tissue, which means that they represent a characteristic of all the tissue included in the voxel and do not distinguish intra- or extracellular metabolite levels.

Future Perspectives

For further adoption in clinical routine, SVS and MRSI acquisition protocols need to be standardized across different scanner manufacturer platforms, as well as automated to allow acquisition of high-quality spectra without assistance of an MR physicist. In clinical routine, an 8- or 10-min measurement during a patient examination should not fail; therefore, the MRS protocol needs to be robust and reliable. Post-processing and quantification also needs to be standardized and preferably automated to remove operator bias, and the output for the clinical end user (the radiologist) should be either a table with metabolite concentrations (for SVS) or maps with metabolite concentrations or ratios superimposed on an anatomical background image (for MRSI). If MRSI data acquisition is standardized across different MR system platforms (but with room for differentiation in spatial resolution, echo time, editing, etc. between different diseases), and post-processing is performed in a standardized fashion, metabolite concentration or ratio maps can become part of the radiologists' toolbox to diagnose and evaluate different neurological diseases.

References

1. Oz G, Alger JR, Barker PB, Bartha R, Bizzi A, Boesch C, Bolan PJ, Brindle KM, Cudalbu C, Dincer A, Dydak U, Emir UE, Frahm J, Gonzalez RG, Gruber S, Gruetter R, Gupta RK, Heerschap A, Henning A, Hetherington HP, Howe FA, Huppi PS, Hurd RE, Kantarci K, Klomp DW, Kreis R, Kruiskamp MJ, Leach MO, Lin AP, Luijten PR, Marjanska M, Maudsley AA, Meyerhoff DJ, Mountford CE, Nelson SJ, Pamir MN, Pan JW, Peet AC, Poptani H, Posse S, Pouwels PJ, Ratai EM, Ross BD, Scheenen TW, Schuster C, Smith IC, Soher BJ, Tkac I, Vigneron DB, Kauppinen RA, Group MRSC (2014) Clinical proton MR spectroscopy in central nervous system disorders. Radiology 270(3):658–679. doi:10.1148/radiol.13130531
2. Andronesi OC, Kim GS, Gerstner E, Batchelor T, Tzika AA, Fantin VR, Vander Heiden MG, Sorensen AG (2012) Detection of 2-hydroxyglutarate in IDH-mutated glioma patients by in vivo spectral-editing and 2D correlation magnetic resonance spectroscopy. Sci Transl Med 4(116):116ra114. doi:10.1126/scitranslmed.3002693
3. Bottomley PA (1987) Spatial localization in NMR spectroscopy in vivo. Ann N Y Acad Sci 508:333–348
4. Scheenen TW, Klomp DW, Wijnen JP, Heerschap A (2008) Short echo time 1H-MRSI of the human brain at 3T with minimal chemical shift displacement errors using adiabatic refocusing pulses. Magn Reson Med 59(1):1–6. doi:10.1002/mrm.21302
5. Garwood M, DelaBarre L (2001) The return of the frequency sweep: designing adiabatic pulses for contemporary NMR. J Magn Reson 153(2):155–177. doi:10.1006/jmre.2001.2340
6. Slotboom J, Mehlkopf AF, Bovee WMMJ (1991) A single-shot localization pulse sequence suited for coils with inhomogeneous Rf fields using adiabatic slice-selective Rf pulses. J Magn Reson 95(2):396–404. doi:10.1016/0022-2364(91)90229-M
7. Oz G, Tkac I (2011) Short-echo, single-shot, full-intensity proton magnetic resonance spectroscopy for neurochemical profiling at 4 T: validation in the cerebellum and brainstem. Magn Reson Med 65(4):901–910. doi:10.1002/mrm.22708
8. Bruhn H, Frahm J, Gyngell ML, Merboldt KD, Hanicke W, Sauter R (1991) Localized proton NMR spectroscopy using stimulated echoes: applications to human skeletal muscle in vivo. Magn Reson Med 17(1):82–94
9. Frahm J, Bruhn H, Gyngell ML, Merboldt KD, Hanicke W, Sauter R (1989) Localized high-resolution proton NMR spectroscopy using stimulated echoes: initial applications to human brain in vivo. Magn Reson Med 9(1):79–93
10. Haase A, Frahm J, Matthaei D, Hanicke W, Bomsdorf H, Kunz D, Tischler R (1986) MR imaging using stimulated echoes (STEAM). Radiology 160(3):787–790. doi:10.1148/radiology.160.3.3737918
11. Ordidge RJ, Bowley RM, McHale G (1988) A general approach to selection of multiple cubic volume elements using the ISIS technique. Magn Reson Med 8(3):323–331
12. Mlynarik V, Gambarota G, Frenkel H, Gruetter R (2006) Localized short-echo-time proton MR spectroscopy with full signal-intensity acquisition. Magn Reson Med 56(5):965–970. doi:10.1002/mrm.21043
13. Adalsteinsson E, Star-Lack J, Meyer CH, Spielman DM (1999) Reduced spatial side lobes in chemical-shift imaging. Magn Reson Med 42(2):314–323
14. Duyn JH, Moonen CT (1993) Fast proton spectroscopic imaging of human brain using multiple spin-echoes. Magn Reson Med 30(4):409–414

15. Adalsteinsson E, Irarrazabal P, Topp S, Meyer C, Macovski A, Spielman DM (1998) Volumetric spectroscopic imaging with spiral-based k-space trajectories. Magn Reson Med 39 (6):889–898
16. Haase A, Frahm J, Hanicke W, Matthaei D (1985) 1H NMR chemical shift selective (CHESS) imaging. Phys Med Biol 30(4):341–344
17. Ogg RJ, Kingsley PB, Taylor JS (1994) WET, a T1- and B1-insensitive water-suppression method for in vivo localized 1H NMR spectroscopy. J Magn Reson B 104(1):1–10
18. Tkac I, Starcuk Z, Choi IY, Gruetter R (1999) In vivo ^{1}H NMR spectroscopy of rat brain at 1 ms echo time. Magn Reson Med 41(4):649–656
19. Li Y, Osorio JA, Ozturk-Isik E, Chen AP, Xu D, Crane JC, Cha S, Chang S, Berger MS, Vigneron DB, Nelson SJ (2006) Considerations in applying 3D PRESS H-1 brain MRSI with an eight-channel phased-array coil at 3 T. Magn Reson Imaging 24(10):1295–1302. doi:10.1016/j.mri.2006.07.012
20. Tran TKC, Vigneron DB, Sailasuta N, Tropp J, Le Roux P, Kurhanewicz J, Nelson S, Hurd R (2000) Very selective suppression pulses for clinical MRSI studies of brain and prostate cancer. Magn Reson Med 43(1):23–33. doi:10.1002/(Sici)1522-2594(200001)43:1<23::Aid-Mrm4>3.0.Co;2-E
21. Henning A, Fuchs A, Murdoch JB, Boesiger P (2009) Slice-selective FID acquisition, localized by outer volume suppression (FIDLOVS) for ^{1}H-MRSI of the human brain at 7 T with minimal signal loss. NMR Biomed 22(7):683–696. doi:10.1002/nbm.1366
22. Gruetter R (1993) Automatic, localized in vivo adjustment of all 1st-order and 2nd-order shim coils. Magn Reson Med 29(6):804–811. doi:10.1002/mrm.1910290613
23. Li SZ, Dardzinski BJ, Collins CM, Yang QX, Smith MB (1996) Three-dimensional mapping of the static magnetic field inside the human head. Magn Reson Med 36(5):705–714. doi:10.1002/mrm.1910360509
24. Gruetter R, Tkac I (2000) Field mapping without reference scan using asymmetric echo-planar techniques. Magn Reson Med 43(2):319–323. doi:10.1002/(Sici)1522-2594(200002)43:2<319::Aid-Mrm22>3.0.Co;2-1
25. Hetherington HP, Chu WJ, Gonen O, Pan JW (2006) Robust fully automated shimming of the human brain for high-field H-1 spectroscopic imaging. Magn Reson Med 56(1):26–33. doi:10.1002/Mrm.20941
26. van Gelderen P, de Zwart JA, Starewicz P, Hinks RS, Duyn JH (2007) Real-time shimming to compensate for respiration-induced B-0 fluctuations. Magn Reson Med 57(2):362–368. doi:10.1002/Mrm.21136
27. Wilm BJ, Duerst Y, Dietrich BE, Wyss M, Vannesjo SJ, Schmid T, Brunner DO, Barmet C, Pruessmann KP (2014) Feedback field control improves linewidths in in vivo magnetic resonance spectroscopy. Magn Reson Med 71(5):1657–1662. doi:10.1002/Mrm.24836
28. Provencher SW (1993) Estimation of metabolite concentrations from localized in vivo proton NMR spectra. Magn Reson Med 30(6):672–679
29. Naressi A, Couturier C, Devos JM, Janssen M, Mangeat C, de Beer R, Graveron-Demilly D (2001) Java-based graphical user interface for the MRUI quantitation package. MAGMA 12 (2–3):141–152
30. Tkac I, Andersen P, Adriany G, Merkle H, Ugurbil K, Gruetter R (2001) In vivo ^{1}H NMR spectroscopy of the human brain at 7 T. Magn Reson Med 46(3):451–456

Chapter 4
Magnetic Resonance Spectroscopy in Dementia

Jonathan Graff-Radford and Kejal Kantarci

Abstract Neuroimaging biomarkers are increasingly being used in clinical practice for early diagnosis and differential diagnosis of dementia and in clinical trials as an outcome measure. Proton magnetic resonance spectroscopy has shown promise in dementia as a diagnostic biomarker with the ability to detect preclinical disease and amnestic mild cognitive impairment and provide ancillary information to distinguish among dementia subtypes. Alzheimer's disease is characterized by decreased *N*-acetylaspartate-to-creatine (NAA/Cr) and elevated *myo*-inositol-to-creatine (mI/Cr) levels. Dementia with Lewy bodies is characterized by normal NAA/Cr levels in the posterior cingulate and elevated choline-to-creatine (Cho/Cr). Vascular dementia demonstrates decreased NAA/Cr but preserved Cho/Cr and mI/Cr in the posterior cingulate. Despite promising studies, MRS is not routinely used in the evaluation for dementia in clinical practice. Improving knowledge of the pathological basis of the metabolite ratio abnormalities, longitudinal studies, and better standardization of the MRS technique may improve the application in dementia.

Keywords Magnetic resonance spectroscopy • Mild cognitive impairment • Alzheimer's disease • Dementia with Lewy bodies • Vascular dementia • Frontotemporal lobar degeneration • Posterior cingulate

Introduction

Neuroimaging biomarkers are increasingly being used in clinical practice for early diagnosis and differential diagnosis of dementia and in clinical trials as an outcome measure. Accurate diagnosis of dementia is not only important for treatment and prognosis but also because neurotherapeutics has moved to targeting underlying protein deposits that characterize the dementia syndrome [1–3]. Because of the

J. Graff-Radford, M.D.
Department of Neurology, Mayo Clinic, Rochester, MN, USA
e-mail: GraffRadford.Jonathan@mayo.edu

K. Kantarci, M.D., M.S. (✉)
Department of Radiology, Mayo Clinic, Rochester, MN, USA
e-mail: kantarci.kejal@mayo.edu

G. Öz (ed.), *Magnetic Resonance Spectroscopy of Degenerative Brain Diseases*,
Contemporary Clinical Neuroscience, DOI 10.1007/978-3-319-33555-1_4

inaccuracies of clinical diagnosis, biomarkers help distinguish underlying pathology and aid clinical diagnosis. For example, while clinical diagnosis matches underlying pathology with adequate sensitivity in the behavioral variant of frontotemporal lobar degeneration (FTLD) [4], sensitivity of clinical diagnosis of dementia with Lewy bodies (DLB) to autopsy confirmation is 32 % and decreases to 12 % for cases of mixed Alzheimer's disease (AD) and DLB pathology [5]. Cortical atrophy on structural brain MRI correlates with the neurofibrillary tangle pathology of AD [6–11]. Specifically, hippocampal atrophy distinguishes clinically diagnosed DLB from AD [12]. In patients with clinically diagnosed DLB, hippocampal atrophy predicts higher Braak-NFT stage [13]. However, structural MRI has limitations. Patients with neurodegenerative disease often have multiple pathologies [14], and patients with mixed AD and DLB pathology may be misclassified as just AD if the diagnosis is made based on structural MRI alone. $[^{18}F]_2$-fluoro-deoxy-D-glucose (FDG) PET may be useful in distinguishing FTLD and AD, especially when diagnostic uncertainty exists, although frontal hypometabolism on PET has been reported in the frontal variant of AD demonstrating functional imaging with PET may not predict underlying pathology in some of the cases [15, 16]. Occipital lobe hypometabolism on FDG PET distinguishes pathologically confirmed DLB from AD [17, 18]. This distinction is useful but limited because occipital metabolism is significantly impaired in posterior cortical atrophy, an atypical form of AD [19], and decreased in advanced AD [20]. Overall, several modalities show promise and are being used in the clinic for early diagnosis and differential diagnosis of dementia including C-11 Pittsburgh compound B (PiB) PET scans and CSF biomarkers such as amyloid beta (Aβ) and tau. Proton magnetic resonance spectroscopy (MRS) may provide additional information and offer an alternative to the less accessible PET techniques that also require radioactive injections.

Background on Disease Pathology

AD is a neurodegenerative disease characterized at autopsy by deposition of tau in neurofibrillary tangles and plaques consisting of extracellular amyloid. Typical AD initially presents with episodic memory impairment consistent with early involvement of the medial temporal lobe with the neurodegenerative pathology. As the neurodegenerative pathology progresses to involve association cortices, semantic memory, executive function, and language become involved. Primary motor and sensory cortices are affected late in the course. Aβ deposition in the brain occurs in a stereotyped pattern initially in the cortex, followed by the hippocampus, before reaching the brain stem and cerebellum [21]. Tau deposition also follows a stereotyped pattern starting in the medial temporal lobes before spreading to the limbic system followed by involvement of the association cortices with the neurofibrillary tangle pathology [22]. Topographic distribution of the neurofibrillary tangle pathology mirrors the topographic distribution of neurodegenerative changes and atrophy in AD [23].

DLB is a neurodegenerative disorder characterized by abnormal deposition of α-synuclein protein in Lewy bodies. The presence of Lewy bodies in limbic and cortical regions differentiates DLB from Parkinson's disease. Significant overlap between AD and DLB pathology occurs with a significant proportion of DLB patients having amyloid plaques. The clinical presentation of DLB includes dementia with hallucinations, parkinsonism, fluctuations, REM sleep behavior disorder, and severe sensitivity to neuroleptics.

Cerebrovascular disease is another common pathology observed in patients with dementia in autopsy series. In most cases however, vascular pathology coexists with the pathology of AD, and pure vascular pathology is relatively uncommon. Vascular lesions, specifically microinfarcts, are more common in patients with dementia than cognitively normal elderly [24]. In a patient with the clinical diagnosis of AD and cerebrovascular disease, the challenge is to determine how much, if any, of the two pathologies contribute to dementia.

Frontotemporal dementia (FTD) refers to clinical syndromes characterized by degeneration of the frontal and temporal lobes. Three clinical variants of FTD exist: behavioral variant FTD (bvFTD), semantic variant primary progressive aphasia (svPPA), and progressive agrammatic/nonfluent aphasia (PNFA). BvFTD is characterized by a progressive change in apathy, disinhibition, and socially inappropriate behavior. SvPPA is a fluent aphasia with severe anomia and loss of word knowledge. PNFA is a nonfluent aphasia characterized by agrammatism. Pathologic diagnoses are based on the primary protein deposited in the brain. The two most common proteins found include the microtubule-associated protein tau and the TAR DNA binding protein of 43 kDa with the fused in sarcoma protein being less common [25].

MRS in Presymptomatic Alzheimer's Disease (AD)

Recent evidence suggests that in patients with familial mutations for AD, Aβ deposition occurs approximately 15 years before symptom onset [26]. This, coupled with the failure of treatment with monoclonal antibodies against Aβ peptides, has shifted the focus of AD research to earlier intervention [27]. To facilitate research targeting earlier identification and treatment of AD, the National Institute on Aging and the Alzheimer's Association developed recommendations of stages of preclinical AD [28]: Stage 1 consists of asymptomatic cerebral amyloidosis (positive PIB or low CSF Aβ 42); stage 2 consists of amyloidosis with evidence of neurodegeneration (hypometabolism on FDG PET, cortical atrophy on MRI, and elevated tau levels in the CSF); and stage 3 pertains to cerebral amyloidosis, evidence of neurodegeneration, and trace cognitive impairment without meeting MCI criteria [29]. The data on MRS in preclinical AD is limited but promising. In patients with normal cognition, those with a higher choline-to-creatine (Cho/Cr) ratio in white matter voxels above the lateral ventricles were at increased risk of a cognitive decline [30]. In asymptomatic Alzheimer mutation carriers,

N-acetylaspartate-to-creatine ratio (NAA/Cr) and NAA-to-*myo*-inositol ratio (NAA/mI) were decreased compared to controls with the size of the difference predicting onset age [31]. In a population-based study of 311 cognitively normal older adults from the Mayo Clinic Study of Aging, increased PiB-PET amyloid deposition correlated with increased mI/Cr and Cho/Cr in the posterior cingulate gyrus (Fig. 4.1). Independent of PIB amyloid load, Cho/Cr elevation correlated

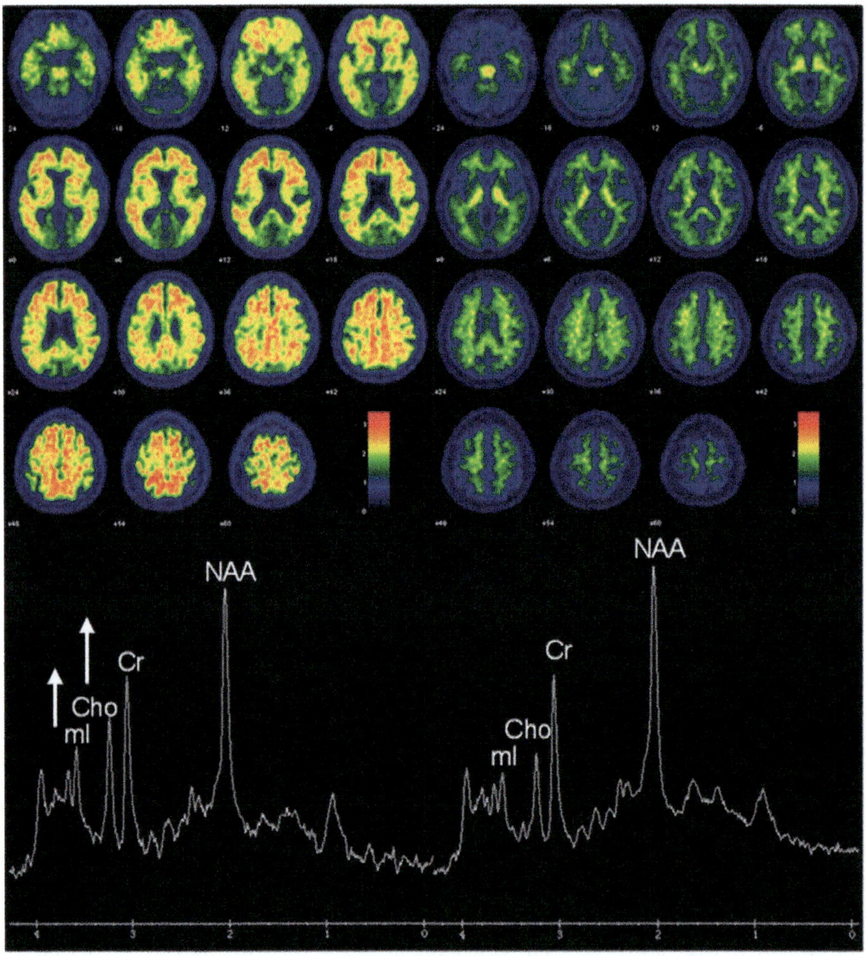

Fig. 4.1 Association between MRS metabolite ratios and cortical Pittsburgh compound-B (PiB) retention ratio on PET. The *left panel* shows the PiB retention ratio PET images in an 82-year-old man with a PiB retention ratio of 2.42, NAA/Cr ratio of 1.51, mI/Cr ratio of 0.58, and Cho/Cr ratio of 0.76. The *right panel* shows the PiB retention ratio in a 78-year-old woman with a PiB retention ratio PET images of 1.27, NAA/Cr ratio of 1.69, mI/Cr ratio of 0.45, and Cho/Cr ratio of 0.54. The spectra are scaled to the Cr peak. High PiB retention is associated with significantly elevated mI/Cr and Cho/Cr ratio and a trend of lower NAA/Cr ratio (With permission from Neuropsychiatric Disease and Treatment) [106]

with worse performance on memory and attention tasks in this cognitively normal cohort. Since Cho/Cr correlated with cognitive performance independent of amyloid in addition to predicting cognitive decline in cognitively normal individuals, MRS shows promise as an additional biomarker for preclinical stages of AD.

MRS in Mild Cognitive Impairment (MCI)

Mild cognitive impairment represents a state of cognitive dysfunction with preserved activities of daily living prior to the onset of dementia. Patients with MCI are at a 10–15 % risk of converting to dementia each year highlighting the importance of studying MCI, particularly for identifying early imaging biomarkers for neurodegenerative dementias [32]. The cognitive impairment associated with MCI can be amnestic, often associated with lower hippocampal volumes, representing prodromal AD, or non-amnestic, associated with preserved hippocampal volumes, which most often represents the prodromal stage of other types of dementia including DLB and vascular dementia (VaD) [33]. In conjunction with lower hippocampal volumes, elevated mI/Cr can distinguish amnestic MCI from non-amnestic MCI [34]. Non-amnestic MCI has normal MRS metabolite ratios, normal hippocampal volumes, but greater proportion of strokes detected on MRI [35]. The MRS changes in amnestic MCI showing decreased NAA/Cr are similar but smaller compared to the MRS findings in AD [36–38]. Compared to controls, subjects with MCI have been shown to have reduced NAA/Cr in the medial temporal lobe, hippocampus, and parietal gray matter with reduced memory performance correlating with lower NAA/Cr in the medial temporal lobe [37]. MRS may also have a role in predicting conversion from MCI to dementia. In a study of 151 MCI subjects, lower posterior cingulate NAA/Cr in addition to hippocampal volumes and infarction predicted conversion to dementia better than any single MRI biomarker alone [34]. A low paratrigonal white matter NAA/Cr also predicted conversion from MCI to dementia [34, 39]. A decline in Cho/Cr predicted stable MCI, while an increase predicted conversion [40]. While most MRS studies performed have been cross-sectional, in a longitudinal study, a decline in NAA/Cr in the posterior cingulate predicted a cognitive decline [40]. In a population-based MRI/MRS study, hippocampal volumes and posterior cingulate NAA/mI independently predicted conversion from MCI to dementia [41]. Since a minority of AD cases may have atypical pathology such as sparing of the hippocampus [42], MRS from the posterior cingulate gyrus may have a complementary role to hippocampal volumetry in predicting progression to dementia in MCI [39, 43–45].

MRS in Alzheimer's Disease (AD)

Interest for MRS as a biomarker for AD began in 1992 when Klunk et al. demonstrated decreased NAA/Cr on autopsied AD brains correlated with Alzheimer's pathology [46]. Subsequently, many studies have shown a consistent AD MRS profile of decreased NAA and elevated mI [38, 47–50]. NAA is considered a marker of neuronal integrity supported by the finding that decreased concentrations of NAA correlated with decreased neuronal density in the cortex of AD subjects [51]. Not surprisingly, the MRS pattern between MCI and AD differs following a similar pattern to the progression of neurofibrillary tangle pathology from limbic to cortical [22]. Compared to MCI subjects, AD subjects had a greater NAA/Cr decline in superior temporal lobe and posterior cingulate voxels [38]. As AD becomes more advanced, decreased NAA/Cr becomes widespread involving the parietal, temporal, frontal, and eventually the occipital lobe [38, 52–55]. While NAA/Cr is consistently decreased in AD compared to normal controls, it is not specific for AD and can be found to be decreased in other neurodegenerative diseases, normal pressure hydrocephalus, and AIDS-related cognitive decline [56–58]. NAA levels in AD can improve with acetylcholinesterase inhibitor treatment [59], but not all studies have confirmed this finding [60].

Shortly after Klunk described decreased NAA/Cr in AD, in 1993 mI/Cr was found to be elevated in AD [48]. In a multivoxel analysis of the superior temporal lobe, posterior cingulate gyri, and medial occipital lobe, elevated mI/Cr occurred in the posterior cingulate [38]. MI is considered a glial marker with elevated levels found in glial proliferation [61, 62].

While some groups have demonstrated an elevation of Cho in AD [49, 63, 64], others have found no difference in Cho between AD and controls [65–68]. Some of these differences may be related to different regions studied. For example, while Cho was found to be elevated in the posterior cingulate and centrum semiovale [63, 64], it was shown to be decreased in the hippocampus [36]. The Cho peak in MRS is related to phosphatidylcholine breakdown products [69]. Therefore, elevations in AD may be the result of greater phospholipid bilayer turnover. Since Cho/Cr ratios decrease with administration of a cholinergic agonist, it has been proposed that increased Cho in AD is a compensatory mechanism to increase acetylcholine, which is depleted in AD [70, 71].

Decreased glutamate (Glu) or (glutamate + glutamine)/Cr ratio occurs in the gray matter of AD patients [72–75]. Hippocampal Glu/Cr ratios and cognitive performance improve with galantamine treatment [76]. Along with decreased NAA levels, decreased Glu levels correlate with frontal amyloid load in transgenic Alzheimer mice [77].

Temporal progression of metabolites in AD is consistent with an increase mI/Cr followed by a decrease in NAA/Cr and then an increase in Cho/Cr [38]. Both NAA/mI and NAA/Cr decline longitudinally in AD compared to controls [78, 79].

MRS metabolite ratios also correlate with cognitive testing and pathologic severity of AD-related changes. NAA/Cr in the medial temporal lobe, primary motor, and sensory cortices correlated with mini-mental state examination [52]. NAA/Cr and mI/Cr were correlated with verbal memory in addition to global cognitive functioning in the posterior cingulate [80]. Decreased NAA/Cr and increased mI/Cr metabolite ratios were associated with higher Braak-NFT stage, higher Aβ plaque burden, and higher chance of AD pathology in the posterior cingulate [81]. NAA/mI ratio proved to be the best predictor of the pathologic likelihood of AD correlating greatest with Braak stage [81].

Transgenic mouse models of AD and ex vivo MRS studies in the human tissue have also provided insight into the pathologic underpinnings of decreased NAA, Glu, and increased mI levels. These findings are consistent with the metabolic abnormalities observed in patients with AD [51, 82–86]. For example, lower NAA and Glu levels were associated with Aβ plaque load in mice with PS2APP mutation [77]. Magic angle spinning ^{1}H MRS in the superior temporal cortex tissue from patients with AD showed a correlation between NAA concentration and neuronal density [51]. An in vivo ^{13}carbon (^{13}C)-MRS and ^{1}H MRS study suggested a link between increased glial or microglial activation and mI elevation in AD [87]. Treatment with anti-β-amyloid therapy reduces mI levels in transgenic mouse models, which suggest a link between mI elevation and β-amyloid deposition [88].

MRS in Dementia with Lewy Bodies (DLB)

MRS findings in DLB are difficult to interpret because AD pathology coexists in over half of DLB cases, and multiple pathologies at autopsy are common [14, 89]. Clinically diagnosed DLB patients have normal NAA/Cr levels [63, 90] but elevated Cho/Cr in the posterior cingulate [57]. Similar to AD patients, low NAA/Cr in the hippocampus and white matter of DLB patients has been reported [91, 92]. The relative contribution of VaD and AD to these findings remains unclear. In a recent multivoxel MRS study, clinically diagnosed DLB patients were characterized by decreased NAA/Cr, increased mI/Cr ratios in the occipital voxel, and elevated Cho/Cr in the posterior cingulate voxel [93]. In an attempt to examine the DLB MRS pattern in the absence of significant coexisting AD pathology, we divided DLB patients into those with preserved hippocampal volumes (low likelihood of coexisting AD) on volumetric MRI and reduced hippocampal volumes (high likelihood of coexisting AD). Patients with low likelihood AD had preserved NAA/Cr in the posterior cingulate. In the subset that came to autopsy, DLB patients with coexisting AD pathology had antemortem MRS patterns of decreased posterior cingulate and frontal NAA/Cr levels. In contrast, patients with pure DLB at autopsy had preserved NAA/Cr in the posterior cingulate. Since NAA is a neuronal marker, these findings are compatible with the fact that DLB has preserved neuronal numbers at autopsy [94]. In contrast, AD is characterized by neuronal loss in the posterior cingulate at autopsy [95].

MI/Cr was elevated in DLB compared to controls in the occipital voxel [93]. This elevation occurred in the absence of coexisting AD pathology. The localization of mI/Cr elevation in DLB is consistent with FDG PET data, which demonstrate greater occipital involvement in DLB compared to AD [96]. The MRS findings in DLB highlight the importance of understanding the relative contributions of multiple pathologies to MRS findings as well as different metabolite patterns in different anatomic locations.

MRS in Vascular Dementia (VaD)

Similar to AD, VaD also demonstrates decreased NAA/Cr compared to controls [57]. In contrast to AD, in the posterior cingulate, Cho/Cr and mI/Cr are not elevated in VaD [57]. Voxel location in gray or white matter plays an important role in VaD as NAA/Cr tends to be lower in white matter [90, 97]. For example, in the frontal lobe, NAA/Cr is lower in VaD than in AD [97].

MRS in Frontotemporal Lobar Degeneration (FTLD)

Using a posterior cingulate voxel, the metabolite ratios in FTLD approximate those in AD (decreased NAA/Cr, increased mI/Cr, and increased Cho/Cr) [57, 98]. However, because frontal lobes are more severely involved in FTLD compared to AD, MRS metabolite levels in the frontal lobe voxel tend to be more abnormal in FTLD compared to AD and successfully distinguish AD from FTLD [98]. In FTLD, the frontal voxel demonstrated decreased NAA/Cr, increased mI/Cr, and decreased Glu plus glutamine, whereas no abnormalities were detected in AD subjects' frontal voxel [98].

Frontotemporal dementia with parkinsonism linked to chromosome 17 (FTDP-17) is an autosomal dominant tauopathy that is linked to mutations in the gene encoding for the microtubule-associated protein tau (*MAPT*) on chromosome 17 [99–102]. Mutations in *MAPT* result in filamentous accumulation of hyperphosphorylated tau in neurons and glia leading to neurodegeneration and atrophy [103]. Progressive accumulation of filamentous tau and subsequent neuronal death is central to the pathogenesis of many neurodegenerative diseases including AD and may begin years before the onset of clinical symptoms. We recently demonstrated ^{1}H MRS metabolite abnormalities in presymptomatic carriers of mutations in the gene encoding for *MAPT* on chromosome 17 [104]. The severity of ^{1}H MRS and MRI abnormalities was associated with the proximity to the estimated age of symptom onset. NAA/mI ratio was fully outside of the control range in presymptomatic *MAPT* mutation carriers who had 5 years to reach estimated age of symptom onset or who were past the estimated age of symptom onset, indicating presence of ^{1}H MRS metabolite abnormalities related to neurodegeneration, years before the onset of symptoms and atrophy in *MAPT* mutation carriers.

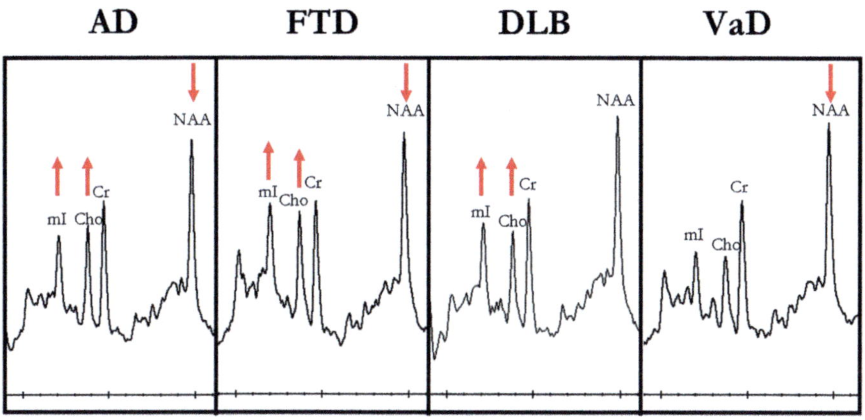

Fig. 4.2 Posterior cingulate gyrus voxel ^{1}H MRS findings in common dementia syndromes: Alzheimer's disease (AD), frontotemporal lobar degeneration (FTLD), dementia with Lewy bodies (DLB), vascular dementia (VaD) (With permission from Neuroimaging Clinics of North America) [107]

Future Perspectives

MRS has shown promise in dementia as a diagnostic biomarker with the ability to detect preclinical disease and amnestic mild cognitive impairment and provide ancillary information to distinguish FTLD, AD, DLB, and VaD (Fig. 4.2). Further, MRS can predict progression early in the disease course from preclinical AD to MCI and from MCI to dementia and correlates with clinical severity. Despite these advantages, MRS is not routinely used in the evaluation for dementia in clinical practice. Better understanding of the pathological basis of the metabolite ratio abnormalities [105] and better standardization of the MRS technique with multisite studies may improve its applications in dementia.

References

1. Farlow M, Arnold SE, van Dyck CH, Aisen PS, Snider BJ, Porsteinsson AP, Friedrich S, Dean RA, Gonzales C, Sethuraman G, DeMattos RB, Mohs R, Paul SM, Siemers ER (2012) Safety and biomarker effects of solanezumab in patients with Alzheimer's disease. Alzheimers Dement 8(4):261–271. doi:10.1016/j.jalz.2011.09.224
2. Gold M, Lorenzl S, Stewart AJ, Morimoto BH, Williams DR, Gozes I (2012) Critical appraisal of the role of davunetide in the treatment of progressive supranuclear palsy. Neuropsychiatr Dis Treat 8:85–93. doi:10.2147/NDT.S12518
3. Salloway S, Sperling R, Fox NC, Blennow K, Klunk W, Raskind M, Sabbagh M, Honig LS, Porsteinsson AP, Ferris S, Reichert M, Ketter N, Nejadnik B, Guenzler V, Miloslavsky M, Wang D, Lu Y, Lull J, Tudor IC, Liu E, Grundman M, Yuen E, Black R, Brashear HR (2014) Two phase 3 trials of bapineuzumab in mild-to-moderate Alzheimer's disease. N Engl J Med 370(4):322–333. doi:10.1056/NEJMoa1304839

4. Harris JM, Gall C, Thompson JC, Richardson AM, Neary D, du Plessis D, Pal P, Mann DM, Snowden JS, Jones M (2013) Sensitivity and specificity of FTDC criteria for behavioral variant frontotemporal dementia. Neurology 80(20):1881–1887. doi:10.1212/WNL.0b013e318292a342
5. Nelson PT, Jicha GA, Kryscio RJ, Abner EL, Schmitt FA, Cooper G, Xu LO, Smith CD, Markesbery WR (2010) Low sensitivity in clinical diagnoses of dementia with Lewy bodies. J Neurol 257(3):359–366. doi:10.1007/s00415-009-5324-y
6. Barkhof F, Polvikoski TM, van Straaten EC, Kalaria RN, Sulkava R, Aronen HJ, Niinisto L, Rastas S, Oinas M, Scheltens P, Erkinjuntti T (2007) The significance of medial temporal lobe atrophy: a postmortem MRI study in the very old. Neurology 69(15):1521–1527
7. Bobinski M, Wegiel J, Tarnawski M, Reisberg B, de Leon MJ, Miller DC, Wisniewski HM (1997) Relationships between regional neuronal loss and neurofibrillary changes in the hippocampal formation and duration and severity of Alzheimer disease. J Neuropathol Exp Neurol 56(4):414–420
8. Gosche KM, Mortimer JA, Smith CD, Markesbery WR, Snowdon DA (2002) Hippocampal volume as an index of Alzheimer neuropathology: findings from the Nun Study. Neurology 58(10):1476–1482
9. Jack CR Jr, Dickson DW, Parisi JE, Xu YC, Cha RH, O'Brien PC, Edland SD, Smith GE, Boeve BF, Tangalos EG, Kokmen E, Petersen RC (2002) Antemortem MRI findings correlate with hippocampal neuropathology in typical aging and dementia. Neurology 58(5):750–757
10. Jagust WJ, Zheng L, Harvey DJ, Mack WJ, Vinters HV, Weiner MW, Ellis WG, Zarow C, Mungas D, Reed BR, Kramer JH, Schuff N, DeCarli C, Chui HC (2008) Neuropathological basis of magnetic resonance images in aging and dementia. Ann Neurol 63(1):72–80
11. Whitwell JL, Shiung MM, Przybelski SA, Weigand SD, Knopman DS, Boeve BF, Petersen RC, Jack CR Jr (2008) MRI patterns of atrophy associated with progression to AD in amnestic mild cognitive impairment. Neurology 70(7):512–520. doi:10.1212/01.wnl.0000280575.77437.a2
12. Whitwell JL, Weigand SD, Shiung MM, Boeve BF, Ferman TJ, Smith GE, Knopman DS, Petersen RC, Benarroch EE, Josephs KA, Jack CR Jr (2007) Focal atrophy in dementia with Lewy bodies on MRI: a distinct pattern from Alzheimer's disease. Brain 130(Pt 3):708–719
13. Kantarci K, Ferman TJ, Boeve BF, Weigand SD, Przybelski S, Vemuri P, Murray MM, Senjem ML, Smith GE, Knopman DS, Petersen RC, Jack CR Jr, Parisi JE, Dickson DW (2012) Focal atrophy on MRI and neuropathologic classification of dementia with Lewy bodies. Neurology 79(6):553–560. doi:10.1212/WNL.0b013e31826357a5, WNL.0b013e31826357a5 [pii]
14. Schneider JA, Arvanitakis Z, Bang W, Bennett DA (2007) Mixed brain pathologies account for most dementia cases in community-dwelling older persons. Neurology 69(24):2197–2204
15. Foster NL, Heidebrink JL, Clark CM, Jagust WJ, Arnold SE, Barbas NR, DeCarli CS, Turner RS, Koeppe RA, Higdon R, Minoshima S (2007) FDG-PET improves accuracy in distinguishing frontotemporal dementia and Alzheimer's disease. Brain 130 (Pt 10):2616–2635. doi:10.1093/brain/awm177
16. Taylor KI, Probst A, Miserez AR, Monsch AU, Tolnay M (2008) Clinical course of neuropathologically confirmed frontal-variant Alzheimer's disease. Nat Clin Pract Neurol 4 (4):226–232. doi:10.1038/ncpneuro0746
17. Minoshima S, Foster NL, Sima AA, Frey KA, Albin RL, Kuhl DE (2001) Alzheimer's disease versus dementia with Lewy bodies: cerebral metabolic distinction with autopsy confirmation. Ann Neurol 50(3):358–365
18. Albin RL, Minoshima S, D'Amato CJ, Frey KA, Kuhl DA, Sima AA (1996) Fluorodeoxyglucose positron emission tomography in diffuse Lewy body disease. Neurology 47 (2):462–466
19. Tang-Wai DF, Graff-Radford NR, Boeve BF, Dickson DW, Parisi JE, Crook R, Caselli RJ, Knopman DS, Petersen RC (2004) Clinical, genetic, and neuropathologic characteristics of posterior cortical atrophy. Neurology 63(7):1168–1174

20. Ishii K, Sasaki M, Kitagaki H, Yamaji S, Sakamoto S, Matsuda K, Mori E (1997) Reduction of cerebellar glucose metabolism in advanced Alzheimer's disease. J Nucl Med 38 (6):925–928
21. Thal DR, Rub U, Orantes M, Braak H (2002) Phases of A beta-deposition in the human brain and its relevance for the development of AD. Neurology 58(12):1791–1800
22. Braak H, Braak E (1991) Neuropathological stageing of Alzheimer-related changes. Acta Neuropathol 82(4):239–259
23. Jack CR Jr, Knopman DS, Jagust WJ, Shaw LM, Aisen PS, Weiner MW, Petersen RC, Trojanowski JQ (2010) Hypothetical model of dynamic biomarkers of the Alzheimer's pathological cascade. Lancet Neurol 9(1):119–128. doi:10.1016/S1474-4422(09)70299-6
24. Arvanitakis Z, Leurgans SE, Barnes LL, Bennett DA, Schneider JA (2011) Microinfarct pathology, dementia, and cognitive systems. Stroke 42(3):722–727. doi:10.1161/STROKEAHA.110.595082
25. Josephs KA, Hodges JR, Snowden JS, Mackenzie IR, Neumann M, Mann DM, Dickson DW (2011) Neuropathological background of phenotypical variability in frontotemporal dementia. Acta Neuropathol 122(2):137–153. doi:10.1007/s00401-011-0839-6
26. Bateman RJ, Xiong C, Benzinger TL, Fagan AM, Goate A, Fox NC, Marcus DS, Cairns NJ, Xie X, Blazey TM, Holtzman DM, Santacruz A, Buckles V, Oliver A, Moulder K, Aisen PS, Ghetti B, Klunk WE, McDade E, Martins RN, Masters CL, Mayeux R, Ringman JM, Rossor MN, Schofield PR, Sperling RA, Salloway S, Morris JC (2012) Clinical and biomarker changes in dominantly inherited Alzheimer's disease. N Engl J Med 367(9):795–804. doi:10.1056/NEJMoa1202753
27. Callaway E (2012) Alzheimer's drugs take a new tack. Nature 489(7414):13–14
28. Sperling RA, Aisen PS, Beckett LA, Bennett DA, Craft S, Fagan AM, Iwatsubo T, Jack CR Jr, Kaye J, Montine TJ, Park DC, Reiman EM, Rowe CC, Siemers E, Stern Y, Yaffe K, Carrillo MC, Thies B, Morrison-Bogorad M, Wagster MV, Phelps CH (2011) Toward defining the preclinical stages of Alzheimer's disease: recommendations from the National Institute on Aging-Alzheimer's Association workgroups on diagnostic guidelines for Alzheimer's disease. Alzheimers Dement 7(3):280–292. doi:10.1016/j.jalz.2011.03.003
29. Albert MS, DeKosky ST, Dickson D, Dubois B, Feldman HH, Fox NC, Gamst A, Holtzman DM, Jagust WJ, Petersen RC, Snyder PJ, Carrillo MC, Thies B, Phelps CH (2011) The diagnosis of mild cognitive impairment due to Alzheimer's disease: recommendations from the National Institute on Aging-Alzheimer's Association workgroups on diagnostic guidelines for Alzheimer's disease. Alzheimers Dement 7(3):270–279. doi:10.1016/j.jalz.2011.03.008
30. den Heijer T, Sijens PE, Prins ND, Hofman A, Koudstaal PJ, Oudkerk M, Breteler MM (2006) MR spectroscopy of brain white matter in the prediction of dementia. Neurology 66 (4):540–544
31. Godbolt AK, Waldman AD, MacManus DG, Schott JM, Frost C, Cipolotti L, Fox NC, Rossor MN (2006) MRS shows abnormalities before symptoms in familial Alzheimer disease. Neurology 66(5):718–722
32. Petersen RC, Roberts RO, Knopman DS, Boeve BF, Geda YE, Ivnik RJ, Smith GE, Jack CR Jr (2009) Mild cognitive impairment: ten years later. Arch Neurol 66(12):1447–1455. doi:10.1001/archneurol.2009.266
33. Petersen RC, Caracciolo B, Brayne C, Gauthier S, Jelic V, Fratiglioni L (2014) Mild cognitive impairment: a concept in evolution. J Intern Med 275(3):214–228. doi:10.1111/joim.12190
34. Kantarci K, Weigand SD, Przybelski SA, Shiung MM, Whitwell JL, Negash S, Knopman DS, Boeve BF, O'Brien PC, Petersen RC, Jack CR Jr (2009) Risk of dementia in MCI: combined effect of cerebrovascular disease, volumetric MRI, and 1H MRS. Neurology 72 (17):1519–1525
35. Kantarci K, Petersen RC, Przybelski SA, Weigand SD, Shiung MM, Whitwell JL, Negash S, Ivnik RJ, Boeve BF, Knopman DS, Smith GE, Jack CR Jr (2008) Hippocampal volumes,

proton magnetic resonance spectroscopy metabolites, and cerebrovascular disease in mild cognitive impairment subtypes. Arch Neurol 65(12):1621–1628
36. Chantal S, Braun CM, Bouchard RW, Labelle M, Boulanger Y (2004) Similar 1H magnetic resonance spectroscopic metabolic pattern in the medial temporal lobes of patients with mild cognitive impairment and Alzheimer disease. Brain Res 1003(1–2):26–35
37. Chao LL, Schuff N, Kramer JH, Du AT, Capizzano AA, O'Neill J, Wolkowitz OM, Jagust WJ, Chui HC, Miller BL, Yaffe K, Weiner MW (2005) Reduced medial temporal lobe N-acetylaspartate in cognitively impaired but nondemented patients. Neurology 64 (2):282–289
38. Kantarci K, Jack CR Jr, Xu YC, Campeau NG, O'Brien PC, Smith GE, Ivnik RJ, Boeve BF, Kokmen E, Tangalos EG, Petersen RC (2000) Regional metabolic patterns in mild cognitive impairment and Alzheimer's disease: a 1H MRS study. Neurology 55(2):210–217
39. Metastasio A, Rinaldi P, Tarducci R, Mariani E, Feliziani FT, Cherubini A, Pelliccioli GP, Gobbi G, Senin U, Mecocci P (2006) Conversion of MCI to dementia: role of proton magnetic resonance spectroscopy. Neurobiol Aging 27(7):926–932. doi:10.1016/j.neurobiolaging.2005.05.002
40. Kantarci K, Weigand SD, Petersen RC, Boeve BF, Knopman DS, Gunter J, Reyes D, Shiung M, O'Brien PC, Smith GE, Ivnik RJ, Tangalos EG, Jack CR Jr (2007) Longitudinal 1H MRS changes in mild cognitive impairment and Alzheimer's disease. Neurobiol Aging 28 (9):1330–1339
41. Kantarci K, Weigand SD, Przybelski SA, Preboske GM, Pankratz VS, Vemuri P, Senjem ML, Murphy MC, Gunter JL, Machulda MM, Ivnik RJ, Roberts RO, Boeve BF, Rocca WA, Knopman DS, Petersen RC, Jack CR Jr (2013) MRI and MRS predictors of mild cognitive impairment in a population-based sample. Neurology 81(2):126–133. doi:10.1212/WNL.0b013e31829a3329
42. Murray ME, Graff-Radford NR, Ross OA, Petersen RC, Duara R, Dickson DW (2011) Neuropathologically defined subtypes of Alzheimer's disease with distinct clinical characteristics: a retrospective study. Lancet Neurol 10(9):785–796. doi:10.1016/S1474-4422(11)70156-9
43. Rami L, Gomez-Anson B, Sanchez-Valle R, Bosch B, Monte GC, Llado A, Molinuevo JL (2007) Longitudinal study of amnesic patients at high risk for Alzheimer's disease: clinical, neuropsychological and magnetic resonance spectroscopy features. Dement Geriatr Cogn Disord 24(5):402–410. doi:10.1159/000109750
44. Fayed N, Davila J, Oliveros A, Castillo J, Medrano JJ (2008) Utility of different MR modalities in mild cognitive impairment and its use as a predictor of conversion to probable dementia. Acad Radiol 15(9):1089–1098. doi:10.1016/j.acra.2008.04.008
45. Burhan AM, Bartha R, Bocti C, Borrie M, Laforce R, Rosa-Neto P, Soucy JP (2013) Role of emerging neuroimaging modalities in patients with cognitive impairment: a review from the Canadian Consensus Conference on the Diagnosis and Treatment of Dementia 2012. Alzheimers Res Ther 5(Suppl 1):S4. doi:10.1186/alzrt200
46. Klunk WE, Panchalingam K, Moossy J, McClure RJ, Pettegrew JW (1992) N-Acetyl-L-aspartate and other amino acid metabolites in Alzheimer's disease brain: a preliminary proton nuclear magnetic resonance study. Neurology 42(8):1578–1585
47. Huang W, Alexander GE, Chang L, Shetty HU, Krasuski JS, Rapoport SI, Schapiro MB (2001) Brain metabolite concentration and dementia severity in Alzheimer's disease: a (1)H MRS study. Neurology 57(4):626–632
48. Miller BL, Moats RA, Shonk T, Ernst T, Woolley S, Ross BD (1993) Alzheimer disease: depiction of increased cerebral myo-inositol with proton MR spectroscopy. Radiology 187 (2):433–437
49. Pfefferbaum A, Adalsteinsson E, Spielman D, Sullivan EV, Lim KO (1999) In vivo spectroscopic quantification of the N-acetyl moiety, creatine, and choline from large volumes of brain gray and white matter: effects of normal aging. Magn Reson Med 41(2):276–284

50. Schuff N, Capizzano AA, Du AT, Amend DL, O'Neill J, Norman D, Kramer J, Jagust W, Miller B, Wolkowitz OM, Yaffe K, Weiner MW (2002) Selective reduction of N-acetylaspartate in medial temporal and parietal lobes in AD. Neurology 58(6):928–935
51. Cheng LL, Newell K, Mallory AE, Hyman BT, Gonzalez RG (2002) Quantification of neurons in Alzheimer and control brains with ex vivo high resolution magic angle spinning proton magnetic resonance spectroscopy and stereology. Magn Reson Imaging 20 (7):527–533
52. Jessen F, Block W, Traber F, Keller E, Flacke S, Papassotiropoulos A, Lamerichs R, Heun R, Schild HH (2000) Proton MR spectroscopy detects a relative decrease of N-acetylaspartate in the medial temporal lobe of patients with AD. Neurology 55(5):684–688
53. Schuff N, Amend DL, Meyerhoff DJ, Tanabe JL, Norman D, Fein G, Weiner MW (1998) Alzheimer disease: quantitative H-1 MR spectroscopic imaging of frontoparietal brain. Radiology 207(1):91–102
54. Tedeschi G, Bertolino A, Lundbom N, Bonavita S, Patronas NJ, Duyn JH, Metman LV, Chase TN, Di Chiro G (1996) Cortical and subcortical chemical pathology in Alzheimer's disease as assessed by multislice proton magnetic resonance spectroscopic imaging. Neurology 47(3):696–704
55. Zhu X, Schuff N, Kornak J, Soher B, Yaffe K, Kramer JH, Ezekiel F, Miller BL, Jagust WJ, Weiner MW (2006) Effects of Alzheimer disease on fronto-parietal brain N-acetyl aspartate and myo-inositol using magnetic resonance spectroscopic imaging. Alzheimer Dis Assoc Disord 20(2):77–85
56. Barker PB, Lee RR, McArthur JC (1995) AIDS dementia complex: evaluation with proton MR spectroscopic imaging. Radiology 195(1):58–64
57. Kantarci K, Petersen RC, Boeve BF, Knopman DS, Tang-Wai DF, O'Brien PC, Weigand SD, Edland SD, Smith GE, Ivnik RJ, Ferman TJ, Tangalos EG, Jack CR Jr (2004) 1H MR spectroscopy in common dementias. Neurology 63(8):1393–1398
58. Shiino A, Matsuda M, Morikawa S, Inubushi T, Akiguchi I, Handa J (1993) Proton magnetic resonance spectroscopy with dementia. Surg Neurol 39(2):143–147
59. Modrego PJ, Pina MA, Fayed N, Diaz M (2006) Changes in metabolite ratios after treatment with rivastigmine in Alzheimer's disease: a nonrandomised controlled trial with magnetic resonance spectroscopy. CNS Drugs 20(10):867–877
60. Modrego PJ, Fayed N, Errea JM, Rios C, Pina MA, Sarasa M (2010) Memantine versus donepezil in mild to moderate Alzheimer's disease: a randomized trial with magnetic resonance spectroscopy. Eur J Neurol 17(3):405–412. doi:10.1111/j.1468-1331.2009.02816.x
61. Bitsch A, Bruhn H, Vougioukas V, Stringaris A, Lassmann H, Frahm J, Bruck W (1999) Inflammatory CNS demyelination: histopathologic correlation with in vivo quantitative proton MR spectroscopy. Am J Neuroradiol 20(9):1619–1627
62. Glanville NT, Byers DM, Cook HW, Spence MW, Palmer FB (1989) Differences in the metabolism of inositol and phosphoinositides by cultured cells of neuronal and glial origin. Biochim Biophys Acta 1004(2):169–179
63. Jones RS, Waldman AD (2004) 1H-MRS evaluation of metabolism in Alzheimer's disease and vascular dementia. Neurol Res 26(5):488–495
64. Meyerhoff DJ, MacKay S, Constans JM, Norman D, Van Dyke C, Fein G, Weiner MW (1994) Axonal injury and membrane alterations in Alzheimer's disease suggested by in vivo proton magnetic resonance spectroscopic imaging. Ann Neurol 36(1):40–47
65. Krishnan KR, Charles HC, Doraiswamy PM, Mintzer J, Weisler R, Yu X, Perdomo C, Ieni JR, Rogers S (2003) Randomized, placebo-controlled trial of the effects of donepezil on neuronal markers and hippocampal volumes in Alzheimer's disease. Am J Psychiatry 160 (11):2003–2011
66. Moats RA, Ernst T, Shonk TK, Ross BD (1994) Abnormal cerebral metabolite concentrations in patients with probable Alzheimer disease. Magn Reson Med 32(1):110–115

67. Rose SE, de Zubicaray GI, Wang D, Galloway GJ, Chalk JB, Eagle SC, Semple J, Doddrell DM (1999) A 1H MRS study of probable Alzheimer's disease and normal aging: implications for longitudinal monitoring of dementia progression. Magn Reson Imaging 17(2):291–299
68. Schuff N, Amend D, Ezekiel F, Steinman SK, Tanabe J, Norman D, Jagust W, Kramer JH, Mastrianni JA, Fein G, Weiner MW (1997) Changes of hippocampal N-acetyl aspartate and volume in Alzheimer's disease. A proton MR spectroscopic imaging and MRI study. Neurology 49(6):1513–1521
69. Klein J (2000) Membrane breakdown in acute and chronic neurodegeneration: focus on choline-containing phospholipids. J Neural Transm 107(8–9):1027–1063
70. Satlin A, Bodick N, Offen WW, Renshaw PF (1997) Brain proton magnetic resonance spectroscopy (1H-MRS) in Alzheimer's disease: changes after treatment with xanomeline, an M1 selective cholinergic agonist. Am J Psychiatry 154(10):1459–1461
71. Wurtman RJ, Blusztajn JK, Marie JC (1985) Autocannibalism of choline-containing membrane phospholipids in the pathogenesis of Alzheimer's disease. Neurochem Int 7:369–372
72. Antuono PG, Jones JL, Wang Y, Li SJ (2001) Decreased glutamate + glutamine in Alzheimer's disease detected in vivo with (1)H-MRS at 0.5 T. Neurology 56(6):737–742
73. Bartha R, Smith M, Rupsingh R, Rylett J, Wells JL, Borrie MJ (2008) High field (1)H MRS of the hippocampus after donepezil treatment in Alzheimer disease. Prog Neuropsychopharmacol Biol Psychiatry 32(3):786–793
74. Hattori N, Abe K, Sakoda S, Sawada T (2002) Proton MR spectroscopic study at 3 Tesla on glutamate/glutamine in Alzheimer's disease. Neuroreport 13(1):183–186
75. Rupsingh R, Borrie M, Smith M, Wells JL, Bartha R (2011) Reduced hippocampal glutamate in Alzheimer disease. Neurobiol Aging 32(5):802–810. doi:10.1016/j.neurobiolaging.2009.05.002
76. Penner J, Rupsingh R, Smith M, Wells JL, Borrie MJ, Bartha R (2010) Increased glutamate in the hippocampus after galantamine treatment for Alzheimer disease. Prog Neuropsychopharmacol Biol Psychiatry 34(1):104–110. doi:10.1016/j.pnpbp.2009.10.007
77. von Kienlin M, Kunnecke B, Metzger F, Steiner G, Richards JG, Ozmen L, Jacobsen H, Loetscher H (2005) Altered metabolic profile in the frontal cortex of PS2APP transgenic mice, monitored throughout their life span. Neurobiol Dis 18(1):32–39. doi:10.1016/j.nbd.2004.09.005
78. Schott JM, Frost C, MacManus DG, Ibrahim F, Waldman AD, Fox NC (2010) Short echo time proton magnetic resonance spectroscopy in Alzheimer's disease: a longitudinal multiple time point study. Brain 133(11):3315–3322. doi:10.1093/brain/awq208
79. Jessen F, Block W, Traber F, Keller E, Flacke S, Lamerichs R, Schild HH, Heun R (2001) Decrease of N-acetylaspartate in the MTL correlates with cognitive decline of AD patients. Neurology 57(5):930–932
80. Kantarci K, Smith GE, Ivnik RJ, Petersen RC, Boeve BF, Knopman DS, Tangalos EG, Jack CR Jr (2002) 1H magnetic resonance spectroscopy, cognitive function, and apolipoprotein E genotype in normal aging, mild cognitive impairment and Alzheimer's disease. J Int Neuropsychol Soc 8(7):934–942
81. Kantarci K, Knopman DS, Dickson DW, Parisi JE, Whitwell JL, Weigand SD, Josephs KA, Boeve BF, Petersen RC, Jack CR Jr (2008) Alzheimer disease: postmortem neuropathologic correlates of antemortem 1H MR spectroscopy metabolite measurements. Radiology 248(1):210–220
82. Chen SQ, Wang PJ, Ten GJ, Zhan W, Li MH, Zang FC (2009) Role of myo-inositol by magnetic resonance spectroscopy in early diagnosis of Alzheimer's disease in APP/PS1 transgenic mice. Dement Geriatr Cogn Disord 28(6):558–566
83. Dedeoglu A, Choi JK, Cormier K, Kowall NW, Jenkins BG (2004) Magnetic resonance spectroscopic analysis of Alzheimer's disease mouse brain that express mutant human APP shows altered neurochemical profile. Brain Res 1012(1-2):60–65. doi:10.1016/j.brainres.2004.02.079

84. Marjanska M, Curran GL, Wengenack TM, Henry PG, Bliss RL, Poduslo JF, Jack CR Jr, Ugurbil K, Garwood M (2005) Monitoring disease progression in transgenic mouse models of Alzheimer's disease with proton magnetic resonance spectroscopy. Proc Natl Acad Sci U S A 102(33):11906–11910
85. Oberg J, Spenger C, Wang FH, Andersson A, Westman E, Skoglund P, Sunnemark D, Norinder U, Klason T, Wahlund LO, Lindberg M (2008) Age related changes in brain metabolites observed by 1H MRS in APP/PS1 mice. Neurobiol Aging 29(9):1423–1433
86. van Duijn S, Nabuurs RJ, van Duinen SG, Natte R, van Buchem MA, Alia A (2013) Longitudinal monitoring of sex-related in vivo metabolic changes in the brain of Alzheimer's disease transgenic mouse using magnetic resonance spectroscopy. J Alzheimers Dis 34 (4):1051–1059. doi:10.3233/JAD-122188
87. Sailasuta N, Harris K, Tran T, Ross B (2011) Minimally invasive biomarker confirms glial activation present in Alzheimer's disease: a preliminary study. Neuropsychiatr Dis Treat 7:495–499
88. Marjanska M, Weigand SD, Preboske G, Wengenack TM, Chamberlain R, Curran GL, Poduslo JF, Garwood M, Kobayashi D, Lin JC, Jack CR Jr (2014) Treatment effects in a transgenic mouse model of Alzheimer's disease: a magnetic resonance spectroscopy study after passive immunization. Neuroscience 259:94–100. doi:10.1016/j.neuroscience.2013.11.052
89. Fujishiro H, Ferman TJ, Boeve BF, Smith GE, Graff-Radford NR, Uitti RJ, Wszolek ZK, Knopman DS, Petersen RC, Parisi JE, Dickson DW (2008) Validation of the neuropathologic criteria of the third consortium for dementia with Lewy bodies for prospectively diagnosed cases. J Neuropathol Exp Neurol 67(7):649–656. doi:10.1097/NEN.0b013e31817d7a1d
90. Kattapong VJ, Brooks WM, Wesley MH, Kodituwakku PW, Rosenberg GA (1996) Proton magnetic resonance spectroscopy of vascular- and Alzheimer-type dementia. Arch Neurol 53 (7):678–680
91. Molina JA, Garcia-Segura JM, Benito-Leon J, Gomez-Escalonilla C, del Ser T, Martinez V, Viano J (2002) Proton magnetic resonance spectroscopy in dementia with Lewy bodies. Eur Neurol 48(3):158–163
92. Xuan X, Ding M, Gong X (2008) Proton magnetic resonance spectroscopy detects a relative decrease of N-acetylaspartate in the hippocampus of patients with dementia with Lewy bodies. J Neuroimaging 18(2):137–141. doi:10.1111/j.1552-6569.2007.00203.x
93. Graff-Radford J, Boeve BF, Murray ME, Ferman TJ, Tosakulwong N, Lesnick TG, Maroney-Smith M, Senjem ML, Gunter J, Smith GE, Knopman DS, Jack CR Jr, Dickson DW, Petersen RC, Kantarci K (2014) Regional proton magnetic resonance spectroscopy patterns in dementia with Lewy bodies. Neurobiol Aging 35(6):1483–1490. doi:10.1016/j.neurobiolaging.2014.01.001
94. Gomez-Isla T, Growdon WB, McNamara M, Newell K, Gomez-Tortosa E, Hedley-Whyte ET, Hyman BT (1999) Clinicopathologic correlates in temporal cortex in dementia with Lewy bodies. Neurology 53(9):2003–2009
95. Brun A, Gustafson L (1976) Distribution of cerebral degeneration in Alzheimer's disease. A clinico-pathological study. Arch Psychiatr Nervenkr 223(1):15–33
96. Kantarci K, Lowe VJ, Boeve BF, Weigand SD, Senjem ML, Przybelski SA, Dickson DW, Parisi JE, Knopman DS, Smith GE, Ferman TJ, Petersen RC, Jack CR Jr (2012) Multimodality imaging characteristics of dementia with Lewy bodies. Neurobiol Aging 33 (9):2091–2105. doi:10.1016/j.neurobiolaging.2011.09.024
97. MacKay S, Meyerhoff DJ, Constans JM, Norman D, Fein G, Weiner MW (1996) Regional gray and white matter metabolite differences in subjects with AD, with subcortical ischemic vascular dementia, and elderly controls with 1H magnetic resonance spectroscopic imaging. Arch Neurol 53(2):167–174
98. Ernst T, Chang L, Melchor R, Mehringer CM (1997) Frontotemporal dementia and early Alzheimer disease: differentiation with frontal lobe H-1 MR spectroscopy. Radiology 203 (3):829–836

99. Boeve BF, Hutton M (2008) Refining frontotemporal dementia with parkinsonism linked to chromosome 17: introducing FTDP-17 (MAPT) and FTDP-17 (PGRN). Arch Neurol 65 (4):460–464
100. Hutton M, Lendon CL, Rizzu P, Baker M, Froelich S, Houlden H, Pickering-Brown S, Chakraverty S, Isaacs A, Grover A, Hackett J, Adamson J, Lincoln S, Dickson D, Davies P, Petersen RC, Stevens M, de Graaff E, Wauters E, van Baren J, Hillebrand M, Joosse M, Kwon JM, Nowotny P, Che LK, Norton J, Morris JC, Reed LA, Trojanowski J, Basun H, Lannfelt L, Neystat M, Fahn S, Dark F, Tannenberg T, Dodd PR, Hayward N, Kwok JB, Schofield PR, Andreadis A, Snowden J, Craufurd D, Neary D, Owen F, Oostra BA, Hardy J, Goate A, van Swieten J, Mann D, Lynch T, Heutink P (1998) Association of missense and 5′-splice-site mutations in tau with the inherited dementia FTDP-17. Nature 393(6686):702–705
101. Rademakers R, Cruts M, van Broeckhoven C (2004) The role of tau (MAPT) in frontotemporal dementia and related tauopathies. Hum Mutat 24(4):277–295
102. Foster NL, Wilhelmsen K, Sima AA, Jones MZ, D'Amato CJ, Gilman S (1997) Frontotemporal dementia and parkinsonism linked to chromosome 17: a consensus conference. Conference Participants. Ann Neurol 41(6):706–715
103. Whitwell JL, Josephs KA, Avula R, Tosakulwong N, Weigand SD, Senjem ML, Vemuri P, Jones DT, Gunter JL, Baker M, Wszolek ZK, Knopman DS, Rademakers R, Petersen RC, Boeve BF, Jack CR Jr (2011) Altered functional connectivity in asymptomatic MAPT subjects: a comparison to bvFTD. Neurology 77(9):866–874. doi:10.1212/WNL.0b013e31822c61f2
104. Kantarci K, Boeve BF, Wszolek ZK, Rademakers R, Whitwell JL, Baker MC, Senjem ML, Samikoglu AR, Knopman DS, Petersen RC, Jack CR Jr (2010) MRS in presymptomatic MAPT mutation carriers: a potential biomarker for tau-mediated pathology. Neurology 75 (9):771–778. doi:10.1212/WNL.0b013e3181f073c7
105. Kantarci K (2013) Proton MRS in mild cognitive impairment. J Magn Reson Imaging 37 (4):770–777. doi:10.1002/jmri.23800
106. Graff-Radford J, Kantarci K (2013) Magnetic resonance spectroscopy in Alzheimer's disease. Neuropsychiatr Dis Treat 9:687–696. doi:10.2147/NDT.S35440
107. Kantarci K (2013) Magnetic resonance spectroscopy in common dementias. Neuroimaging Clin N Am 23(3):393–406. doi:10.1016/j.nic.2012.10.004

Chapter 5
Magnetic Resonance Spectroscopy in Parkinsonian Disorders

Ulrike Dydak, David A. Edmondson, and S. Elizabeth Zauber

Abstract Parkinsonian disorders are a group of mostly neurodegenerative diseases that affect the basal ganglia. Symptoms include slowness of movement, increased muscle tone, and changes in walking and balance. The most common form of parkinsonism is idiopathic Parkinson's disease. Currently, only symptomatic treatments are available. These treatments can improve symptoms and quality of life, but do not prevent neurodegeneration or slow the course of disease.

By detecting metabolic changes in specific brain regions, magnetic resonance spectroscopy (MRS) shows promise in helping to distinguish between different types of parkinsonian disorders, to aid in early diagnosis, and potentially to track disease progression. However, heterogeneous results are common in the literature to date, in part due to differences in patient characteristics, brain regions examined, evolving MRS techniques, and the effects of medications.

This chapter reviews the clinical and pathological features of parkinsonian disorders and MRS studies in these disorders and describes the use of animal models to further investigate the connection between MRS findings and pathological brain changes.

Keywords Parkinson's disease • Parkinsonism • Alpha-synuclein • Progressive supranuclear palsy • Multiple system atrophy • Dementia with Lewy bodies • Corticobasal degeneration

---

U. Dydak, Ph.D. (✉) • D.A. Edmondson, M.S.
School of Health Sciences, Purdue University, West Lafayette, IN 47907, USA

Department of Radiology and Imaging Sciences, Indiana University School of Medicine, Indianapolis, IN 46202, USA
e-mail: udydak@purdue.edu; edmondsd@purdue.edu

S.E. Zauber, M.D.
Department of Neurology, Indiana University School of Medicine, Indianapolis, IN 46202, USA
e-mail: szauber@iupui.edu

G. Öz (ed.), *Magnetic Resonance Spectroscopy of Degenerative Brain Diseases*, Contemporary Clinical Neuroscience, DOI 10.1007/978-3-319-33555-1_5

Symptoms, Pathology, and Classification of Parkinsonian Disorders

The term parkinsonian disorders, further called PDS, encompasses any disorder that causes parkinsonism. Parkinsonism is a clinical syndrome characterized by a combination of clinical features: bradykinesia (slowness of movement), rigidity (muscle stiffness), tremor, and postural instability. Patients with parkinsonism may experience a variety of symptoms including soft or slurred speech, slow shuffling gait, stooped posture, loss of balance, and falls. Most patients with parkinsonism meet clinical criteria for idiopathic Parkinson's disease (IPD); however, other pathological causes of parkinsonism do occur and may be difficult to distinguish clinically from IPD, especially in the first few years of the disease (see below). Atypical refers to any other cause of parkinsonism besides IPD. Clinical recognition of atypical parkinsonism is important, since most atypical forms of parkinsonism have a more rapidly progressive course and poor prognosis. The distinction is also critical for research, since atypical forms of parkinsonism have distinct pathophysiology and would not be expected to respond to IPD-targeted therapies. This chapter focuses on the role of magnetic resonance spectroscopy (MRS) in diagnosis, monitoring, and research of the most common neurodegenerative causes of parkinsonism.

Idiopathic Parkinson's Disease

Idiopathic Parkinson's disease (IPD) is a common neurodegenerative disease affecting seven million people globally. Pathological changes in IPD include the loss of dopaminergic neurons in the substantia nigra pars compacta and accumulation of alpha-synuclein-containing Lewy bodies. Because these neurons project to the striatum (caudate and putamen), cell loss in the substantia nigra pars compacta is the likely cause of bradykinesia (slowness of movement) and rigidity (increase muscle tone) in IPD. The resulting reduced striatal dopamine causes decreased activity in the direct basal ganglia pathway, leaving the indirect pathway to inhibit the cortex and therefore overall movement. Neuronal cell loss is not limited to the dopaminergic system. It also occurs in the locus coeruleus, nucleus basalis of Meynert, pedunculopontine nucleus, and raphe nucleus, which leads to reductions in neurotransmitters such as serotonin and norepinephrine. Lewy bodies, which are pathological neuronal inclusions of abnormally aggregated protein, are also widespread and recently have been reported to occur in the peripheral nervous system, the autonomic nervous system, and the enteric nervous system as well. It is hypothesized that pathological changes occur first in the periphery, followed by the brainstem, and later in the cortex in stages called Braak stages 1–6 [1].

IPD produces a wide spectrum of clinical symptoms including motor symptoms such as slowness, stiffness, tremor, changes in gait, balance, speech, and posture.

Non-motor symptoms include cognitive impairment, depression, anxiety, apathy, pain, and changes in bowel and bladder function. These non-motor symptoms are common, may be present decades before diagnosis, worsen with disease progression, and have a significant impact on quality of life [2].

Symptomatic treatments (see below) can improve motor and non-motor symptoms; however, no treatment to date has been shown to slow disease progression. The search for disease-modifying treatments, such as neuroprotective or restorative agents, is a major focus of current research. One reason why disease-modifying trials have failed to date may be that, at the time of diagnosis, over 50 % of dopaminergic neurons have already been lost [3] and it is too late for intervention. Pathological changes in IPD can be detected up to two decades prior to diagnosis resulting in non-motor symptoms such as constipation, reduced sense of smell, sleep disorders, and mood disorders [4]. It is hypothesized that initiating a neuroprotective therapy would be most effective in the pre-motor or prodromal phase of disease. This earliest phase of "preclinical" IPD refers to people who have no clinical symptoms but in whom imaging or colonic biopsy can detect pathological changes. "Pre-motor" IPD refers to people with subtle non-motor symptoms such as changes in sleep, mood, and autonomic function, who do not yet have any motor features of the disease. "Pre-diagnostic" patients have subtle motor symptoms that are not sufficient to meet full diagnostic criteria for IPD [5]. Accurately diagnosing patients in these pre-motor phases will require improved diagnostic tools such as advanced MRI methods.

In the research setting, the diagnosis of IPD is made using United Kingdom Parkinson's disease brain bank criteria [6]. Clinical diagnostic accuracy improves with disease duration, such that the clinical diagnosis of IPD strongly predicts pathological findings late in the disease course, but can be substantially less accurate in the first 5 years of disease. Most clinical trials in IPD use a standardized motor rating scale to determine response to treatment called the Unified Parkinson Disease Rating Scale (UPDRS). The scale was updated in 2007; the newest version is called MDS-UDPRS [7]. Another common measure of disease severity is the Hoehn and Yahr scale which broadly rates disease severity into 1 of 5 stages. In stage 1, patients have symptoms on one side of the body. In stage 2, symptoms are bilateral. Stage 3 is bilateral symptoms with impaired balance, stage 4 is impairment in balance and walking such that an assistive device is required, and in stage 5 patients are unable to walk [8]. Not every patient will progress through each Hoehn and Yahr stage.

Drugs that act on the dopaminergic system are the first line of treatment in IPD. Since the motor symptoms of IPD result from low striatal dopamine, replacing or enhancing brain dopamine pharmacologically can lessen these motor symptoms. The most commonly used drugs are therefore dopamine agonists and carbidopa/levodopa. There are four dopamine agonist medications: rotigotine, ropinirole, pramipexole, and apomorphine. Medications usually provide good symptom control early in the course of IPD. As the disease progresses, patients may experience more troublesome side effects from medications such as levodopa-induced dyskinesias (involuntary twisting movements) or hallucinations, among others.

Advanced IPD patients also experience symptoms that are resistant to dopaminergic therapy such as dementia, difficulty with walking, and loss of balance. These dopamine-resistant symptoms are likely related to more widespread pathological changes. Nondrug treatments are available such as physical, occupational, and speech therapy, which can be useful throughout the course of disease, but are especially useful in patients with more advanced disease.

A small subset of IPD patients are treated with a surgical procedure called deep brain stimulation surgery (DBS). These are typically younger onset patients with good cognitive function who have significant motor symptoms not entirely relieved with dopaminergic medications. The most common brain target for the surgery is the subthalamic nucleus (STN), although other brain regions may be targeted. The precise mechanism by which DBS improves PD symptoms is unknown, but broadly it is understood to allow for more physiological neuronal cell firing patterns in the basal ganglia. DBS is an effective treatment of symptoms for selected patients; however, it is not a cure and does not slow disease progression.

The search for disease-modifying treatments that can slow disease progression is a major focus of research. Finding such a therapy requires improved understanding of the pathogenesis of IPD. The exact cause of cell death in IPD is still unknown; however, both genetic and environmental factors are thought to play a role. The most well-accepted environmental risk factor is pesticide exposure, either through farming, rural living, or drinking well water. Known genetic mutations account for only a small percentage of IPD cases; however, the process of identifying causative genes has significantly improved the understanding of IPD pathogenesis. It is now understood that changes in mitochondrial and lysosomal function and abnormalities in intracellular protein trafficking contribute to neuronal dysfunction and cell death [9].

Atypical Parkinsonian Disorders

The possibilities for differential diagnosis for parkinsonism are broad and include toxic, metabolic, and genetic diseases. This review focuses on the neurodegenerative diseases that most closely mimic IPD. In the first few years of symptoms, atypical parkinsonian disorders (APD) may closely resemble IPD. However, atypical forms of parkinsonism have a more rapid course and some unique clinical features. In recent years, advanced neuroimaging techniques, including MRS, have helped increase differential diagnostic accuracy.

Dementia with Lewy Bodies

Dementia with Lewy bodies (DLB) is the second most common form of dementia after Alzheimer's disease. DLB can be distinguished from IPD by its clinical time course. In DLB, motor features of parkinsonism and dementia occur within the first

2 years, while in IPD dementia occurs late in the disease course (10–20 years from onset). Patients with DLB experience vivid visual hallucinations and marked fluctuations in cognitive function. Pathological changes resemble those of the late stage IPD with widespread Lewy bodies and neurodegeneration. Reduced dopamine in the nigrostriatal system is responsible for motor features, while cholinergic denervation is responsible for cognitive deficits.

Multiple System Atrophy

Multiple system atrophy (MSA) is also a disorder of alpha-synuclein. In contrast to IPD and DLB, which have intraneuronal inclusions, MSA is pathologically characterized by alpha-synuclein inclusions in oligodendroglia, as well as neurons. As the name suggests, multiple areas of the central nervous system are involved. Dopamine depletion occurs due to atrophy of the nigrostriatal pathway. However, neurodegeneration also occurs in brainstem, cerebellum, and corticospinal tracts. Clinically, MSA is characterized by early involvement of the autonomic nervous system, limited response to levodopa, early impairment in gait, and the presence of cerebellar and corticospinal tract signs. Most patients have a subtype of MSA called MSA-P (parkinsonian MSA), which closely resembles IPD with additional symptoms of rigidity and bradykinesia. Individuals with MSA who have prominent cerebellar features are classified as MSA-C, while those who present initially with autonomic failure (loss of control of bowel and bladder or severe orthostatic hypotension) are termed MSA-A. For diagnosis, standard MRI can be used to detect volume loss in the putamen, pons, and cerebellum. In the clinical research setting, the Unified Multiple System Atrophy Rating Scale (UMSARS) is used to quantify disease severity [10].

Progressive Supranuclear Palsy

Progressive supranuclear palsy (PSP) is clinically characterized by severe disturbance of balance (with falls typically in the first year), axial rigidity (rather than appendicular rigidity in parkinsonism), limitation of voluntary down-gaze, and subcortical dementia. Poor balance is likely caused by neurofibrillary tangles containing abnormal phosphorylated tau protein in the cholinergic pedunculopontine nucleus [11]. Tangles also occur in the basal ganglia, brainstem, and frontal cortex. The most distinctive MRI finding is atrophy of the midbrain. A standard rating scale has been developed to quantify the clinical features, called the Progressive Supranuclear Palsy Rating Scale or PSPRS [12].

Corticobasal Degeneration

Corticobasal degeneration is closely related to progressive supranuclear palsy, with neurofibrillary tangles; however, patients with this disease have more pathological involvement of the cortex than the brainstem, and as a result, clinical symptoms include cortical sensory loss, aphasia, and apraxia, in addition to motor features of parkinsonism. Standard MRI may show asymmetric cortical atrophy.

Manganism

Manganism is a form of parkinsonism characterized by excessive and chronic accumulation of manganese in the basal ganglia. The clinical expression of manganism was first described in cases of occupational exposure [13] and continues to be most commonly studied in welders, who are exposed to manganese-containing fumes [14–18]. However, manganism can also occur in patients with liver failure due to reduced excretion of dietary manganese via the hepatobiliary system [19, 20]. Recently, intravenous drug users of ephedrone have been found to develop manganism due to high levels of manganese contaminants in the drug [21].

The clinical features of manganism include bradykinesia, rigidity, and tremor; however, there are some clinical differences from IPD. Patients with manganism are less likely than IPD patients to have resting tremor but usually have postural tremor. They may have a difference in their walking described as a "cock gait" and are more likely to have dystonia. Importantly, in contrast to patients with IPD, manganism patients typically do not improve with levodopa treatment [22] nor has any other form of symptomatic treatment been established to date, making early and correct diagnosis of manganism highly important. An active area of research is the use of MRI and MRS to find accurate biomarkers for early diagnosis, as well as research into the mechanism of manganese toxicity. Brain MRI of subjects recently exposed to Manganese shows increased T1 signal intensity in the basal ganglia, reflecting manganese accumulation, a finding not observed in IPD patients [23, 24]. Pathological studies of humans with manganism are limited, but they do report changes in the globus pallidus. The substantia nigra, where the bulk of cell death occurs in IPD, is not affected in manganism [25]. Recent studies in nonhuman primates show a marked inhibition of striatal dopamine release in the absence of nigrostriatal dopamine terminal degeneration typical for IPD [26]. With increasing sensitivity and accuracy to measure gamma-aminobutyric acid (GABA) and glutamate (Glu), MRS is a promising tool to further study the differential metabolic imbalances in the basal ganglia pathways in IPD, manganism, and other APDs.

Potential Clinical Utility of Magnetic Resonance Spectroscopy

Magnetic resonance spectroscopy (MRS) allows for a noninvasive assessment of metabolic abnormalities, and more recently the technique has become sensitive enough to detect changes in neurotransmitter concentrations in vivo. It is hypothesized that in most diseases metabolic changes occur prior to structural changes. Thus, assessments of brain biochemistry have the potential to be more sensitive than structural imaging to early pathological changes and to subtle changes in disease progression. Because of its sensitivity, MRS may be useful both at detecting biomarkers of disease progression and for aiding early diagnosis. This is particularly important in disorders like PDS, of which some have a slow disease progression. The slow disease progression may result first in neurochemical and functional changes, before atrophy becomes a visible structural marker of neurodegeneration. It is during the early stages of disease when differential diagnosis using MRS may be crucial for the development of disease-modifying therapies, monitoring of disease progression, or assessment of treatment efficacy.

MRS does not make use of ionizing radiation, does not require the administration of contrast agents, and is thus truly noninvasive. Compared to other molecular imaging techniques, such as PET or SPECT, it is less expensive and is a more straightforward way of assessing biochemical information. Yet, in order to compare results across multiple studies and sites, rigorous data quality control and standardized protocols should be employed (see below).

Standard proton MRS (^{1}H MRS) techniques allow for the quantification of several metabolites providing information on neuronal integrity, myelin and glial cells, energy metabolism, and other metabolic compounds [27]. The largest peak in a healthy brain ^{1}H spectrum originates from *N*-acetylaspartate (NAA), a marker of neuronal viability and function. NAA is reduced in many brain disorders, due to neuronal or axonal loss. As such, reduction of NAA is a characteristic feature of neurodegenerative diseases. Early studies probing the utility of MRS in IPD diagnosis investigated basal ganglia and cortical structures involved in motor dysfunction. Yet, since NAA is also involved in several neuronal processes such as mitochondrial function, osmoregulation, and lipid and protein synthesis, its depletion may equally be interpreted as loss of neuronal function and does not immediately indicate neuronal cell death.

Total creatine (tCr) is a combined peak at 3.03 ppm, which is composed of creatine (Cr) and phosphocreatine (PCr) and serves as an energy buffer. In earlier literature it was highly common to use tCr as internal reference and provide metabolite concentrations as ratio to tCr. While the concentration of tCr is assumed to be relatively stable due to it being the sum of Cr and PCr, this assumption requires careful testing and has often been shown as unjustified, particularly in pathological conditions.

Total choline (tCho) stands for the combination of several choline-containing compounds, including free choline, phosphorylcholine, and glycerophosphorylcholine, which is a precursor to acetylcholine. Changes in tCho are often associated with abnormalities in membrane synthesis or degradation or neuro-inflammation.

Myo-inositol (mI), another detectable metabolite, is suggested as a marker for gliosis. Elevated mI levels have been associated with several neurodegenerative diseases [28, 29] and were found very early in dementia [30, 31]. An increase in another metabolite, lactate (Lac), is usually interpreted as marker of anaerobic metabolism such as in hypoxia, ischemia, or cancer. In neurodegenerative disorders and in MRS studies of IPD in particular, Lac plays a minor role.

The pathophysiology of parkinsonian disorders involves the indirect and direct pathways of movement in the basal ganglia, which are neuronal circuits facilitating the initiation and execution of voluntary movement. These pathways depend on excitatory and inhibitory signaling and well-balanced regulation of the neurotransmitters dopamine (DA), GABA, and Glu (Fig. 5.1). Thus, exploring these neurotransmitters by MRS is of great interest to understand where and how these neuronal pathways are disrupted in PDS. Glu is readily detectable by short-TE ^{1}H MRS, but its differentiation from the glutamine MRS signal is only considered unambiguous at field strengths of 3T and higher. The detection and quantification of GABA require special editing techniques at clinical field strengths of 1.5T and 3T, the most common being the MEGA-PRESS technique [32]. While accurate detection of GABA using regular short-TE MRS becomes feasible at 4T and higher [33], and even under special experimental conditions at 3T [34], reports on GABA concentrations measured without editing techniques at 1.5T and 3T have to be viewed with caution. Studying GABA levels in parkinsonism is of interest in assessing the motor aspects of the disease: non-MRS, invasive, and ex vivo techniques have described increased GABA levels in the basal ganglia of patients with movement disorders [35]. Both in IPD patients and in animal IPD models, loss of dopaminergic striatal neurons has been found to be accompanied by increased striatal GABA content [36–39]. In postmortem human studies, these increased GABA levels were found to be inversely correlated with the severity of DA loss [37]. DA itself is not visible by MRS due to its low concentration in the order of nM.

Phosphorus (^{31}P) MRS allows for measurement of compounds related to cellular energy metabolism of the brain, including both high-energy phosphates such as adenosine triphosphate (ATP) and phosphocreatine (PCr) and also the low-energy metabolites adenosine diphosphate (ADP) and inorganic phosphate (Pi). Since abnormalities in mitochondrial function contribute to neuronal dysfunction and cell death [9], studying brain energy metabolism by noninvasive ^{31}P MRS techniques can prove a useful tool for both diagnosis and monitoring of disease in PDS.

Being able to measure in vivo concentrations of these metabolites and neurotransmitters, MRS is a promising tool to assess metabolic biomarkers of early disease states, disruptions of metabolic pathways, or biochemical response to treatment in PDS. Later parts of this chapter will summarize and discuss studies in which MRS was used to find metabolic differences between IPD and controls, to find a marker of disease progression, to monitor the effects of treatment, as well as to differentiate between different forms of APDs.

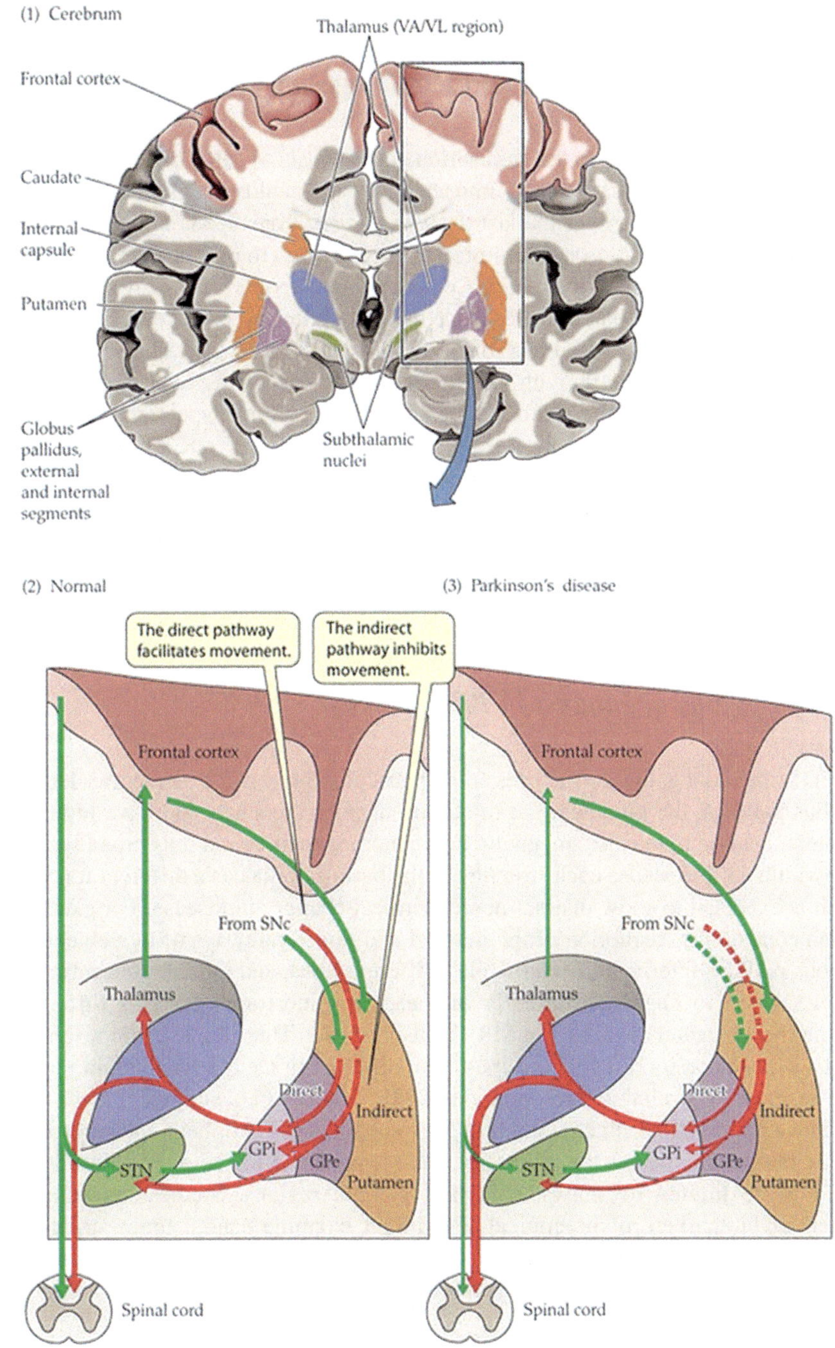

Fig. 5.1 Normal and abnormal circuitry of the basal ganglia. (1) Location of the basal ganglia in a cross section of the human brain. (2) Basic circuit in the normal brain. Excitatory input is shown in *green* and inhibitory input in *red*. (3) In Parkinson's disease the dopaminergic neurons of the substantia nigra pars compacta (SNc) degenerate, unbalancing their modulatory effects on the striatum (putamen and caudate nucleus). Reproduced with permission from Hill et al: Animal Physiology, Third Edition Companion Website. Box Extension 19.2

Technical Considerations and Challenges of MRS in PDS

Technical improvements in the acquisition and analysis of MRS data, as well as in study designs, have made huge impacts on data quality, reproducibility, and reliability of MRS data over the last decades. As such one needs to be cautious in comparing study results from literature 10–20 years ago to more current literature. The test-retest coefficient of variance for the main metabolites (NAA, tCr, tCho, mI, Glu) has become less than 6 % when one uses an optimized protocol and sequence [40]. The widespread availability of high-field MRS at 3T and higher magnetic fields has yielded spectra with better signal-to-noise ratios and high spectral resolution, thus making the identification and quantification of spectral peaks easier and more reliable. At the same time, the use of PRESS, one of the standard localization strategies, at higher field strength, also introduces larger variance in the volume of origin for each metabolite peak (chemical shift displacement). This limitation can be overcome by using localization schemes less prone to this artifact such as STEAM and sLASER.

Brain Regions of Interest in PDS

In PDS, the basal ganglia structures are of particular interest. Yet, structures like the globus pallidus, the putamen, the midbrain, or pons are small and have high iron content, leading to intrinsic magnetic field inhomogeneities and thus broad spectral linewidths [33]. Because each structure in the basal ganglia has a different function, each is expected to show distinct neurochemical changes in disease. For example, when considering disruption of the indirect and direct pathways of movement, the globus pallidus interna and externa play different roles, and GABA concentrations are expected to change differently in these two substructures. This difference cannot be adequately resolved in MRS measurements. Thus, high spatial resolution and adequate corrections for contribution of different tissue types and brain regions are necessary for reliable interpretation of MRS data in PDS studies.

Since pathological changes are thought to begin in the periphery and spread first to the brainstem, then to the midbrain, and eventually to the cortex, studies do not need to be limited to analysis of basal ganglia regions. Studies investigating potential biomarkers of preclinical IPD might examine caudal brain structures, such as the pons [41], while studies focused on understanding metabolic correlates of particular motor or non-motor symptoms may examine the motor cortex or other cortical areas [42].

Choice of MRS Technique and Correction Factors

MR spectroscopic imaging (MRSI) is a technique that can provide metabolic information with a somewhat higher spatial resolution over a larger anatomical coverage in about the same scan time as single voxel MRS (SVS). Yet good quality MRSI data can only be obtained if the whole MRSI region to be covered has high magnetic field homogeneity. Thus, cortical regions close to the skull, as well as other brain regions such as the midbrain and pons, are more amenable to single voxel MRS techniques than to MRSI. However, if the main focus is on studying the metabolism of the basal ganglia structures, one or more axial slices of MRSI may nicely cover most basal ganglia structures in one scan (Fig. 5.2).

The minimal size for volumes of interest (VOIs) in single voxel MRS on clinical scanners is restricted to at least 1 ml, implying that an MRS VOI always contains more than one type of brain tissue (gray matter, white matter, cerebrospinal fluid (CSF)). Because it is well known that gray and white matter contain different metabolite concentrations, it is important to consider the contributions of different brain structures and brain tissue types to the VOI. Even the smaller MRSI voxels still contain mixtures of CSF, gray matter, and white matter and seldom encompass a single brain structure. Therefore, different voxel compositions can easily mask or confound metabolic findings and may contribute to the inconsistent results found in the IPD literature. In the past, CSF contribution to the VOI was corrected for by reporting metabolic concentrations as ratios, often using tCr as the reference metabolite, e.g., NAA/tCr. Studies that use tCr as the reference do not account for the differences in [tCr] between gray and white matter, [tCr] regional variation across the brain, and age dependence of [tCr]. In 2002, O'Neill et al. were one of the first investigators to use quantitative MRS approaches to study IPD, reporting absolute concentrations and accounting for voxel tissue composition [45]. They

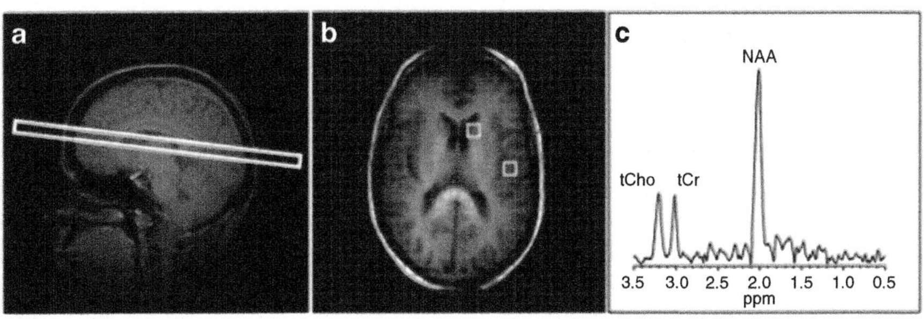

Fig. 5.2 Proton MRSI of a subject with idiopathic Parkinson's disease at 3.0T. (**a**) Prescribed central slice of interest for ^{1}H MRSI scan. (**b**) Sample ^{1}H MRSI grid overlaid on a corresponding localizer MR image with illustrations of $10 \times 10 \times 15$ mm^3 striatal and gray matter voxels of interest. (**c**) A representative proton spectrum showing resonances for *N*-acetylaspartate (NAA), total creatine (tCr), and total choline (tCho). Reproduced with permission from Weiduschat et al. (2015). Usefulness of proton and phosphorus MR spectroscopic imaging for early diagnosis of Parkinson's disease. [43]

not only reported significantly lower tCr in the substantia nigra of IPD patients compared to matched controls, but also substantial atrophy of the putamen, the globus pallidus, and the prefrontal cortex, corrected for total brain volume. While results of volumetric changes in IPD do vary in the literature, atrophy is fairly consistently reported in basal ganglia structures. This highlights how important the correction of metabolite concentrations for age- and disease-related atrophy is. Further, using the relative contribution of a particular brain structure of interest (e.g., the often studied lentiform nucleus) to the MRS VOI as covariate for statistical analysis will help to account for atrophy of a particular structure.

Increased Brain Iron Content in IPD

It has been suggested that brain iron content of basal ganglia nuclei is increased in IPD and possibly also in other forms of parkinsonian disorders [46, 47]. The iron level of the substantia nigra, as measured by T2* MRI, has even been suggested as a marker of disease progression in IPD [48]. The very same mechanism that allows MRI techniques to quantify brain iron content, namely, the fact that iron is paramagnetic and thus decreases the T2 relaxation time, also causes the linewidth of spectral peaks to widen. This increases the challenge of obtaining well-resolved, high-quality spectra from high-iron brain regions. It also may be a confounder in the quantification of MRS visible metabolites, depending on the quantification technique used. Since T2 itself is different for each metabolite of interest and depends on the field strength [49], the influence of iron on T2 may depend on the metabolite as well. Thus, using metabolite ratios may yield confounded results.

Study Population Heterogeneity

Results vary and are sometimes even conflicting in the human MRS literature in PDS. The heterogeneous results may be due to technical differences as well as the different inclusion and exclusion criteria used. Because it appears that a decrease of cortical NAA is primarily a feature of dementia and thus appears in IPD patients with cognitive decline, but not in patients with pure motor dysfunction (see Sect. on APDs), it is important that IPD patients, with and without dementia, are differentiated [50, 51]. Furthermore, the concentration of NAA—as well as many other metabolites—is age dependent [52, 53], and thus age needs to be rigorously accounted for in all statistical models.

Overview of Clinical MRS in Parkinsonian Disorders

Metabolite Changes in IPD Versus Controls

When using MRS to study a particular pathology, the first step is to identify a metabolic signature of the disease. One way is to search for spectroscopic differences between the patient population and age-matched healthy controls. Since the hallmark of IPD is the neurodegeneration of nigrostriatal dopaminergic neurons, NAA (as a marker for neuronal viability) is a classic candidate to be examined in such studies. In fact, most cross-sectional MRS studies on PDS report on NAA findings; however, the range of findings is wide and needs to be carefully differentiated with respect to patient classification, brain region, and data acquisition and analysis methods. Early studies, using 1.5T MRI, mostly report on ratios of NAA to either tCho or tCr. An early multicenter study in 1995 included 151 IPD patients and 97 controls and reported lower NAA/tCr in the striatum in an elderly subset of patients (age >51 years), as well as lower NAA/tCho in a subset of drug-naïve patients [54]. In general, over the past 20 years, the majority of MRS studies on IPD patients reported decreases of NAA/tCr or NAA/tCho. Since neuropathological changes in IPD are known to start in the brainstem and midbrain and later spread to the cortex, we will discuss studies investigating basal ganglia structures separately from studies investigating the cerebral cortex.

^{1}H MRS Studies of the Basal Ganglia

Since reliably obtaining good quality MRS data from small regions of interest only emerged with the advent of high-field MRI scanners, earlier studies often reported on larger, composite basal ganglia regions such as the lentiform nucleus or the striatum. The only early study that reported a significantly lower NAA/tCr in the striatum found it in a subset of elderly patients, as well as in drug-naïve patients, as noted above, probably due to a large sample size [54]. Another study found no differences in NAA/tCr but instead decreases in both, NAA and tCr, and a decrease in NAA/tCho, in the contralateral putamen [55]. Several other early studies investigating the lentiform nucleus, putamen, or thalamus did not find significant changes of NAA, tCr, or tCho [56–59], except for an increase of tCho in the lentiform nucleus reported later by Clarke et al. in 2000 [60]. A recent MRSI study analyzing a region of interest in the striatal area also reports unchanged metabolite levels, potentially due to the short duration of disease in their patient sample compared to other studies [43].

Since nigrostriatal dopaminergic degeneration is the hallmark of IPD, the overwhelming majority of MRS studies of IPD over the past decade have targeted the substantia nigra (SN). As mentioned above, this region is small and lies in the midbrain, which makes it difficult to obtain a narrow spectral linewidth. The SN is also known to have increased iron content in IPD, proportional with the severity of the disease [48]. Thus, conflicting findings are not surprising.

Decreases in NAA/tCr in the SN, in line with diminished neuronal viability, have been reported by several investigators [33, 44, 61–63]. Some of these studies investigated the laterality of their findings and showed asymmetry of NAA/tCr [61, 62], consistent with the known pathological asymmetry in early IPD. Using the improved sensitivity of a 3T high-resolution MRSI technique (point spread function corrected voxel size of 2 ml), Seraji-Bozorgzad et al. followed drug-naïve patients over 3 months and found a 4.4 % decrease of NAA/tCr in IPD patients versus controls [64], in line with the percentage of neuronal loss per decade found ex vivo [65]. Together with an asymmetry index of 16.1 % in IPD versus only 1.6 % in the control group, this study suggests that a decrease in NAA/tCr in the SN may serve as a marker of disease progression in unmedicated patients.

Medication status of patients likely plays a confounding role in clinical studies, as NAA levels have been shown to depend on the medication status (see Sect. on monitoring disease progression). These findings suggest that at least part of the decrease in NAA should be interpreted as inhibition of the neuronal function rather than as a marker of cell loss. Since NAA is synthesized by neuronal mitochondria, lower NAA levels have been hypothesized to reflect impaired mitochondrial function, in line with mitochondrial electron transport chain dysfunction being a primary or secondary event in parkinsonian pathogenesis [66].

Even though they also used high-resolution MRS at high-field strength and scanned the patients after at least 12 h of withdrawal from their respective medications, some studies did not find differences in NAA in the SN [41, 45]. Potentially, these findings might be explained by the positioning of the single voxel or the MRSI grid and differentiation of the rostral and caudal parts of the SN. According to two studies by Gröger et al., only the rostral part of the SN showed a decrease in NAA/tCr and mI/tCr, while both metabolite ratios increased in the caudal part of the SN compared to controls (Fig. 5.3) [44, 63]. Later, they used an improved quantification method using internal water as reference instead of tCr. They found higher NAA and lower tCho in the rostral SN compared to the caudal SN in both IPD patients and controls, while higher tCr in the rostral compared to the caudal part was only found in IPD patients. NAA was found to be significantly reduced in both regions in IPD with respect to the control group [67]. According to these results, which agree with postmortem results by Gerlach et al. [68], the differential changes in NAA/tCr ratios could be due to spatially different concentrations of tCr found in the caudal and rostral parts of the SN in IPD patients. Further findings suggesting changes in tCr instead of (or in addition to) NAA are reported in the 2002 study by O'Neill et al. who also used internal water as reference instead of reporting ratios. This group did not detect any decreases in NAA, tCho, or mI but instead found a decrease in tCr in the substantia nigra, accompanied by atrophy of putamen, globus pallidus, and prefrontal cortex [45].

In summary, changes in NAA/tCr, primarily found in the SN and in agreement with the neuropathology of IPD, could only be detected with high resolution (~1–2 ml), with careful voxel placement, and with sensitivity only achieved by

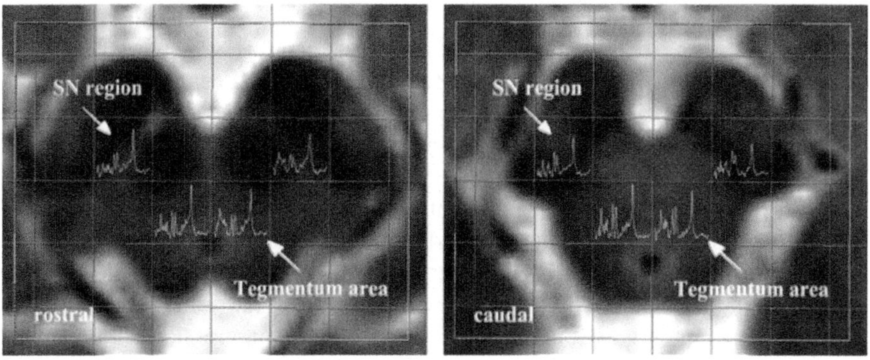

Fig. 5.3 Rostral (*left*) and caudal (*right*) voxels investigated by ^{1}H MRSI in the regions of the substantia nigra as well as in the midbrain tegmentum areas. Reproduced with permission from Gröger et al. (2011) Three-dimensional magnetic resonance spectroscopic imaging in the substantia nigra of healthy controls and patients with Parkinson's disease. Eur Radiol 21:1962–1969 [44]

3T or higher magnetic fields. Since most changes are reported as a ratio to tCr, reflecting energy status or glial activation [64, 69], a decrease in NAA/tCr may be interpreted as a loss of neuronal viability, a deficit in energy metabolism, gliosis, or any combination thereof. Therefore, quantification methods using internal water as is done by LCModel [70]—though requiring correction for CSF content of the volume of interest—are of advantage for interpretation of the results.

^{1}H MRS Studies of Cortical Structures

A large variety of cortical brain regions have been probed by MRS in IPD. Fortunately, the cortical regions associated with the basal ganglia-thalamocortical networks require less stringent resolution than MRS of the substantia nigra; however, other confounders, such as the existence of dementia, may strongly affect cortical measurements. A reduction in NAA/tCr in the temporoparietal cortex was found in IPD patients by Hu et al. [71], as well as by Taylor et al. [72], who also reported a decrease in tCho/tCr for the same region. Furthermore decreased NAA/tCr ratios were found in IPD patients in the motor cortex [73], presupplementary motor cortex [74], posterior cingulate cortex (PCC) [75], and anterior cingulate cortex (ACC) [42] and in temporal gray matter [76]. Importantly, Lewis et al. found an association between reduced NAA/tCr in the ACC and poor executive function and increased psychotic symptoms. Another study took the presence of mild cognitive impairment (MCI) into account and found a decrease in NAA/tCr in the occipital cortex and an increase of tCho/tCr in the PCC only in IPD patients with MCI [51]. Finally, several single voxel and MRSI studies report no significant metabolite changes at all between IPD patients and controls in the investigated cortical regions [43, 59, 77].

In summary, when interpreting these results, one needs to consider that most studies only sample between 10 and 20 patients and controls, very few studies exist on medication-naïve patients, and very few studies control for, or even report on, the medications used by their subject pool. Only few studies differentiate between IPD patients with or without dementia or cognitive impairment, and many early studies did not correct for gray matter, white matter, and CSF content of the volumes of interest. In fact, taking into account that dementia and cognitive impairment are associated with reduced cortical NAA [78], decreased NAA in cortical regions might be a differentiator between IPD without dementia and IPD with dementia. Progressive cognitive symptoms and the emergence of other non-motor symptoms commonly occur with disease progression in IPD. One hypothesis is that the progression of clinical symptoms relates to cortical changes in specific brain regions. This would explain why no metabolic changes are found in cortical regions in early stages of IPD but are found in more advanced disease when more clinical symptoms, such as cognitive impairment, occur. Cortical changes detected in early stage parkinsonism may indicate the presence of an atypical parkinsonian disorder (see Sect. 4.3).

Alterations in Glutamate and GABA

As described above, the balance of the neurotransmitters Glu and GABA plays an essential role in the direct and indirect pathways of movement. However, few studies have examined the roles of these neurotransmitters in the basal ganglia. In 2007, Kicker et al. used TE averaging for unambiguous detection of Glu [79] in the lentiform nucleus, but did not find any changes in Glu in IPD patients (neither on nor 12 h off medication) compared to controls [80]. In a study on metabolic differences between IPD with and without dementia and Alzheimer's disease, Griffith et al. found lower levels of Glu/tCr ratios in the PCC in IPD patients with dementia compared both to controls and to Alzheimer's disease patients [81]. Emir et al. found elevated GABA in the pons (+64 %) and in the putamen (+32 %) using a short-TE STEAM sequence at 7T (Fig. 5.4) [41]. No further GABA MRS studies have been published in human IPD; however, some evidence for increased levels of GABA in the thalamus in parkinsonism comes from studies on manganese-exposed metal workers. Using MEGA-PRESS spectral editing, Dydak et al. showed significantly increased GABA/tCr levels (+80 %) in a relatively large volume centered on the thalamus in highly exposed smelter workers in China [82]. A few years later, the same group showed that increasing thalamic GABA/tCr in manganese-exposed workers predicted reduced motor performance, consistent with increased thalamic inhibition of excitatory glutamate to the motor cortex resulting in hypoactive motor symptoms [83]. Similar studies still need to be done in IPD to determine whether GABA level changes are a potential marker of early motor impairment or only specific to the mechanism of manganese-induced parkinsonism. To date only pilot data exists that suggests that increased thalamic GABA changes in IPD correlate with disease severity, as measured by UPDRS-III [84, 85].

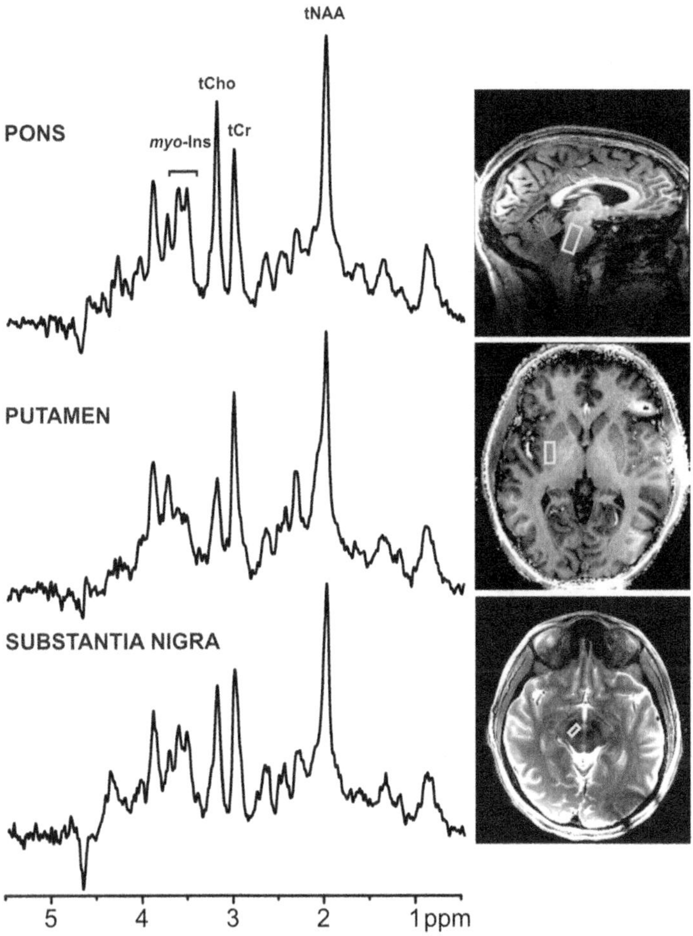

Fig. 5.4 Examples of ^{1}H MR spectra obtained in one patient with PD with STEAM (TR = 5 s, TE = 8 ms) from three VOIs. Reproduced under the Creative Commons Attribution license from Emir et al. (2012) Elevated Pontine and Putamenal GABA Levels in Mild-Moderate Parkinson Disease Detected by 7 Tesla Proton MRS (MRS of Brainstem and Striatum in Parkinson Disease). PLoS One 7:e30918 [41]

^{31}P MRS in PD

Involvement of mitochondrial dysfunction is suggested in the early pathophysiology of IPD and may be assessed by investigating energy metabolism using phosphorus (^{31}P) MRS. A combined ^{31}P and ^{1}H MRSI study at 3T was performed in a group of early IPD and a group of advanced IPD patients, as compared to matched control subjects [86]. The putamen and midbrain were explored. While no significant changes in ^{1}H metabolites, such as NAA, tCr, or tCho, were found, the high-energy phosphates ATP and PCr were found to be significantly lower in both IPD patient groups compared to controls. Low-energy phosphate ADP and inorganic phosphate were unchanged. These results suggest early involvement of mitochondrial dysfunction. In contrast, investigating a striatal and parietal gray matter brain region, a recent study reported unaltered levels of ^{31}P metabolites [43] in early PD. The same group also reported a sex difference in ^{31}P metabolites in these same brain regions, with male PD patients having lower ATP and PCr levels than female PD patients [87]. This indicates that the gender distribution within a study population may need to be considered when comparing results across studies. For example, the Hattingen study [86] had a study sample with 11 males and 5 females, whereas the Weiduschat study [43] had 10 male and 10 female patients in their sample, which could explain why the first study was able to detect changes in the high-energy phosphates in contrast to the second.

Differential MRS Findings in APD

Clinical differentiation between IPD and other forms of atypical parkinsonism (APD), such as dementia with Lewy bodies (DLB), multiple system atrophy (MSA), progressive supranuclear palsy (PSP), or corticobasal degeneration (CBD), is often challenging, particularly in early stages of the disease. This leads to a high rate of misclassifications. Yet, distinguishing between these forms of APD is important for clinical prognosis and will become essential once disease-modifying treatments are available. Assuming that chemical/metabolic changes underlie all neurological symptoms and may be present prior to the clinical manifestation of such symptoms, MRS is expected to play an important role in differentiating IPD from ADPs. In particular, some ADPs like DLB are clinically differentiated from IPD more by their time course than by the early symptoms. Since DLB is a form of dementia, MRS has the potential to detect metabolic changes in brain areas that are affected in dementia rather than in motor disorders.

This concept of differentiation through MRS has been tested in several studies. Lower NAA and elevated mI levels in cortical regions such as the occipital, temporal, parietal, and frontal cortices are well established in dementing disease such as mild cognitive impairment (MCI) and Alzheimer's disease [78]. In comparison to nondemented IPD patients, lower NAA levels in the occipital cortex were found both in patients with Parkinson disease dementia (PDD) [50] and in IPD patients with MCI [51]. Nie et al. also reported a higher tCho/tCr ratio in the posterior cingulate cortex in the MCI group [51]. Another study looking at the posterior cingulate gyrus showed significantly reduced NAA/tCr in PDD compared to nondemented IPD as well as age-matched controls [88]. In that same study, NAA/tCr correlated with neuropsychological performance in both patient groups, but not with severity of motor impairment, suggesting NAA/tCr is a marker for dementia in patients with IPD. Lewis et al. also report a similar correlation between decreased cortical NAA with neuropsychological results [42]. They found that lower NAA/tCr in the anterior cingulate cortex was associated with poor executive function and increased psychotic symptoms. Pagonabarraga et al. [89] report similar findings in their MRS study to evaluate temporal lobe dysfunction in IPD as it relates to progression to dementia. Studying IPD patients without cognitive impairment, with MCI and with dementia (PDD), they found that cognitive impairment was related to dorsolateral prefrontal dysfunction (reduced NAA) even in early stages such as in MCI, while progression to PDD was linked to the additional impairment of temporal lobe structures (reduced NAA in hippocampus in PDD patients).

In line with the notion that reductions in *cortical* NAA reflect the additional neuronal dysfunction in APDs, but not of IPD, Abe et al. studied patients with PSP, CBD, MSA, and IPD and found reduced NAA/tCr in the putamen in all four patient groups compared to healthy controls, as well as in CBD patients compared to MSA and IPD patients. However, only PDP, CBD, and MSA patients, but not IPD patients, also displayed lowered NAA/tCr in the frontal cortex [90].

Another group of studies established that in the basal ganglia, in particular in striatal regions, reductions of NAA are more prominent or more widely distributed in APDs than in IPD: early MRS studies at 1.5T could find lower NAA/tCr in the lentiform nucleus in MSA-P (parkinsonian MSA), but not in IPD [56], or lower NAA/tCr in MSA as compared to IPD patients [91]. Guevara et al. (2010) reported lower NAA concentrations in the pallidum, putamen, and lentiform nucleus in patients with PSP and MSA-P compared to healthy controls and patients with IPD [92]. As discussed above, the detection limit of alterations in NAA in the basal ganglia is strongly dependent on the sensitivity and specificity of the MRS exam, i.e., it is influenced by magnetic field strength and spatial resolution. Thus, the fact that these studies at 1.5T detected changes in APDs but not in IPD might simply indicate that NAA reductions in basal ganglia areas are expected to be higher in APD compared to IPD. This would also explain why some studies also reported decreases in NAA/tCr or NAA/tCho in the lentiform nucleus in IPD, as outlined in the review by Firbank [93]. At 3T, using the NAA/tCr levels in putamen and pontine basis was suggested as discriminator between MSA and IPD by Watanabe et al. (2004), with NAA/tCr being lower in the putamen only in MSA-P patients but in the pontine basis also in MSA-C patients. Finally, another 3T MRS study demonstrated a differentiation between IPD and APDs by comparing NAA/tCr within different regions of the substantia nigra (SN): NAA/tCr was found higher in caudal voxels of the SN than in rostral voxels only in IPD patients. In contrast, the opposite relationship was found for APDs and healthy controls [94].

In summary, many of these studies still need to be verified by reproduction of the results, which are all based on statistical group differences but do not yet allow for individual diagnosis. Without a clear establishment of "normal" NAA values in the brain structures of interest, or those "typical" for IPD, it will remain difficult to use MRS *in the basal ganglia* to differentiate IPD from APDs. However, while these results await further confirmation, the presence of metabolic alterations *in cortical structures* seems to be a strong indicator for the presence of an APD and/or additional dementia, especially in earlier stages of the disease.

MRS to Monitor Disease Progression or Treatment Effect

Since the motor symptoms of IPD result from low striatal DA, good symptom control early in the course of IPD is achieved by medications replacing or enhancing brain DA, such as levodopa. Already in 1995 and 1997, two studies reported significantly lower NAA/tCho in drug-naïve IPD patients compared to controls in striatum and putamen, respectively, but normal levels in levodopa-treated patients [54, 95]. However, these studies were cross-sectional and thus could not yet prove that NAA would be restored in the same patient after treatment. A decade later a prospective study design was able to show that dopaminergic treatment increased NAA in IPD patients to normal levels in the motor cortex: Lucetti et al. [96] reported initially lower levels in NAA/tCr and tCho/tCr in de novo IPD patients. After 6 months of treatment with the dopamine agonist pergolide, an increase to normal levels both in NAA/tCr and tCho/tCr, together with an improvement of motor symptoms as measured by UPDRS-III, was seen. A recent MRS study at 3T used a randomized permuted block study in drug-on and drug-off condition in 20 patients with parkinsonian syndromes [97]. Significantly lower NAA, tCr, and mI in the putamen were reported in drug-off conditions compared to healthy controls. Levodopa therapy nearly restored NAA as well as tCr to normal levels, i.e., levels were still slightly lower but without significant difference to normal levels anymore. Yet, mI remained low in the patients even in drug-on condition. No correlations between putaminal metabolite levels and motor function (UPDRS-III scores) were found.

In cases where pharmacological treatments do not adequately control severe motor symptoms, patients may be eligible for deep brain stimulation (DBS). Due to the challenges of performing MRI on patients with an implanted neurostimulator, MRI studies on the effects of DBS at higher field strength do not exist. There is however one MRS study performed at 1T comparing brain metabolism in left and right globus pallidus, as well as left fronto-basal cortex before and after bilateral DBS of the subthalamic nucleus. The authors reported no changes in globus pallidus but increased *cortical* NAA/tCho and NAA/tCr ratios after DBS, correlating with improvement of motor performances (UPDRS-III) [98].

NAA is known to reflect neuronal viability and thus neuronal dysfunction, as well as neurodegeneration. Dopaminergic neurodegeneration is a slowly progressive and nonreversible feature of IPD starting in the substantia nigra and spreading through the basal ganglia and eventually to the cortex with an accompanying increase in clinical symptoms. Both dopaminergic treatment and DBS alleviate motor symptoms, but neither is a cure, nor do they slow down disease progression. With this in mind, the interpretation of decreased NAA levels found in MRS studies of PDS seems to reflect both components: Only *cortical* NAA levels have been associated with motor or cognitive impairment depending on the brain region studied and seem to be reversible as far as symptoms are reversible through treatment. Here, NAA levels seem to primarily reflect the severity of symptoms and as such the reversible impairment of neuronal function in these regions, until neurodegenerative processes also reach the cortex in later stages of IPD. Lower NAA levels in basal ganglia have not been associated with either motor or cognitive deficits but in contrast have been shown to follow the known rate of neurodegeneration [64] and to reflect postmortem findings of neuronal loss [64]. Yet, the Mazual study still found reversible NAA levels in the putamen. Thus, NAA levels in basal ganglia structures may be reflective of both neuronal dysfunction and early degeneration—having both a reversible and a nonreversible component. Long-term prospective studies would be needed to differentiate dysfunction from degeneration. Of most interest would be a metabolite marker of disease progression that is independent of medication status. The "neurodegenerative component" of NAA levels could be such a marker but is confounded by the "neuronal function" component, which is partially alleviated by medication. Potentially, results from MRS studies performed in medicated subjects might be more indicative of underlying disease progression, but more studies are needed to understand the effects of medication on metabolite levels. Only few studies reported on significantly lower mI levels in IPD patients [44, 97], yet the finding that decreased mI in the putamen was not altered by medication status in the Mazual study [97] makes mI an interesting candidate for a marker of disease progression.

Overview of MRS in Animal Models of Parkinsonian Disorders

As with other diseases, animal models have been developed to probe into the mechanistic properties of IPD. Because lesions occur within the brain, the models incorporate chemicals that either can cross the blood-brain barrier or are directly applied to applicable regions of the brain, such as the substantia nigra and striatum. Historically, there are two commonly used neurotoxic animal models that mimic the depletion of DA neurons in IPD: 6-hydroxydopamine (6-OHDA) and 1-methyl-4-phenyl-1,2,3,6-tetrahydropyridine (MPTP). Both have been used on a variety of animals, large and small, including mice, dogs, cats, and nonhuman primates. While these models have been favorites in the research community, the suspicion that environmental factors may contribute to IPD has led to the development of additional models in recent years. These include using pesticides, as well as the metal manganese, to produce symptoms of parkinsonism. There have been many reviews and in-depth discussion on animal model designs for parkinsonism that the interested reader is referred to [99–101]. The following will give a short introduction to the most common animal models, as well as an overview on MRS studies and their results performed in these models.

6-OHDA

6-OHDA is a potent neurotoxin that causes discrete damage to dopaminergic neurons. 6-OHDA is very similar in shape to dopamine with the exception of an additional hydroxyl group. The additional hydroxyl group renders the molecule toxic, as it can autooxidize and induce oxidative stress to dopaminergic neurons. Consequently, this leads to apoptotic cell death and the loss of neuronal activity. Much like dopamine, 6-OHDA cannot cross the blood-brain barrier and therefore needs to be applied directly to the desired area of the brain. In the past, the most commonly administered location in rat models was the medial forebrain bundle, but recently other areas have been used to better represent human IPD, such as the striatum and substantia nigra. Often 6-OHDA is only injected into one side of the brain, leaving the other side as a reference. This is beneficial because it allows for each animal to be its own control, which eliminates some sources of variability between animals. Articles reviewing the 6-OHDA model and its applications are abundant [99, 102–104].

While 6-OHDA has provided around 50 years of IPD research [99], very few studies have used MRS as an instrument for measuring the impact of this chemical on the modeled brain. Those using MRS to study the effects of 6-OHDA on metabolite concentrations lack consistency in their approaches and results, even though they were all performed on rats. In one study, 6-OHDA was applied to the left striatum of rats resulting in a decrease in tCho/tCr and an increase in Lac/tCr in the afflicted striatum [105]. While the interpretation of changes in tCho is largely ambiguous, increased lactate could indicate hypoxic conditions, as is seen in ischemia [27]. However, no studies have reported similar findings on either tCho or Lac upon 6-OHDA administration since.

Some indirect evidence of neurodegeneration was provided by Hou et al. in 2010 when they showed that 6-OHDA injection into the SN correlated with a decrease in NAA/tCr in the ipsilateral frontal cortex. This decrease in NAA/tCr corresponded to a loss of dopaminergic neurons in the SN analyzed postmortem [106]. NAA/tCr also decreased in the injected striatum in another study, correlating with a significant impairment of the rats' left forepaw performance [107]. More recently, Coune et al.'s study using 14.1T in vivo MRS showed significant decreases in NAA similar to Hou et al. While a decline in NAA is generally attributed to neurodegeneration in a particular region [108], a decrease in NAA in a downstream location might indicate loss of neuronal function, rather than simply neurodegeneration, as discussed above in the section on clinical findings, complicating interpretation. More recently, GABA has been found to increase in the striata after 6-OHDA injection [109, 110], consistent with recent findings in humans [41]. Other metabolites were reported to decrease in an ex vivo study, including Glu, Cr, glycine, alanine, and taurine [109]. Gao et al. also found differences in the contralateral striatum of increasing Glu and decreasing GABA, compared to baseline scans prior to 6-OHDA injection [109]. This suggests that there might be compensatory changes in the contralateral striatum resulting from DA cell loss in the affected striatum.

MPTP

The MPTP model was unintentionally developed after MPTP was accidentally synthesized during the production of an illicit opioid analgesic drug called MPPP. After clusters of intravenous drug users unexpectedly developed a sudden onset of parkinsonism, investigations found an impurity, MPTP, to be the direct cause. In short, MPTP crosses the blood-brain barrier where it is quickly taken up by astrocytes and metabolized into 1-methyl-4-phenyl-dihydropyridine (MPP+) (Fig. 5.5). MPP+ is then able to enter dopaminergic neurons through the dopamine transporter (DAT) where it concentrates in the mitochondria of the neurons and inhibits the function of a protein in the electron transport chain (called complex I), which interrupts a significant portion of the neuron's ATP synthesis. In regions replete with dopaminergic neurons, such as the SN and striatum, this inhibition causes reactive oxidative species to form while also limiting the available ATP in the neurons, leading inevitably to cell death [99, 101, 104].

Various animals, including nonhuman primates, cats, rats, and mice, have been made parkinsonian with MPTP. One of the hallmarks of this model is that results more closely resemble the gradual onset of IPD when MPTP is chronically administered to animals, rather than giving large doses at once [108]. In a 1998 study of chronically administered MPTP to nonhuman primates, a single voxel PRESS sequence was used to detect increases in striatal Lac/macromolecules and tCho. This increase remained until 10 months after the last administration and was accompanied by a significant increase in the tCho/tCr ratio and a significant

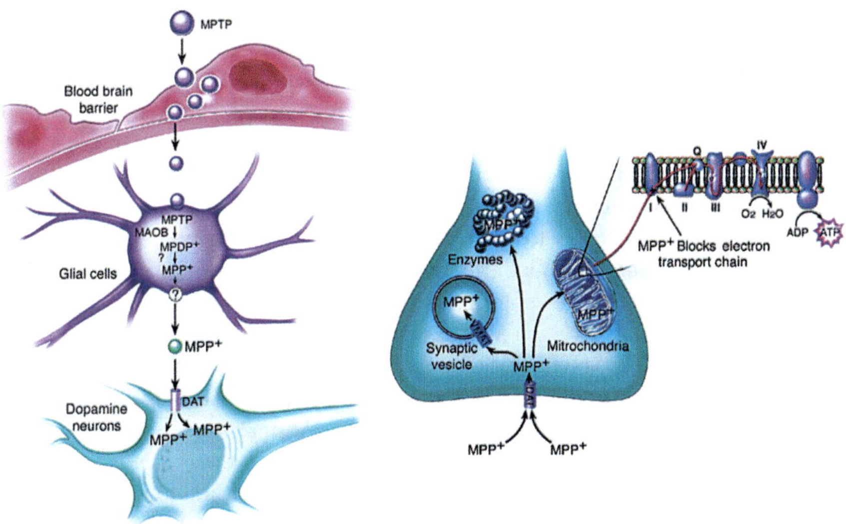

Fig. 5.5 Schematic of MPTP metabolism and toxicity in dopaminergic neurons. Explanation in text. Reproduced with permission from Dauer W, Przedborski S (2003) Parkinson's Disease: Mechanisms and Models. Neuron 39:889–909 [104]

decrease in the NAA/tCr ratio [108]. A later study using acute MPTP exposure over 10 days reported a similar decrease in NAA/tCr ratio along with evidence of increased Lac in cat striata [111].

A benefit of the MPTP model over 6-OHDA is the ability to test the interactions of different drugs concurrent with MPTP exposure. In one study using mice, exposure to MPTP caused an increase in striatal Lac/tCr 2 h after exposure, but in a second round using a DA uptake inhibitor or monoamine oxidase inhibitor prior to exposure, the Lac/tCr increase did not occur [112]. In a comparative study using marmosets, NAA/tCr ratios reduced significantly after 1 week and remained low after 3.5 weeks in marmosets that were exposed to MPTP. However, for the marmosets that had taken a neuroprotective agent prior to MPTP exposure, NAA/tCr ratios slightly decreased 1 week later but then significantly increased 3.5 weeks after exposure [113]. In two separate studies, Chassain et al. found increased glutamine, Glu, and GABA in the striatum of MPTP treated mice, but this effect could be attenuated if pretreated with levodopa [114, 115].

Other Novel Models

There are a few other neurotoxic animal models for IPD. While none of these models have been used in studies where MRS has been employed, they hold promise for future studies of environmental risk factors of IPD, particularly exposure to pesticides. Several pesticides have been shown to produce PD-like symptoms including paraquat and rotenone. Paraquat is a commonly used herbicide that has a structure similar to MPP+ but a different toxic mechanism. Paraquat creates reactive oxygen species (ROS) by redox cycling, meaning it continues to produce these ROS even while undergoing mechanisms intended to make it less reactive. This induces significant lipid peroxidation and eventual degeneration [99–101, 116]. Rotenone is a naturally occurring compound from the roots of some plant species that is used as an insecticide and to kill fish. Rotenone inhibits complex I of the electron transport chain, and due to nigrostriatal dopaminergic neurons' comparatively higher sensitivity to complex I inhibition, small chronic doses can cause parkinsonian symptoms without affecting other regions of the brain [99, 100, 117].

Because some genes have been found to contribute to IPD (mutations are responsible for about 10 % of all IPD cases) [101], genetic animal models are also being developed. Some genes known to cause IPD are LRRK2, parkin, DJ1, SNCA, and PINK1. Mutations in parkin account for about half of the familial cases of IPD and 20 % of the early onset IPD cases. Animal models based on these mutations are newer than the neurotoxic models, but their use is increasing, particularly the models that result in alpha-synuclein overproduction. There are a few ways to create these genetic models. One is to breed the specific transgenic animal model desired, and the other is to induce mutation of the specific gene with a viral vector. For example, Coune et al. injected the SN of rats unilaterally with an adeno-associated virus (AAV) that had been prepared to express human alpha-synuclein in adult nigral neurons. Twelve weeks after injection, the affected side showed a significant increase in GABA compared to the unaffected side using 14.1T MRS [110]. While MRS is yet to be employed with any consistency with these relatively newer models, it is likely that these models will become more common in the near future for studying IPD.

Future Perspectives

IPD is a neurodegenerative disease, characterized among others by slowly progressive dopaminergic neurodegeneration that spreads from brainstem to basal ganglia and ultimately to the cortex. Concurrently, clinical symptoms, that at first primarily include motor dysfunction, progress and start to include additional dysfunctions, such as impaired cognition. The exact type of additional dysfunction depends on the affected cortical regions. Development of treatments that can slow or even stop the neurodegenerative process in IPD strongly depends on early and accurate discrimination of IPD from other parkinsonian disorders to optimize the therapy, as well as on biomarkers of disease progression for monitoring treatment efficacy. Furthermore, such a biomarker at best should be sensitive enough to detect changes in the preclinical or pre-diagnostic phase of the disease.

MRS measures underlying metabolic changes in vivo, potentially before they result in structural brain changes (such as volume loss due to neurodegeneration) or overt clinical symptoms. The ability to noninvasively measure changes in metabolite and/or neurotransmitter concentrations with MRS holds promise to help understand the underlying mechanism and may serve a biomarker of disease progression. At first glance the results of MRS studies in PDS reported in this chapter may seem inconsistent. Yet, both human and animal MRS have made huge technical advances since researchers started using it in IPD research: With increasing magnetic field strength as well as developments of improved acquisition and analysis techniques, the spatial resolution as well as the sensitivity of MRS has increased, and quantification of metabolites has significantly better reproducibility today compared to 20 years ago. Thus, achieving accurate and reproducible insights into the metabolism of even small basal ganglia regions such as caudal and rostral parts of the substantia nigra, or achieving increased sensitivity to identify disease-related metabolic changes even earlier in the disease, will become more feasible.

What remains challenging is to understand confounders that mask reproducibility and correct interpretation of results. Accuracy and reproducibility will remain crucially dependent on controlling for potential confounders, such as brain tissue composition, patient group homogeneity (controlling for disease severity, additional symptoms such as cognitive impairment or depression, APDs), brain iron content and its possible effects on T2 of the studied metabolites, and, last but not least, medication status. At minimum, medication status should be reported in every study.

Future research should focus on ways to differentiate between metabolic changes that correlate with reversible symptoms and those that reflect neurodegenerative progression of the disease. As discussed in this chapter, changes in basal ganglia NAA levels seem to contain a component reflecting neuronal function, which may be reversible with medication, as well as a component reflecting neuronal cell death, which is not reversible. With advancing disease stage, the latter component is assumed to be increasing. The nonreversible component of NAA may also be interpreted as reflecting mitochondrial dysfunction, eventually leading to degeneration. Furthermore, cortical changes in NAA seem to be related more to cognitive deficits as seen in later stages of IPD or in APDs. This interpretation might explain many of the seemingly conflicting results on NAA presented throughout the chapter. In the search for a biomarker of disease progression, future studies should focus on whether metabolic signals are reversible with medication. The current literature suggests that changes in mI and/or GABA might be interesting candidates for nonreversible or early changes in IPD. A differentiation between cortical regions and basal ganglia is of interest for interpretation of results and possibly for identifying APDs early in the disease. Furthermore, longitudinal studies are needed to better understand the dependence on disease severity and symptoms. Once the dependence of metabolite signals on current medication is better understood, such longitudinal studies can be performed in medicated status, as this will highlight the nonreversible metabolic changes reflecting the true disease progression. This nonreversible component of the metabolic response is of most interest when assessing the efficacy of any *novel* therapy that aims to slow or stop progression of IPD instead of alleviating symptoms. Finally, with novel animal models of IPD emerging, which more closely resemble the time course and the pathogenesis of the human disease, back-translation of human MRS findings to the animal model for systematic studies and validation of reversible versus nonreversible metabolic changes or dependence of metabolic changes on particular symptoms will be more readily feasible.

References

1. Braak H, Torch S, Lambeng N (2003) Staging of brain pathology related to sporadic Parkinson's disease. Neurobiol Aging 24:197–211
2. Duncan GW, Khoo TK, Yarnall AJ et al (2014) Health-related quality of life in early Parkinson's disease: the impact of nonmotor symptoms. Mov Disord 29:195–202
3. Guttman JM, Burkholder JJ, Kish JS et al (1997) [11C]RTI-32 PET studies of the dopamine transporter in early dopa-naive Parkinson's disease: Implications for the symptomatic threshold. Neurology 48:1578–1583
4. Postuma RB, Aarsland D, Barone P et al (2012) Identifying prodromal Parkinson's disease: pre-motor disorders in Parkinson's disease. Mov Disord 27:617–626. doi:10.1002/mds.24996
5. Siderowf A, Lang AE (2012) Premotor Parkinson's disease: concepts and definitions. Mov Disord 27:608–614

6. Hughes A, Daniel S, Kilford L, Lees A (1992) Accuracy of clinical diagnosis of idiopathic Parkinson's disease: a clinico-pathological study of 100 cases. J Neurol Neurosurg Psychiatry 55:181–184
7. Goetz CG, Fahn S, Matrinez-Martin P et al (2007) Movement disorder society-sponsored revision of the unified Parkinson's disease rating scale (MDS-UPDRS): process, format, and clinimetric testing plan. Mov Disord 22:41–47
8. Hoehn M, Yahr M (1967) Parkinsonism: onset, progression and mortality. Neurology 17:427–442
9. Kalia LV, Lang AE (2015) Parkinson's disease. Lancet 386:896–912
10. Wenning GK, Tison F, Seppi K et al (2004) Development and validation of the unified multiple system atrophy rating scale (UMSARS). Mov Disord 19:1391–1402
11. Dickson DW (2007) Neuropathology of parkinsonian disease. In: Jankovic J, Tolosa E (eds) Park Dis Mov Disord, 5th edn. Lippincott Williams & Wilkins, Philadelphia, pp 271–294
12. Golbe LI, Ohman-Strickland PA (2007) A clinical rating scale for progressive supranuclear palsy. Brain 130:1552–1565
13. Couper J (1837) On the effects of black oxide of manganese when inhaled into the lungs. Br Ann Med Pharm, Vital Stat Gen Sci 1:41–42
14. Mena I, Marin O, Fuenzalida S, Cotzias G (1967) Chronic manganese poisoning. Clinical picture and manganese turnover. Neurology 17:128–136
15. Cook D, Fahn S, Brait K (1974) Chronic manganese intoxication. Arch Neurol 30:59–64
16. Bowler RM, Gysens S, Diamond E et al (2006) Manganese exposure: neuropsychological and neurological symptoms and effects in welders. Neurotoxicology 27:315–326. doi:10.1016/j.neuro.2005.10.007
17. Racette BA, Aschner M, Guilarte TR et al (2012) Pathophysiology of manganese-associated neurotoxicity. Neurotoxicology 33:881–886. doi:10.1016/j.neuro.2011.12.010
18. Long Z, Jiang Y-M, Li X-R et al (2014) Vulnerability of welders to manganese exposure—a neuroimaging study. Neurotoxicology 45:285–292. doi:10.1016/j.neuro.2014.03.007
19. Hauser R, Zesiewicz T, Martinez C et al (1996) Blood manganese correlates with brain magnetic resonance imaging changes in patients with liver disease. Can J Neurol Sci 23:95–98
20. Rose C, Butterworth RF, Zayed J et al (1999) Manganese deposition in basal ganglia structures results from both portal-systemic shunting and liver dysfunction. Gastroenterology 117:640–644. doi:10.1016/S0016-5085(99)70457-9
21. Sikk K, Haldre S, Aquilonius S-M, Taba P (2011) Manganese-induced parkinsonism due to ephedrone abuse. Parkinsons Dis 2011:1–8. doi:10.4061/2011/865319
22. Perl P, Olanow C (2007) The neuropathology of manganese-induced parkinsonism. J Neuropathol Exp Neurol 66:675–682
23. Eriksson H, Tedroff J, Thuomas K-Å et al (1992) Manganese induced brain lesions in Macaca fascicularis as revealed by positron emission tomography and magnetic resonance imaging. Arch Toxicol 66:403–407
24. Kim SH, Chang KH, Chi JG et al (1999) Sequential change of MR signal intensity of the brain after manganese administration in rabbits: Correlation with manganese concentration and histopathologic findings. Invest Radiol 34:383–393
25. Olanow CW (2004) Manganese-induced parkinsonism and Parkinson's disease. Ann N Y Acad Sci 1012:209–223. doi:10.1196/annals.1306.018
26. Guilarte TR (2010) Manganese and Parkinson's disease: a critical review and new findings. Environ Health Perspect 118:1071–1080. doi:10.1289/ehp.0901748
27. De Graff RA (2007) In vivo NMR spectroscopy: principles and techniques, 2nd edn. Wiley, Hoboken
28. Öz G, Iltis I, Hutter D et al (2011) Distinct neurochemical profiles of spinocerebellar ataxias 1, 2, 6, and cerebellar multiple system atrophy. Cerebellum 10:208–217
29. Miller BL, Moats RA, Shonk T et al (1993) Alzheimer disease: depiction of increased cerebral myo-inositol with proton MR spectroscopy. Radiology 187:433–437

30. Godbolt AK, Waldman AD, Macmanus DG et al (2006) MRS shows abnormalities before symptoms in familial Alzheimer disease. Neurology 66:718–722
31. Kantarci K, Boeve BF, Wszolek ZK et al (2010) MRS in presymptomatic MAPT mutation carriers: a potential biomarker for tau-mediated pathology. Neurology 75:771–778
32. Mullins PG, Mcgonigle DJ, O'Gorman RL et al (2014) Current practice in the use of MEGA-PRESS spectroscopy for the detection of GABA. Neuroimage 86:43–52
33. Öz G, Terpstra M, Tkáč I et al (2006) Proton MRS of the unilateral substantia nigra in the human brain at 4 tesla: detection of high GABA concentrations. Magn Reson Med 55:296–301
34. Near J, Andersson J, Maron E et al (2013) Unedited in vivo detection and quantification of [gamma]-aminobutyric acid in the occipital cortex using short-TE MRS at 3T. NMR Biomed 26:1353–1362
35. Galvan A, Wichmann T (2007) GABAergic circuits in the basal ganglia and movement disorders. Prog Brain Res 160:287–312
36. Hornykiewicz O (2001) Chemical neuroanatomy of the basal ganglia—normal and in Parkinson's disease. J Chem Neuroanat 22:3–12
37. Kish S, Rajput A, Gilbert J et al (1987) GABA-dopamine relationship in Parkinson's disease striatum. Adv Neurol 45:75–77
38. Perry TL, Javoy-Agid F, Agid Y, Fibiger HC (1983) Striatal GABAergic neuronal activity is not reduced in Parkinson's disease. J Neurochem 40:1120–1123
39. Tanaka Y, Niijima K, Mizuno Y, Yoshida M (1986) Changes in gamma-aminobutyrate, glutamate, aspartate, glycine, and taurine contents in the striatum after unilateral nigrostriatal lesions in rats. Exp Neurol 91:259–268
40. Terpstra M, Cheong I, Lyu T et al (2015) Test-retest reproducibility of neurochemical profiles with short-echo, single-voxel MR spectroscopy at 3T and 7T. Magn Reson Med. doi:10.1002/mrm.26022
41. Emir UE, Tuite PJ, Öz G (2012) Elevated pontine and putamenal GABA levels in mild-moderate Parkinson disease detected by 7 tesla proton MRS. PLoS One 7, e30918. doi:10.1371/journal.pone.0030918
42. Lewis S, Shine J, Duffy S et al (2012) Anterior cingulate integrity: executive and neuropsychiatric features in Parkinson's disease. Mov Disord 27:1262–1267
43. Weiduschat N, Mao X, Beal M et al (2015) Usefulness of proton and phosphorus MR spectroscopic imaging for early diagnosis of Parkinson's disease. J Neuroimaging 25:105–110
44. Gröger A, Chadzynski G, Godau J et al (2011) Three-dimensional magnetic resonance spectroscopic imaging in the substantia nigra of healthy controls and patients with Parkinson's disease. Eur Radiol 21:1962–1969. doi:10.1007/s00330-011-2123-5
45. O'Neill J, Schuff N, Marks WJ et al (2002) Quantitative ^{1}H magnetic resonance spectroscopy and MRI of Parkinson's disease. Mov Disord 17:917–927. doi:10.1002/mds.10214
46. Dexter DT, Wells FR, Lees AJ et al (1989) Increased nigral iron content and alterations in other metal ions occurring in brain in Parkinson's disease. J Neurochem 52:1830–1836. doi:10.1111/j.1471-4159.1989.tb07264.x
47. Berg D, Hochstrasser H, Schweitzer K, Riess O (2006) Disturbance of iron metabolism in Parkinson's disease—ultrasonography as a biomarker. Neurotox Res 9:1–13
48. Ulla M, Bonny J-M, Ouchchane L et al (2013) Is R-2* a new MRI biomarker for the progression of Parkinson's disease? A longitudinal follow-up. PLoS One 8, e57904
49. Mlynárik V, Gruber S, Moser E (2001) Proton T_1 and T_2 relaxation times of human brain metabolites at 3 Tesla. NMR Biomed 14:325–331. doi:10.1002/nbm.713
50. Summerfield C, Gómez-Ansón B, Tolosa E et al (2002) Dementia in Parkinson disease. Arch Neurol. doi:10.1001/archneur.59.9.1415
51. Nie K, Zhang Y, Huang B et al (2013) Marked N-acetylaspartate and choline metabolite changes in Parkinson's disease patients with mild cognitive impairment. Parkinsonism Relat Disord 19:329–334

52. Kadota T, Horinouchi T, Kuroda C (2001) Development and aging of the cerebrum: assessment with proton MR spectroscopy. Am J Neuroradiol 22:128–135
53. Angelie E, Bonmartin A, Boudraa A et al (2001) Regional differences and metabolic changes in normal aging of the human brain: proton MR spectroscopic imaging study. Am J Neuroradiol 22:119–127
54. Holshouser BA, Komu M, Möller HE et al (1995) Localized proton NMR spectroscopy in the striatum of patients with idiopathic parkinson's disease: a multicenter pilot study. Magn Reson Med 33:589–594
55. Chaudhuri KR, Lemmens GM, Williams SCR et al (1996) Proton magnetic resonance spectroscopy of the striatum in Parkinson's disease patients with motor response fluctuations. Parkinsonism Relat Disord 2:63–67
56. Davie C, Wenning G, Barker G et al (1995) Differentiation of multiple system atrophy from idiopathic Parkinson's disease using proton magnetic resonance spectroscopy. Ann Neurol 37:204–210
57. Tedeschi G, Litvan I, Bonavita S et al (1997) Proton magnetic resonance spectroscopic imaging in progressive supranuclear palsy, Parkinson's disease and corticobasal degeneration. Brain 120:1541–1552
58. Cruz C, Aminoff M, Meyerhoff D et al (1997) Proton MR spectroscopic imaging of the striatum in Parkinson's disease. Magn Reson Imaging 15:619–624
59. Clarke CE, Lowry M, Horsman A (1997) Unchanged basal ganglia N-acetylaspartate and glutamate in idiopathic Parkinson's disease measured by proton magnetic resonance spectroscopy. Mov Disord 12:297–301
60. Clarke CE, Lowry M (2000) Basal ganglia metabolite concentrations in idiopathic Parkinson's disease and multiple system atrophy measured by proton magnetic resonance spectroscopy. Eur J Neurol 7:661–665
61. Choe BY, Park JW, Lee KS et al (1998) Neuronal laterality in Parkinson's disease with unilateral symptom by in vivo ^{1}H magnetic resonance spectroscopy. Invest Radiol 33:450–455
62. Seraji-Bozorgzad N, Bao F, Shneyder N et al (2014) Quantitative MRI biomarker study in Parkinson's disease: high-field 1H-MR spectroscopic and multi-modal MRI longitudinal study of the substantia nigra. Neurology 82(10 Supplement): P6.069
63. Gröger A, Bender B, Wurster I et al (2013) Differentiation between idiopathic and atypical parkinsonian syndromes using three-dimensional magnetic resonance spectroscopic imaging. J Neurol 84:644–649
64. Seraji-Bozorgzad N, Bao F, George E, et al. (2015) Longitudinal study of the substantia nigra in Parkinson disease: A high-field ^{1} H-MR spectroscopy imaging study. Mov Disord 30:1400–1404. doi: 10.1002/mds.26323
65. Fearnley JM, Lees AJ (1991) Ageing and Parkinson's disease: substantia nigra regional selectivity. Brain 114:2283–2301
66. Schapira A, Gu M, Taanman J et al (1998) Mitochondria in the etiology and pathogenesis of Parkinson's disease. Ann Neurol 44:S89–S98
67. Gröger A, Kolb R, Schäfer R, Klose U (2014) Dopamine reduction in the substantia nigra of Parkinson's disease patients confirmed by in vivo magnetic resonance spectroscopic imaging. PLoS One 9, e84081
68. Gerlach M, Gsell W, Kornhuber J et al (1996) A post mortem study on neurochemical markers of dopaminergic, GABA-ergic and glutamatergic neurons in basal ganglia-thalamocortical circuits in Parkinson syndrome. Brain Res 741:142–152
69. Öz G, Alger JRJ, Barker PBP et al (2014) Clinical proton MR spectroscopy in central nervous system disorders. Radiology 270:658–679. doi:10.1148/radiol.13130531
70. Provencher SW (1993) Estimation of metabolite concentrations from localized in vivo proton NMR spectra. Magn Reson Med 30:672–679. doi:10.1002/mrm.1910300604

71. Hu M, Taylor-Robinson S, Chaudhuri K et al (1999) Evidence for cortical dysfunction in clinically non-demented patients with Parkinson's disease: a proton MR spectroscopy study. J Neurol Neurosurg Psychiatry 67:20–27
72. Taylor-Robinson S, Turjanski N, Bhattacharya S et al (1999) A proton magnetic resonance spectroscopy study of the striatum and cerebral cortex in Parkinson's disease. Metab Brain Dis 14:45–55
73. Lucetti C, Del Dotto P, Gambaccini G et al (2001) Proton magnetic resonance spectroscopy (^{1}H-MRS) of motor cortex and basal ganglia in de novo Parkinson's disease patients. Neurol Sci 22:69–70
74. Camicioli R, Hanstock C, Bouchard T et al (2007) Magnetic resonance spectroscopic evidence for presupplementary motor area neuronal dysfunction in Parkinson's disease. Mov Disord 22:382–386
75. Camicioli R, Korzan J, Foster S et al (2004) Posterior cingulate metabolic changes occur in Parkinson's disease patients without dementia. Neurosci Lett 354:177–180
76. Levin BE, Katzen HL, Maudsley A et al (2014) Whole-brain proton MR spectroscopic imaging in Parkinson's disease. J Neuroimaging 24:39–44
77. Bowen BC, Block RE, Sanchez-Ramos J et al (1995) Proton MR spectroscopy of the brain in 14 patients with Parkinson disease. Am J Neuroradiol 16:61–68
78. Tumati S, Martens S, Aleman A (2013) Magnetic resonance spectroscopy in mild cognitive impairment: Systematic review and meta-analysis. Neurosci Biobehav Rev 37:2571–2586. doi:10.1016/j.neubiorev.2013.08.004
79. Hurd R, Sailasuta N, Srinivasan R et al (2004) Measurement of brain glutamate using TE-averaged PRESS at 3T. Magn Reson Med 51:435–440. doi:10.1002/mrm.20007
80. Kickler N, Krack P, Fraix V et al (2007) Glutamate measurement in Parkinson's disease using MRS at 3 T field strength. NMR Biomed 20:757–762
81. Griffith H, Okonkwo O, O'Brien T, Hollander J (2008) Reduced brain glutamate in patients with Parkinson's disease. NMR Biomed 21:381–387
82. Dydak U, Jiang Y, Long L et al (2011) In vivo measurement of brain GABA concentrations by magnetic resonance spectroscopy in smelters occupationally exposed to manganese. Environ Health Perspect 119:219–224
83. Long Z, Li X, Xu J et al (2014) Thalamic GABA predicts fine motor performance in manganese-exposed smelter workers. PLoS One 9, e88220
84. Dharmadhikari S, Ma R, Yeh C-L, et al. (2015) MRS of basal-ganglia in Parkinson's Disease reveals higher GABA levels. In: 23rd Annual Meet Exhibition International Society for Magnetic Resonance in Medicine Toronto, Canada, p 23:2209
85. Dydak U, Dharmadhikari S, Snyder S, Zauber SE (2015) Increased Thalamic GABA Levels Correlate with Parkinson Disease Severity. AD/PD Conference, Nice, France; March 18–21 2015
86. Hattingen E, Magerkurth J, Pilatus U et al (2009) Phosphorus and proton magnetic resonance spectroscopy demonstrates mitochondrial dysfunction in early and advanced Parkinson's disease. Brain 132:3285–3297. doi:10.1093/brain/awp293
87. Weiduschat N, Mao X, Beal MF et al (2014) Sex differences in cerebral energy metabolism in Parkinson's disease: a phosphorus magnetic resonance spectroscopic imaging study. Parkinsonism Relat Disord 20:545–548. doi:10.1016/j.parkreldis.2014.02.003
88. Griffith HR, den Hollander JA, Okonkwo OC et al (2008) Brain N-acetylaspartate is reduced in Parkinson disease with dementia. Alzheimer Dis Assoc Disord 22:54–60. doi:10.1097/WAD.0b013e3181611011
89. Pagonabarraga J, Gómez-Ansón B, Rotger R et al (2012) Spectroscopic changes associated with mild cognitive impairment and dementia in Parkinson's disease. Dement Geriatr Cogn Disord 34:312–318. doi:10.1159/000345537
90. Abe K, Terakawa H, Takanashi M et al (2000) Proton magnetic resonance spectroscopy of patients with parkinsonism. Brain Res Bull 52:589–595

91. Frederico F, Simone I, Lucivero V et al (1999) Usefulness of proton magnetic resonance spectroscopy in differentiating parkinsonian syndromes. Ital J Neurol Sci 20:223–229
92. Guevara CA, Blain CR, Stahl D et al (2010) Quantitative magnetic resonance spectroscopic imaging in Parkinson's disease, progressive supranuclear palsy and multiple system atrophy. Eur J Neurol 17:1193–1202
93. Firbank M, Harrison R, O'Brien J (2002) A comprehensive review of proton magnetic resonance spectroscopy studies in dementia and Parkinson's disease. Dement Geriatr Cogn Disord 14:64–76
94. Watanabe H, Fukatsu H, Katsuno M et al (2004) Multiple regional ^{1}H-MR spectroscopy in multiple system atrophy: NAA/Cr reduction in pontine base as a valuable diagnostic marker. J Neurol 75:103–109
95. Ellis CM, Lemmens G, Williams SC et al (1997) Changes in putamen N-acetylaspartate and choline ratios in untreated and levodopa-treated Parkinson's disease: A proton magnetic resonance spectroscopy study. Neurology 49:438–444
96. Lucetti C, Del Dotto P, Gambaccini G et al (2007) Influences of dopaminergic treatment on motor cortex in Parkinson disease: a MRI/MRS study. Mov Disord 22:2170–2175
97. Mazuel L, Chassain C, Jean B et al (2016) Proton MR spectroscopy for diagnosis and evaluation of treatment efficacy in Parkinson disease. Radiology 278:505–513
98. Llumiguano C, Kovacs N, Usprung Z et al (2008) ^{1}H-MRS experiences after bilateral DBS of the STN in Parkinson's disease. Parkinsonism Relat Disord 14:229–232
99. Cannon JR, Greenamyre JT (2010) Neurotoxic in vivo models of Parkinson's disease. Recent advances. Prog Brain Res. 184:17–33. doi:10.1016/S0079-6123(10)84002-6
100. Cannon JR, Greenamyre JT (2011) The role of environmental exposures in neurodegeneration and neurodegenerative diseases. Toxicol Sci 124:225–250. doi:10.1093/toxsci/kfr239
101. Blesa J, Phani S, Jackson-Lewis V, Przedborski S (2012) Classic and new animal models of Parkinson's disease. J Biomed Biotechnol. 2012:845618. doi:10.1155/2012/845618
102. Deumens R, Blokland A, Prickaerts J (2002) Modeling Parkinson's disease in rats: an evaluation of 6-OHDA lesions of the nigrostriatal pathway. Exp Neurol 175:303–317. doi:10.1006/exnr.2002.7891
103. Blum D, Torch S, Lambeng N (2001) Molecular pathways involved in the neurotoxicity of 6-OHDA, dopamine and MPTP: contribution to the apoptotic theory in Parkinson's disease. Prog Neurobiol 65:135–172
104. Dauer W, Przedborski S (2003) Parkinson's disease: mechanisms and models. Neuron 39:889–909. doi:10.1016/S0896-6273(03)00568-3
105. Kickler N, Lacombe E, Chassain C et al (2009) Assessment of metabolic changes in the striatum of a rat model of parkinsonism: an in vivo ^{1}H MRS study. NMR Biomed 22:207–212. doi:10.1002/nbm.1305
106. Hou Z, Lei H, Hong S et al (2010) Functional changes in the frontal cortex in Parkinson's disease using a rat model. J Clin Neurosci 17:628–633. doi:10.1016/j.jocn.2009.07.101
107. Kim SY, Choe BY, Lee HS et al (2011) Forelimb akinesia and metabolic alteration in the striatum following unilateral 6-hydroxydopamine lesion in rats: An in vivo proton magnetic resonance spectroscopy study. Neurochem J 5:270–277. doi:10.1134/S1819712411040088
108. Brownell AL, Jenkins BG, Elmaleh DR et al (1998) Combined PET/MRS brain studies show dynamic and long-term physiological changes in a primate model of Parkinson disease. Nat Med 4:1308–1312. doi:10.1038/3300
109. Gao HC, Zhu H, Song CY et al (2013) Metabolic changes detected by ex vivo high resolution 1H NMR spectroscopy in the striatum of 6-OHDA-induced Parkinson's rat. Mol Neurobiol 47:123–130. doi:10.1007/s12035-012-8336-z
110. Coune PG, Craveiro M, Gaugler MN et al (2013) An in vivo ultrahigh field 14.1T ^{1}H-MRS study on 6-OHDA and α-synuclein-based rat models of Parkinson's disease: GABA as an early disease marker. NMR Biomed 26:43–50. doi:10.1002/nbm.2817

111. Podell M, Hadjiconstantinou M, Smith MA, Neff NH (2003) Proton magnetic resonance imaging and spectroscopy identify metabolic changes in the striatum in the MPTP feline model of parkinsonism. Exp Neurol 179:159–166. doi:10.1016/S0014-4886(02)00015-8
112. Koga K, Mori A, Ohashi S et al (2006) ^{1}H MRS identifies lactate rise in the striatum of MPTP-treated C57BL/6 mice. Eur J Neurosci 23:1077–1081. doi:10.1111/j.1460-9568.2006.04610.x
113. van Vlieta SAM, Blezer ELA, Jongsma MJ et al (2008) Exploring the neuroprotective effects of modafinil in a marmoset Parkinson model with immunohistochemistry, magnetic resonance imaging and spectroscopy. Brain Res 1189:219–228. doi:10.1016/j.brainres.2007.10.059
114. Chassain C, Bielicki G, Carcenac C et al (2013) Does MPTP intoxication in mice induce metabolite changes in the nucleus accumbens? A ^{1}H nuclear MRS study. NMR Biomed 26:336–347. doi:10.1002/nbm.2853
115. Chassain C, Bielicki G, Keller C et al (2010) Metabolic changes detected in vivo by ^{1}H MRS in the MPTP-intoxicated mouse. NMR Biomed 23:547–553. doi:10.1002/nbm.1504
116. Wu B, Song B, Tian S et al (2012) Central nervous system damage due to acute paraquat poisoning: A neuroimaging study with 3.0T MRI. Neurotoxicology 33:1330–1337. doi:10.1016/j.neuro.2012.08.007
117. Cannon JR, Tapias VM, Na HM et al (2009) A highly reproducible rotenone model of Parkinson's disease. Neurobiol Dis 34:279–290. doi:10.1016/j.nbd.2009.01.016

Chapter 6
Magnetic Resonance Spectroscopy in Huntington's Disease

Fanny Mochel, Janet M. Dubinsky, and Pierre-Gilles Henry

Abstract Huntington's disease (HD) is an autosomal dominant neurodegenerative disease with complete penetrance. Although the understanding of the cellular mechanisms that drive neurodegeneration in HD and account for the characteristic pattern of neuronal vulnerability is incomplete, defects in energy metabolism, particularly mitochondrial function, represent a common thread in studies of HD pathogenesis in animal models and humans. Here we review the metabolic dysfunction captured by in vivo proton and phosphorus magnetic resonance spectroscopy (MRS) in animal models of HD and human carriers of the mutated huntingtin protein. Having access to a presymptomatic population of individuals gives a unique possibility of approaching early pathophysiological changes in HD. Although longitudinal studies are needed to determine more precisely the time course of these metabolic changes in humans, MRS tools are already used in clinical trials to obtain proof of concepts of the ability of disease-modifying drugs to impact on disease progression in HD.

Keywords Huntington's disease • Caudate • Putamen • Energy metabolism

F. Mochel, M.D., Ph.D. (✉)
Institut du Cerveau et de la Moelle épinière, ICM, Inserm U 1127, CNRS UMR 7225, Sorbonne Universités, UPMC Univ Paris 06 UMR S 1127, 75013 Paris, France

Département de Génétique, Salpêtrière University Hospital, 75013 Paris, France

Groupe de Recherche Clinique Neurométabolique, Université Pierre and Marie Curie, Paris, France
e-mail: fanny.mochel@upmc.fr

J.M. Dubinsky, Ph.D.
Department of Neuroscience, University of Minnesota, Minneapolis, MN 55455, USA
e-mail: dubin001@umn.edu

P.-G. Henry, Ph.D.
Department of Radiology, Center for Magnetic Resonance Research, University of Minnesota, Minneapolis, MN 55455, USA
e-mail: henry@cmrr.umn.edu

G. Öz (ed.), *Magnetic Resonance Spectroscopy of Degenerative Brain Diseases*, Contemporary Clinical Neuroscience, DOI 10.1007/978-3-319-33555-1_6

Huntington's Disease: A Metabolic Disease?

Huntington's disease (HD) has a prevalence of 5–10 per 100,000 in the general population of the Western hemisphere. The disorder is reported in all ethnicities, although it is much more common in Scotland and Venezuela and less common in Finland, China, Japan, and black South Africans [1]. HD affects at least 40,000 people living in Europe, and an estimated additional 80,000 individuals carry the HD mutation but remain as yet unaffected. In the United States, there are about 30,000 affected individuals, and another 150,000 Americans have a genetic risk for developing the disease [2].

Molecular Diagnosis and Presymptomatic Testing

HD is inherited as an autosomal dominant trait. In individuals with HD, a polymorphic trinucleotide repeat sequence (CAGn), near the 5′ end of the *HTT* gene, is expanded beyond the normal repeat range, leading to the translation of an expanded polyglutamine sequence in the huntingtin protein (Htt) [3]. In the normal population, the number of CAG repeats varies from 17 to 29. Individuals with adult-onset HD usually have a CAG expansion from 40 to 55, whereas those with juvenile onset have CAG expansions greater than 60 that are often inherited from the father. However, penetrance of alleles with a CAG repeat in the range of 36–39 is reduced. Although people with intermediate alleles (27–35 repeats) will not experience HD symptoms themselves, the repeat inherited by their children can be longer than their own. The CAG repeat length is inversely correlated with the age at onset; the longer the CAG repeat stretches within the *HTT* gene, the earlier is the onset and the more severe the clinical manifestations of HD [4].

The availability of the HD genetic test makes possible the identification of mutant gene carriers long before they become symptomatic [5]. Presymptomatic testing has been offered since 1993 within a multidisciplinary framework [6]. Having access to a presymptomatic population of individuals also gives a unique possibility of approaching early pathophysiological changes in HD. This feature is pivotal in identifying therapies that will be efficient not only at slowing but also stopping or even reversing the pathological process in HD. When symptoms develop in patients, atrophy of the underlying region is already pronounced [7]. Therefore, the presymptomatic phase in HD provides a unique window for therapeutic intervention.

Clinical Manifestations

Individuals who have 36 CAG repeats or more may develop the clinical symptoms and signs of HD including motor, cognitive, and neuropsychiatric abnormalities that cause a progressive loss of functional capacity and shortened lifespan. The clinical features of HD usually emerge in adulthood between 30 and 50 years of age, after which illness progresses steadily over a period of 15–25 years. Patients with HD may present with motor signs (over 50 %), with behavioural signs, or with both motor and behavioural signs [8]. Patients themselves may be unaware or unconcerned about early cognitive and motor changes. The motor disorder usually begins with clumsiness and fidgetiness that evolve into chorea. The presence and severity of chorea vary markedly from subject to subject and over time. In addition to chorea, patients with HD have bradykinesia and motor impersistence, with difficulty sustaining ongoing movement. With advancing disease, there is progression of bradykinesia and dystonic movements appear. The gait disorder of HD is complex, with chorea, parkinsonism, lapses in tone of antigravity muscles, and ataxia. Ultimately, progressive bradykinesia and intractable falls lead to the wheelchair- or bed-bound state. Dysarthria and dysphagia progressively impair communication and nutrition.

Behavioural changes contribute largely to disability in HD [9]. The most common changes in early disease are irritability, anxiety, and mood disturbance. Depressed mood is very common with a lifetime risk of about 50 % [10]. The critical periods for suicide risk are prior to receiving a diagnosis and when patients experience a loss of independence [11]. Psychosis is not common but may be difficult to treat. Apathy increases in concert with disease severity and is a nearly universal feature of advanced disease [12]. Behavioural and psychiatric disorders may predate the onset of overt HD by as long as a decade, reflecting early pathological changes in the non-motor areas of the striatum [8]. Cognitive changes are frequent in HD [8]. The dementia of HD fits the description of subcortical dementia with disordered attention, concentration, motivation, insight, judgement, and problem solving rather than traditional cortical signs such as aphasia and apraxia [13].

Neuropathology

The initial detailed description of HD was that of George Huntington, a medical practitioner of Pomeroy, Ohio, in 1872. The pathology of HD includes prominent neuronal loss and gliosis in the caudate nucleus and putamen along with regional and more diffuse atrophy. On gross examination of the brain, the atrophy may be diffuse or confined to the frontal lobe or the striatum [14]. With the progression of the disease, neostriatal degeneration appears to simultaneously move in a caudo-rostral direction, in a dorsoventral direction, and in a medio-lateral direction [14]. Large cortical neurons in layer VI are also involved, as are neurons in the thalamus, pars reticulata of the substantia nigra, superior olive, lateral tuberal nucleus of the hypothalamus, and deep cerebellar nuclei [15]. Premature death of medium spiny gamma-aminobutyric acid-ergic (GABAergic) projection neurons of the neostriatum with gradual atrophy of the caudate nucleus, putamen, and external segment of the globus pallidus is the neuropathologic hallmark of HD [14]. Interneurons of the striatum are generally spared. Early, there is preferential loss of GABAergic neurons that co-localize enkephalin, dynorphin, and substance P [16]. These neurons are thought to predominate in the indirect pathway, accounting for difficulties suppressing abnormal movement early in the disease course. With disease progression, all GABAergic medium spiny neurons are affected, including those in the direct pathway, explaining the emergence of parkinsonism in later disease.

Pathophysiology and Energy Metabolism in HD

In HD, both normal and mutant alleles are expressed. Gain-of-function alterations in which the mutant Htt has protein-level toxicity, as well as loss of function of normal Htt, have been proposed to contribute to HD [17]. Like many neurodegenerative diseases, HD is characterized by a cascade of events leading to neuronal cell death. The main pathophysiological mechanisms proposed to underlie HD are (1) neuronal aggregates, (2) transcriptional dysregulation, (3) excitotoxicity, (4) mitochondrial dysfunction and altered energy metabolism, (5) changes in axonal transport, and (6) signalling dysfunction [18].

Energy metabolism has been under the scope of HD research for many years due to several observations in both patients and models of the disease. A reduction in ATP production was shown in the brain of HD mice, including presymptomatic mice [19]. A significant reduction in ATP levels and mitochondrial respiration was measured in striatal cells of HD mice, although the respiratory chain complexes were not impaired [20]. PET studies conducted in HD patients showed that glucose consumption is reduced in the brain, providing strong evidence for hypometabolism, especially in the basal ganglia, even in presymptomatic mutation carriers [21–23]. The underlying cause of this early energy deficit in HD brain is currently unknown, but impaired glycolysis [24], tricarboxylic acid (TCA) cycle [25], and/or oxidative phosphorylation [20] may be involved. Mutant Htt was shown to decrease the expression of peroxisome proliferator-activated receptor coactivator-1alpha (PGC-1α) in the striatum of HD mice and patients, through a CREB-dependent transcriptional inhibition [26]. PGC-1α is a transcriptional co-activator that regulates key energetic metabolic pathways, both in the brain and peripheral tissues [27]. The possible role of PGC-1α in HD was initially suspected from the observation of selective striatal lesions in the PGC-1α knockout mouse [28]. Downregulation of PGC-1α in HD striatum was then shown to affect mitochondrial energy metabolism, possibly by impairing oxidative phosphorylation [26]. In addition, the inhibition of succinate dehydrogenase, by 3-nitropropionate

(3NP) or malonate, mimicking HD neuropathology in baboons [29] and mice [30], indicates that a lack of substrates for the TCA and the respiratory chain is implicated in the energy deficit of HD brain.

Clinical Utility of MRS in HD

By tracking several brain regions longitudinally in presymptomatic individuals that were predicted to be far or close to motor onset according to the individual's age and *HTT* CAG repeat length [4], the TRACK-HD study has shown that progressive striatal atrophy, and especially caudate atrophy, is a component of the disease that can be observed decades before predicted motor onset [31]. Several brain regions besides the caudate are also involved early in the disease process, but the neural networks that underlie such alterations are unknown [31]. Metabolic changes are likely to occur prior to brain atrophy and could potentially be reversed with early therapeutic intervention. A real challenge in HD is therefore to identify and capture the metabolic dysfunction that leads to structural changes. The sensitivity of magnetic resonance spectroscopy (MRS) should allow the characterization of such metabolic dysfunction. Because of its ability to study brain metabolism in vivo, including energy metabolism using ^{31}P MRS is of particular clinical relevance in HD, and especially at the presymptomatic stage of the disease.

As a monogenic disease with full penetrance (for alleles with CAG repeats $>$39)—carriers develop the disease in an age-dependent manner; HD is a model to understand mechanisms of neurodegeneration that may be common to more frequent neurodegenerative diseases such as Parkinson's or Alzheimer's diseases. Major discoveries in HD are likely to impact largely on the field of neurodegeneration. Consequently, promising therapeutic strategies have been developed in HD among which are (1) the lowering of mutated *HTT* using RNA interference, (2) the correction of toxic post-translational modifications of the mutated *HTT*, and (3) the compensation of key downstream cellular alterations like vesicular transport or mitochondrial functions. However, due to the small effect sizes of any clinical parameter, it is useful to identify robust and functional biomarkers of HD for proof-of-concept studies testing these exciting treatments before they can be launched in large clinical trials. The detection by MRS of dynamic metabolic changes linked to the progression of the disease could lead to a better understanding of the nature of the metabolic changes in HD and identification of novel brain biomarkers. In addition, these biomarkers could be used to assess the efficacy of therapeutic strategies in HD carriers, and especially those targeting brain energy metabolism [32].

Overview of MRS in Animal Models of HD

Proton MRS

Since the earliest development of animal models of HD, MRS has been utilized to biochemically characterize the disease state and longitudinally follow disease progression. Moreover, these metabolite profiles have proved useful in assessing the impact of proposed treatments and gaining insight into underlying disease mechanisms. MRS has also been applied to biochemical preparations of synthetic polyglutamine molecules to assess the structure of monomers, fibrils, and aggregates [33, 34]. While these experiments have shed light on the molecular nature of the toxic structure, they will not be addressed in the current discussion of in vivo MRS applications.

One of the earliest animal models of HD involved systemic injections of metabolic inhibitors, including 3NP, to produce striatal lesions. Decreased total N-acetylaspartate (tNAA) and increased lactate concentrations accompanied these lesions [35, 36]. Low-dose, chronic administration of 3NP, an inhibitor of succinate dehydrogenase, an enzyme involved in both the TCA cycle and the mitochondrial electron transport chain, produces striatal degeneration and motor symptoms reminiscent of HD, demonstrating that disruption of energy metabolism reproduces the main features of HD phenotypes [35, 37, 38]. In rats receiving a dietary creatine supplementation, the striatal lactate/tNAA ratio detected by ^{1}H MRS was reduced compared to 3NP alone [38]. As the effects of 3NP administration developed slowly in baboon, absolute concentrations of striatal tNAA, creatine, and choline decreased initially followed many weeks later by an increased lactate concentration [37]. When ^{13}C MRS was used to measure TCA cycle activity in the striatum of 3NP-treated rats, the TCA cycling rate was lower than controls, confirming mitochondrial inhibition [39]. More recent analytical measurements of striatal ATP and phosphocreatine (PCr) concentrations from 3NP-treated rats reported that 3NP resulted in a decrease in both ATP and PCr concentrations, whereas PCr was shown to increase in genetic mouse models [40]. Directly lesioning the striatum with quinolinic acid (QA), an NMDA receptor agonist that produced excitotoxic damage and produced ^{1}H MRS detectable decreases in glutathione (GSH) and glutamate (Glu), with corresponding increases in glutamine (Gln), a different pattern of changes than those observed in the 3NP model [41]. In addition, decreases in tNAA, total creatine (tCr, Cr+PCr), and taurine were observed [41]. While both of these biochemical approaches (3NP and QA) to modelling HD mimicked the striatal lesion, the mechanisms producing these lesions did not necessarily mimic the disease. Nonetheless, characterizing these models demonstrated the feasibility of using MRS to monitor changes in metabolite concentrations in vivo in animal models of HD.

With the development of genetic mouse models of HD, MRS provided insightful characterization of metabolic changes associated with the onset and progress of disease. ^{1}H MRS at 4.7T was first applied to the R6/2 mouse, an early HD model with rapidly progressive motor dysfunction and a short lifespan [42, 43]. The observation of a decreased tNAA/Cr ratio and increased Gln+Glu level was confirmed by ^{1}H MRS of in vitro samples of striatal and cortical brain extracts and interpreted to indicate disruption of the regulation of the neurotransmitter glutamate [42]. Similarly, decreases in tNAA and increases in taurine and choline were individually quantified in a subsequent study including two other more slowly progressing genetic models, the N171-82Q and YAC72 mice, supporting the idea that similar metabolite changes are related to HD and not to the effects of a specific genetic construct [44]. In the R6/1 HD mouse, which exhibits a slower disease progression than the R6/2 mouse, the loss of tNAA was observed at an age in which striatal DARPP32 expression had also declined [45], aligning this metabolite change to other phenotypic markers. The decline in tNAA levels in HD mice corroborated similar measurements from ^{1}H MRS studies in humans [46].

The most precise characterization of both striatal and cortical metabolite changes has been reported in longitudinal studies of the R6/2 mouse model at 9.4T [47, 48]. Initial reports of decline in the absolute concentrations of striatal NAA and increases in Cr, glycerophosphocholine (GPC), GSH, and Gln, among others, at both 8 and 12 weeks of age [47] were followed up with a more detailed longitudinal progression of both cortical and striatal changes at 4, 8, 12, and 15 weeks [48]. The increased spatial resolution from the higher field strength provided measurements localized to dorsal cortex or striatum without interference from surrounding structures. The asynchronous changes reported among 18 small molecules indicated that as disease progressed, multiple mechanisms altered different metabolites. Striatal PCr increased, and PE decreased stepwise at 8 weeks without further changes, while both striatal and cortical Gln, Cr, Cho, Gln, and Lac continuously increased, and NAA steadily decreased. To capture this multidimensional landscape, all metabolite concentrations were combined using a Partial Least Squares Discriminant Analysis (PLSDA). The two derived principal components, which accounted for 75 % of the variability, described a trajectory that separated the more diseased mice from early onset and controls [48]. Longitudinal, behavioural, and MRI-derived anatomical measurements were compared to the metabolite profiles to determine which would be the most sensitive indicator of disease. As a biomarker, the first principal component was more sensitive and better able to recognize and predict disease stage than striatal volume, climbing behaviour, body weight, or any single metabolite [48].

Longitudinal assessment of metabolite changes in the Q140 knock-in mouse model demonstrated many changes comparable to the R6/2 mouse [49]. One notable difference was in total choline (tCho), increasing in R6/2 while remaining stable in Q140. Interestingly, in the more slowly progressing Q140 model, striatal metabolite changes preceded cortical ones [49]. Surprisingly, in the BACHD mouse, a model expressing the full-length human mutant *HTT* gene, the only metabolite observed to change over the course of 24 months was an increase in lactate, suggesting this model progresses more slowly than previously thought [50]. The non-invasive assessment of brain metabolites via ^{1}H MRS has become so valued that it was one of the variables measured in an initial characterization of a novel knock-in mouse model, zQ175 [51]. Comparable to the previously reported metabolite profiles, the zQ175 striatum at one year is characterized by decreased NAA and increased concentrations of glutamine, taurine, and total creatine [51].

This approach was also applied to ^{1}H MRS data from very young Q111 mice, a model of the earliest stages of disease [52, 53]. Increases in Cr and decreases in PCr at 6 weeks were homeostatically restored to normal by 13 weeks [54]. Other changes in the Q111 striatal metabolite profile included increases in Gln, as observed in R6/2, but also increased glucose and decreased taurine; changes are not encountered previously. At these time points during mouse adolescence, the PLSDA analysis could distinguish genetic carriers from wild-type littermates, even though the metabolite changes at these early ages largely contrasted with those encountered during overt disease in the R6/2 (Fig. 6.1) [48, 54]. This study illustrates the fluid nature of the systemic response to the consequences of abnormal *HTT* gene expression. Early in life, homeostatic changes can compensate for perturbations, producing metabolite shifts that are different from those encountered during manifest disease. The increases in PCr and/or Cr values detected by ^{1}H MRS in the above studies were confirmed independently by analytical HPLC analysis of brain regions harvested from R6/2 and Q111 mice [40]. Thus ^{1}H MRS provides a reliable, non-invasive method for detecting and following the dynamics of disease progression in vivo.

Phosphorus MRS

While ^{1}H MRS provides information on multiple metabolites, suitable for metabolomic characterization, ^{31}P MRS yields information on phosphorus-containing metabolites providing potential mechanistic information about energy status. The absence of changes in human occipital measures of Pi/ATP in response to visual stimulation among individuals with early-stage HD illustrates the ability of ^{31}P MRS to provide information regarding energy utilization during functional challenges [55]. Due to the smaller brains of HD mice and the lower sensitivity of ^{31} P MRS compared with ^{1}H MRS, the spatial resolution of ^{31}P MRS in mice at 9.4T is currently limited to relatively large regions, e.g. dorsal brain [54]. Functional challenges are also difficult to present to anaesthetized animals, restricting measurements to steady state examination of phosphorus-containing metabolites for both R6/2 and Q111 mice [54]. In 11-week-old R6/2 brain, GPC increased and PCr and NADP trended upwards. Measured ATP levels appeared normal, illustrating ATP's strong homeostatic basal regulation even in advanced disease. Analytical measurements of ATP concentration in brain extracts detect an ~10 % decrease [40], below the resolution of the ^{31}P MRS for the sample size studied. The calculated relative rate of the creatine kinase reaction increased as a consequence of the overall increase in total creatine [54] but inconsistent with reports of downregulation of that enzyme [56]. In young Q111 mice, the homeostatic ATP regulation was observed clearly in calculations using values from both ^{1}H and ^{31}P MRS [54]. The phosphorylation potential

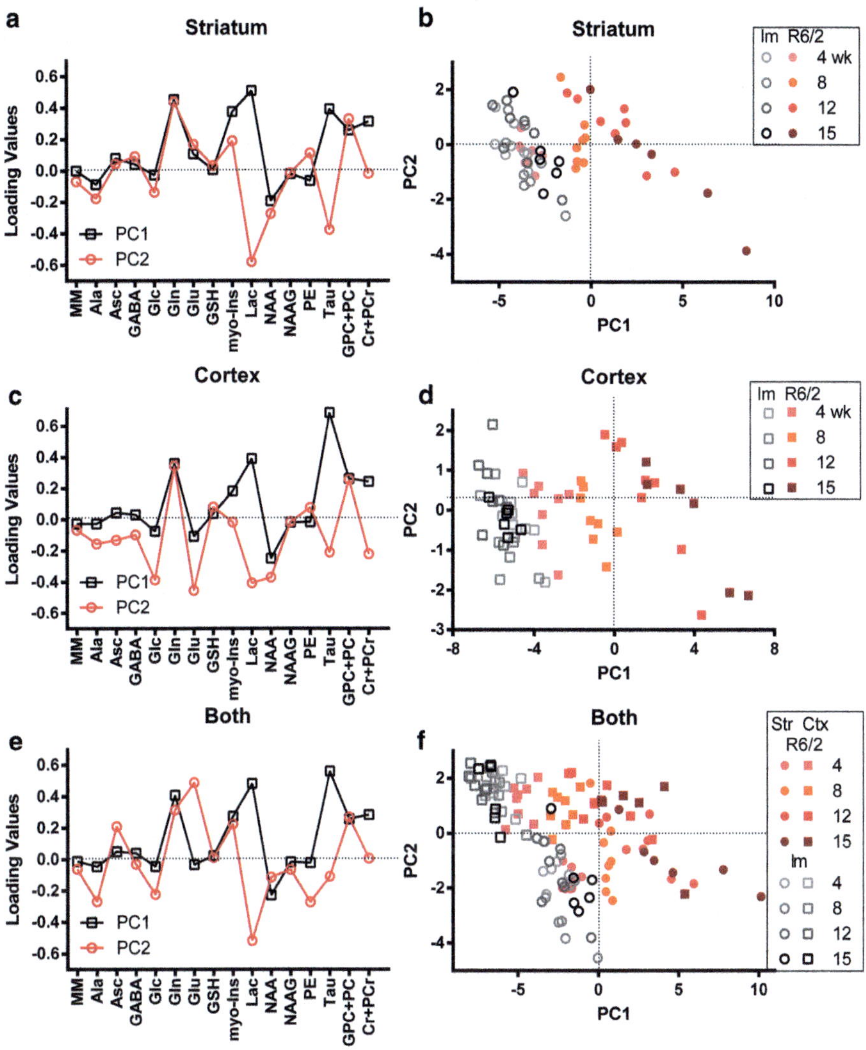

Fig. 6.1 Partial Least Squares Discriminant Analysis (PLSDA) of striatal (**a**, **b**) and cortical (**c**, **d**) metabolites in separate principal component (PC) spaces or combined in one space (**e**, **f**) [48]. The loading plots provide the relative weights of each metabolite concentration used in computing PC1 and PC2 (**a**, **c**, **e**). *Solid symbols* represent R6/2, *open symbols* represent littermate controls, *squares* represent cortex and *triangles* represent striatum. *Grey scale* differs depending on age

declined, and the concentration of ADP and the relative rate of oxidative phosphorylation were transiently lower at 6 weeks, recovering by 13 weeks [54], illustrating the resilience of brain tissue just beginning to accumulate Htt aggregates [53, 57].

In addition to energy compounds, ^{31}P MRS yields measures of intracellular pH, derived from the Pi peak. A cytoplasmic alkalinization was encountered both in rat striatum lesioned with 3NP and occipital cortex of HD patients [58], again demonstrating the ability of MRS to detect parallel changes in both preclinical models and human disease. However, the ^{31}P MRS studies of young Q111 and diseased R6/2 mice did not detect any changes in intracellular pH [54].

^{31}P MRS has yielded insights into the shifts in energy metabolism that occur in both human disease and animal HD models. As the resolution improves with higher field magnets, functional challenges are expected to yield greater insights into the metabolic shifts that accrue as disease progresses. One example of this is the recent report of the ability to measure the cerebral metabolic rate of oxygen ($CMRO_2$) in mouse brain using a 16.4T magnet [59]. When applied to transgenic mouse models of HD, this technique may be able to directly determine if HD limits energy capacity.

Biomarker Considerations

^{1}H MRS can yield anywhere from 4 to 19 metabolites depending upon field strength, magnetic field homogeneity (shim), and signal to noise ratio. Including all of this metabolite information, rather than just a few favourite metabolites, in a biomarker calculation should in principle increase the accuracy and usefulness of the biomarker. Other mathematical combinatorial techniques have been proposed beyond simple metabolite ratioing. An ROC analysis of R6/2 ^{1}H MRS data suggested that PCr or Cr+PCr alone might be sufficient to identify disease onset and track progression [60]. The problem with this analysis is that it derives from the most severe and rapidly progressive animal model and may be insensitive to early-onset metabolite changes. PLSDA has been successfully applied to longitudinal MRS analysis in multiple mouse HD models to yield combined pseudo-variables that track disease progression [48, 54]. A PLSDA mathematical model derived from the R6/2 mouse can be used to assess disease progression in the Q140 mouse and vice versa [49]. More complicated mathematical processes [61] have been proposed but lack validation with data from multiple mouse models. As various MRS measures are demonstrated to have validity in both preclinical models and human disease, consensus upon how to best utilize the rich metabolomic information needs to be developed.

Overview of Clinical MRS in HD

Methodological Considerations

MRS studies of subjects with HD present a number of challenges specific to this disease. First, while motion is always a concern in MRS, it is particularly problematic in subjects with HD. Motor symptoms may result in involuntary movements during scanning in subjects with HD. Even in presymptomatic individuals or subjects, in the early stages of HD, small voxels that match structures of interest (e.g. in caudate or putamen) can make MRS acquisitions particularly sensitive to movement (a 5-mm movement has potentially much more impact on measured metabolite concentrations when using a 2-cm^3 voxel than when using a 15-cm^3 voxel). Recent progress in real-time motion and shim correction for MRS could be used in the future to alleviate this problem [62, 63].

In addition, cohorts of HD subjects differ widely between studies with respect to disease severity, disease duration, age, etc. This makes it difficult to compare results from different studies and may contribute to higher standard deviations across groups. Even when presymptomatic and patient groups are clearly distinguished, significant heterogeneity within each group is difficult to avoid. Correcting for disease burden or disease duration may help in reducing this heterogeneity.

Data acquisition techniques vary widely across studies, sometimes making comparison of results difficult. While many early studies used long-echo time (TE) and focused on singlet resonances (tNAA, tCr, tCho), recent studies generally used shorter TE, which gives access to additional metabolites such as *myo*-inositol (mIns) and Glu. Higher magnetic fields (3T and 7T) offer the potential of additional gains in quantification precision [64].

Accurate quantification is another challenge. Early studies often reported metabolite ratios, which can be ambiguous. For example, an unchanged tNAA/tCr ratio can mask a simultaneous decrease in both tNAA and tCr. More recent studies have used LCModel or similar techniques to quantify short-TE spectra. The use of water as a concentration reference allows determination of "absolute" metabolite concentrations without using concentration ratios. Handling of the macromolecule (MM) baseline is particularly important, and inclusion of an experimentally measured MM spectrum is recommended. Without inclusion of such an MM spectrum, the baseline can be estimated, but care is required with potential bias due to incorrect baseline estimation. Insufficient outer volume suppression may also cause lipid contamination artefacts and have a strong effect on quantification. Improvements in the localization performance of MRS pulse sequences in the past 15 years probably explain why lactate increases reported in HD in earlier studies were not observed in more recent studies.

Finally, avoiding or correcting for partial volume effects is crucial in HD studies, particularly for cerebrospinal fluid (CSF). The most studied structures (caudate and putamen) display pronounced atrophy in HD and are therefore prone to partial volume effects from white matter and, in the case of caudate, from nearby CSF. The choice of smaller voxels and careful voxel positioning help minimize these effects. Alternatively, image segmentation can be used to estimate the amount of white matter or CSF in the voxel. Estimation of CSF content can also be performed using a bi-exponential fit of unsuppressed water signal as a function of TE.

Clinical Studies Using Proton MRS

^{1}H MRS studies in subjects with HD were reported as early as 1993 [35]. In the 1990s, several studies were performed at 1.5T using long-echo times [46, 65–68] and one at shorter TE (30 ms) [69]. Most of these studies reported only metabolite ratios, with a few exceptions [69]. In the 2000s, studies at short-echo time became more widespread, giving access to J-coupled metabolites such as mIns and Glu. Finally, in the 2010s, studies appeared at higher magnetic field (3T and 7T). The largest study to date included 84 subjects—30 controls, 25 presymptomatic carriers, and 29 patients with early HD (Fig. 6.2) [70]. Most studies have focused on caudate and putamen, but several studies have also looked at other cortical and subcortical regions. In the following, we discuss the most significant ^{1}H MRS findings in individuals affected by HD and then in presymptomatic individuals.

Most MRS studies to date have been performed in symptomatic HD subjects—or in mixed symptomatic/presymptomatic cohorts. In the putamen, at least four studies have been published [69–72]. Most but not all found decreased tNAA and decreased tCr in symptomatic HD subjects compared to controls [69, 70, 72]. Decreased Glu [70, 72], increased mIns, and increased tCho compared to controls [69, 70] were also reported. These observations are consistent with

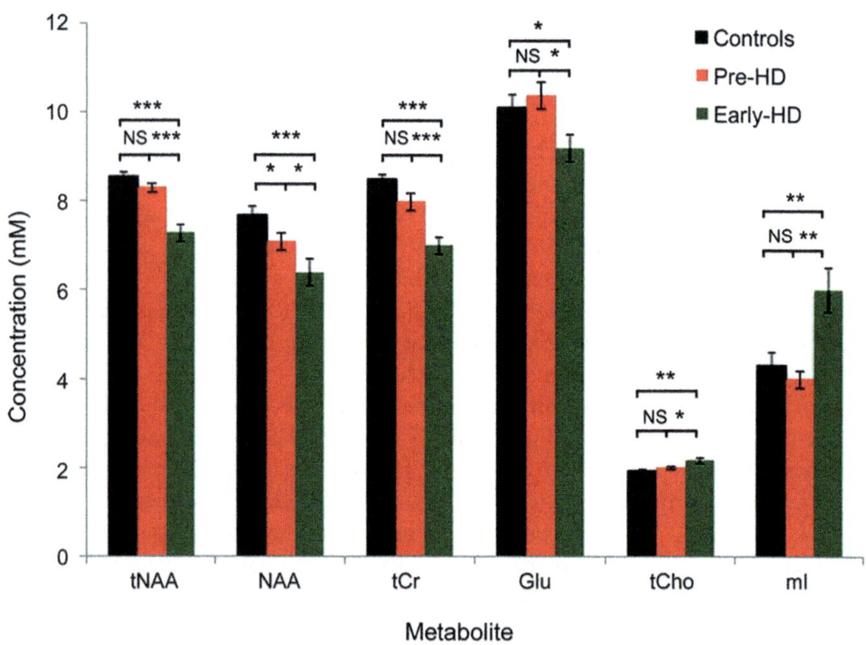

Fig. 6.2 ^{1}H MRS in 30 controls, 25 presymptomatic individuals, and 29 subjects with HD [70]. Error bars denote SEM for each group. *Asterisks* indicate level of significance: $*p < 0.05$, $**p < 0.01$, and $***p < 0.001$. Group comparisons were made using one-way analysis of variance with post hoc Newman-Keuls Multiple Comparison Tests

neurodegeneration and gliosis. In the caudate, two studies reported decreased tNAA and decreased tCr [67, 72], while another reported only a decrease in Glu, although only when considering concentration ratios [71]. Observations of increased lactate in the striatum in early studies [46, 66] were not confirmed in subsequent studies. Other brain regions have been studied more sparsely and generally showed fewer changes, consistent with striatum being the most affected structure in HD. In occipital cortex, two studies reported no change in the absolute concentration of any metabolites [68, 69], while the ratio tNAA/tCr was significantly decreased [69], and ratios Lac/tNAA and tCho/tNAA were significantly increased [46]. In prefrontal cortex, two studies reported no change in any metabolite [71, 72], while one study reported decreased tNAA and increased Lac [65]. Decreased tNAA and Glu were reported in posterior cingulate cortex [73]. Measurements in thalamus and hypothalamus at 7T reported no change [73], although decreased tNAA/tCr was reported in an earlier study [74]. Overall, keeping in mind the great variety of brain regions studied and MRS methodologies used, changes in brain regions other than striatum mostly show either a small change in tNAA or no change at all. Many of these studies, however, have relatively small numbers of subjects and may not have sufficient power to detect small changes, which may explain apparent inconsistencies.

Just a handful of studies have reported results in cohorts of presymptomatic individuals that included no subjects with diagnosed HD. The largest study to date reported a significant 8 % decrease in NAA—but not tNAA—in the putamen of presymptomatic individuals relative to controls [70]. In contrast, other studies reported no change in putamen [72, 75], caudate [72], or striatum [76]. In prefrontal cortex, there was either no change [72] or an increase in tCho [76], although the absence of shown spectra in the latter study makes it impossible to assess spectral quality. No change was reported in the hypothalamus [72] or thalamus [72, 75]. Again, these studies have small numbers of subjects for these specific structures and may not have sufficient power to detect small changes. Overall, it seems that there is a small (<10 %) decrease in tNAA in the putamen in presymptomatic HD carriers, which can be detected in studies with sufficiently large (~25) numbers of subjects. Across presymptomatic individuals and HD subjects, changes in tNAA correlated with performance on a motor function task and with disease burden [70].

Clinical Studies Using Phosphorus MRS

Only a few studies have been reported using ^{31}P MRS in HD subjects. The first one reported no change in posterior parietal region, except for a small increase in Pi that was attributed to increased CSF [69]. Another study reported an increase in pH in the brain of HD patients [58].

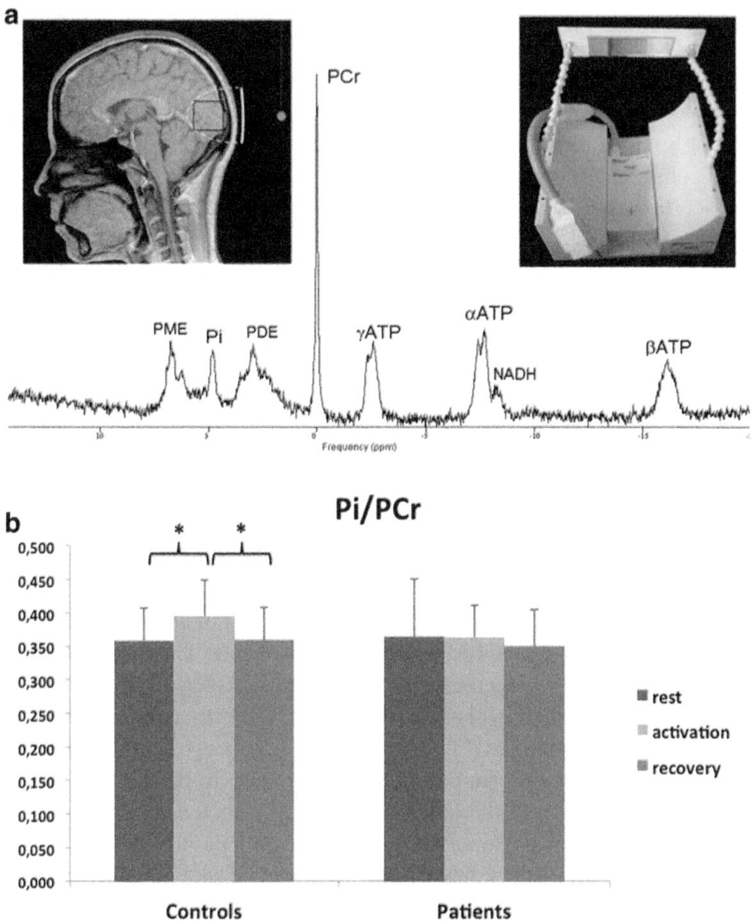

Fig. 6.3 (**a**) Representative ^{31}P NMR spectrum from the occipital cortex of a control subject at 3T (pulse-acquire, 240 repetitions, TR = 2 s, 8 min total acquisition time) [55]. In addition, an ^{1}H image shows the position of the sphere at the centre of the ^{31}P surface coil, aligned with the visual cortex (*upper left*). The surface coil is drawn on the image to show the size and position of the coil relative to the head. The *dashed white line* indicates the sensitive volume of the coil encompassing most of the visual cortex. The square on the ^{1}H image shows the region used for localized ^{1}H shimming. A picture of the ^{31}P coil is also shown (*upper right*). (**b**) Pi/PCr ratio before, during, and after visual stimulation in 15 early HD patients and 15 age- and sex-matched controls. In controls, the Friedman test was significant for Pi/PCr ($p = 0.006$), and the Bonferroni-corrected Wilcoxon-signed crank test indicated increased Pi/PCr between rest and activation ($p = 0.024$) followed by a decrease between activation and recovery ($p = 0.012$). These changes were absent in HD patients; $*p < 0.05$

More recently, a study in occipital cortex during functional visual activation reported an increase in Pi/PCr and Pi/ATP in controls during activation compared to rest, but this functional increase was not seen in subjects with HD suggesting compromised energy metabolism (Fig. 6.3) [55].

Perspectives

The above-mentioned studies demonstrate that MRS permits detection of metabolic changes in HD. Longitudinal studies are needed in humans to determine more precisely the time course of these metabolic changes and compare them with other measures such as the rate of caudate atrophy or Unified HD Rating Scale (UHDRS) scores. Because metabolic changes are thought to precede structural changes, measurements of metabolic parameters could potentially detect a positive response to treatment earlier than structural imaging.

From a technical standpoint, future studies could benefit from higher magnetic fields, which are expected to provide better quantification precision, especially for J-coupled metabolites such as glutamate and glutamine [64]. Real-time motion and shim correction could also be beneficial to mitigate the effect of movement. Other modalities such as ^{31}P magnetization transfer or ^{17}O MRS could be used to further investigate energy metabolism in HD. A recent study in mice using ^{17}O MRS also showed promise to study energy metabolism [59].

References

1. Squitieri F et al (1994) DNA haplotype analysis of Huntington disease reveals clues to the origins and mechanisms of CAG expansion and reasons for geographic variations of prevalence. Hum Mol Genet 3(12):2103–2114
2. Pringsheim T et al (2012) The incidence and prevalence of Huntington's disease: a systematic review and meta-analysis. Mov Disord 27(9):1083–1091
3. A novel gene containing a trinucleotide repeat that is expanded and unstable on Huntington's disease chromosomes. The Huntington's Disease Collaborative Research Group. (1993) Cell, 72(6):971–983
4. Langbehn DR et al (2004) A new model for prediction of the age of onset and penetrance for Huntington's disease based on CAG length. Clin Genet 65(4):267–277
5. Tibben A (2007) Predictive testing for Huntington's disease. Brain Res Bull 72(2-3):165–171
6. MacLeod R et al (2013) Recommendations for the predictive genetic test in Huntington's disease. Clin Genet 83(3):221–231
7. Tabrizi SJ et al (2009) Biological and clinical manifestations of Huntington's disease in the longitudinal TRACK-HD study: cross-sectional analysis of baseline data. Lancet Neurol 8 (9):791–801
8. Paulsen JS et al (2008) Detection of Huntington's disease decades before diagnosis: the Predict-HD study. J Neurol Neurosurg Psychiatry 79(8):874–880
9. Duff K et al (2007) Psychiatric symptoms in Huntington's disease before diagnosis: the predict-HD study. Biol Psychiatry 62(12):1341–1346
10. Paulsen JS et al (2005) Depression and stages of Huntington's disease. J Neuropsychiatry Clin Neurosci 17(4):496–502
11. Paulsen JS et al (2005) Critical periods of suicide risk in Huntington's disease. Am J Psychiatry 162(4):725–731
12. Reedeker N et al (2011) Incidence, course, and predictors of apathy in Huntington's disease: a two-year prospective study. J Neuropsychiatry Clin Neurosci 23(4):434–441
13. Montoya A et al (2006) Brain imaging and cognitive dysfunctions in Huntington's disease. J Psychiatry Neurosci 31(1):21–29

14. Vonsattel JPG, Keller C, Amaya MDP (2008) Neuropathology of Huntington's disease. In: Litvan CDAI (ed) Handbook of clinical neurology. Elsevier B. V, Amsterdam
15. Vonsattel JP, DiFiglia M (1998) Huntington disease. J Neuropathol Exp Neurol 57(5):369–384
16. Mitchell IJ, Cooper AJ, Griffiths MR (1999) The selective vulnerability of striatopallidal neurons. Prog Neurobiol 59(6):691–719
17. Zuccato C, Cattaneo E (2014) Huntington's disease. Handb Exp Pharmacol 220:357–409
18. Ross CA et al (2014) Huntington disease: natural history, biomarkers and prospects for therapeutics. Nat Rev Neurol 10(4):204–216
19. Gines S et al (2003) Specific progressive cAMP reduction implicates energy deficit in presymptomatic Huntington's disease knock-in mice. Hum Mol Genet 12(5):497–508
20. Milakovic T, Johnson GV (2005) Mitochondrial respiration and ATP production are significantly impaired in striatal cells expressing mutant huntingtin. J Biol Chem 280 (35):30773–30782
21. Antonini A et al (1996) Striatal glucose metabolism and dopamine D2 receptor binding in asymptomatic gene carriers and patients with Huntington's disease. Brain 119 (Pt 6):2085–2095
22. Grafton ST et al (1992) Serial changes of cerebral glucose metabolism and caudate size in persons at risk for Huntington's disease. Arch Neurol 49(11):1161–1167
23. Kuwert T et al (1993) Striatal glucose consumption in chorea-free subjects at risk of Huntington's disease. J Neurol 241(1):31–36
24. Browne SE, Beal MF (2004) The energetics of Huntington's disease. Neurochem Res 29 (3):531–546
25. Tabrizi SJ et al (1999) Biochemical abnormalities and excitotoxicity in Huntington's disease brain. Ann Neurol 45(1):25–32
26. Cui L et al (2006) Transcriptional repression of PGC-1alpha by mutant huntingtin leads to mitochondrial dysfunction and neurodegeneration. Cell 127(1):59–69
27. Lin J, Handschin C, Spiegelman BM (2005) Metabolic control through the PGC-1 family of transcription coactivators. Cell Metab 1(6):361–370
28. Lin J et al (2004) Defects in adaptive energy metabolism with CNS-linked hyperactivity in PGC-1alpha null mice. Cell 119(1):121–135
29. Palfi S et al (1996) Chronic 3-nitropropionic acid treatment in baboons replicates the cognitive and motor deficits of Huntington's disease. J Neurosci 16(9):3019–3025
30. Brouillet E et al (1998) Partial inhibition of brain succinate dehydrogenase is sufficient to initiate striatal degeneration in rat. J Neurochem 70(2):794–805
31. Tabrizi SJ et al (2013) Predictors of phenotypic progression and disease onset in premanifest and early-stage Huntington's disease in the TRACK-HD study: analysis of 36-month observational data. Lancet Neurol 12(7):637–649
32. Mochel F et al (2010) Dietary anaplerotic therapy improves peripheral tissue energy metabolism in patients with Huntington's disease. Eur J Hum Genet 18(9):1057–1060
33. Kar K et al (2013) beta-hairpin-mediated nucleation of polyglutamine amyloid formation. J Mol Biol 425(7):1183–1197
34. Pedersen JT, Heegaard NH (2013) Analysis of protein aggregation in neurodegenerative disease. Anal Chem 85(9):4215–4227
35. Beal MF et al (1993) Neurochemical and histologic characterization of striatal excitotoxic lesions produced by the mitochondrial toxin 3-nitropropionic acid. J Neurosci 13 (10):4181–4192
36. Jenkins BG et al (1996) Non-invasive neurochemical analysis of focal excitotoxic lesions in models of neurodegenerative illness using spectroscopic imaging. J Cereb Blood Flow Metab 16(3):450–461
37. Dautry C et al (1999) Serial ^{1}H-NMR spectroscopy study of metabolic impairment in primates chronically treated with the succinate dehydrogenase inhibitor 3-nitropropionic acid. Neurobiol Dis 6(4):259–268

38. Matthews RT et al (1998) Neuroprotective effects of creatine and cyclocreatine in animal models of Huntington's disease. J Neurosci 18(1):156–163
39. Henry PG et al (2002) Decreased TCA cycle rate in the rat brain after acute 3-NP treatment measured by in vivo ^{1}H-[^{13}C] NMR spectroscopy. J Neurochem 82(4):857–866
40. Mochel F et al (2012) Early alterations of brain cellular energy homeostasis in Huntington disease models. J Biol Chem 287(2):1361–1370
41. Tkac I et al (2001) Metabolic changes in quinolinic acid-lesioned rat striatum detected non-invasively by in vivo ^{1}H NMR spectroscopy. J Neurosci Res 66(5):891–898
42. Jenkins BG et al (2000) Nonlinear decrease over time in N-acetyl aspartate levels in the absence of neuronal loss and increases in glutamine and glucose in transgenic Huntington's disease mice. J Neurochem 74(5):2108–2119
43. Mangiarini L et al (1996) Exon 1 of the HD gene with an expanded CAG repeat is sufficient to cause a progressive neurological phenotype in transgenic mice. Cell 87(3):493–506
44. Jenkins BG et al (2005) Effects of CAG repeat length, HTT protein length and protein context on cerebral metabolism measured using magnetic resonance spectroscopy in transgenic mouse models of Huntington's disease. J Neurochem 95(2):553–562
45. van Dellen A et al (2000) N-Acetylaspartate and DARPP-32 levels decrease in the corpus striatum of Huntington's disease mice. Neuroreport 11(17):3751–3757
46. Jenkins BG et al (1998) 1H NMR spectroscopy studies of Huntington's disease: correlations with CAG repeat numbers. Neurology 50(5):1357–1365
47. Tkac I et al (2007) Neurochemical changes in Huntington R6/2 mouse striatum detected by in vivo 1H NMR spectroscopy. J Neurochem 100(5):1397–1406
48. Zacharoff L et al (2012) Cortical metabolites as biomarkers in the R6/2 model of Huntington's disease. J Cereb Blood Flow Metab 32(3):502–514
49. Zacharoff L et al. (2010) Biochemical changes in Q140 striatum precedes progressive volume loss. In: Proceedings of the society for neuroscience. San Diego, CA
50. Zacharoff L et al. (2011) Striatum specific lactate change in BACHD model of Huntington's disease. In: Proceedings of the society for neuroscience. Washington, DC.
51. Heikkinen T et al (2012) Characterization of neurophysiological and behavioral changes, MRI brain volumetry and ^{1}H MRS in zQ175 knock-in mouse model of Huntington's disease. PLoS One 7(12), e50717
52. Wheeler VC et al (1999) Length-dependent gametic CAG repeat instability in the Huntington's disease knock-in mouse. Hum Mol Genet 8(1):115–122
53. Wheeler VC et al (2000) Long glutamine tracts cause nuclear localization of a novel form of huntingtin in medium spiny striatal neurons in HdhQ92 and HdhQ111 knock-in mice. Hum Mol Genet 9(4):503–513
54. Tkac I et al (2012) Homeostatic adaptations in brain energy metabolism in mouse models of Huntington disease. J Cereb Blood Flow Metab 32(11):1977–1988
55. Mochel F et al (2012) Abnormal response to cortical activation in early stages of Huntington disease. Mov Disord 27(7):907–910
56. Kim J et al (2010) Reduced creatine kinase as a central and peripheral biomarker in Huntington's disease. Biochim Biophys Acta 1802(7-8):673–681
57. Wheeler VC et al (2002) Early phenotypes that presage late-onset neurodegenerative disease allow testing of modifiers in Hdh CAG knock-in mice. Hum Mol Genet 11(6):633–640
58. Chaumeil MM et al (2012) pH as a biomarker of neurodegeneration in Huntington's disease: a translational rodent-human MRS study. J Cereb Blood Flow Metab 32(5):771–779
59. Cui W et al (2013) Non-invasive measurement of cerebral oxygen metabolism in the mouse brain by ultra-high field ^{17}O MR spectroscopy. J Cereb Blood Flow Metab 33(12):1846–1849
60. Nikas JB, Keene CD, Low WC (2010) Comparison of analytical mathematical approaches for identifying key nuclear magnetic resonance spectroscopy biomarkers in the diagnosis and assessment of clinical change of diseases. J Comp Neurol 518(20):4091–4112

61. Nikas JB, Low WC (2011) Application of clustering analyses to the diagnosis of Huntington disease in mice and other diseases with well-defined group boundaries. Comput Methods Programs Biomed 104(3):e133–e147
62. Hess AT et al (2011) Real-time motion and B0 corrected single voxel spectroscopy using volumetric navigators. Magn Reson Med 66(2):314–323
63. Keating B, Ernst T (2012) Real-time dynamic frequency and shim correction for single-voxel magnetic resonance spectroscopy. Magn Reson Med 68(5):1339–1345
64. Deelchand DK, Iltis I, Henry PG (2014) Improved quantification precision of human brain short echo-time ^{1}H magnetic resonance spectroscopy at high magnetic field: a simulation study. Magn Reson Med 72(1):20–25
65. Harms L et al (1997) Decreased N-acetyl-aspartate/choline ratio and increased lactate in the frontal lobe of patients with Huntington's disease: a proton magnetic resonance spectroscopy study. J Neurol Neurosurg Psychiatry 62(1):27–30
66. Jenkins BG et al (1993) Evidence for impairment of energy metabolism in vivo in Huntington's disease using localized ^{1}H NMR spectroscopy. Neurology 43(12):2689–2695
67. Sanchez-Pernaute R et al (1999) Clinical correlation of striatal ^{1}H MRS changes in Huntington's disease. Neurology 53(4):806–812
68. Taylor-Robinson SD et al (1996) Proton magnetic resonance spectroscopy in Huntington's disease: evidence in favour of the glutamate excitotoxic theory. Mov Disord 11(2):167–173
69. Hoang TQ et al (1998) Quantitative proton-decoupled ^{31}P MRS and ^{1}H MRS in the evaluation of Huntington's and Parkinson's diseases. Neurology 50(4):1033–1040
70. Sturrock A et al (2010) Magnetic resonance spectroscopy biomarkers in premanifest and early Huntington disease. Neurology 75(19):1702–1710
71. Padowski JM et al (2014) Neurochemical correlates of caudate atrophy in Huntington's disease. Mov Disord 29(3):327–335
72. van den Bogaard SJ et al (2011) Exploratory 7-Tesla magnetic resonance spectroscopy in Huntington's disease provides in vivo evidence for impaired energy metabolism. J Neurol 258 (12):2230–2239
73. Unschuld PG et al (2012) Brain metabolite alterations and cognitive dysfunction in early Huntington's disease. Mov Disord 27(7):895–902
74. Ruocco HH et al (2007) Evidence of thalamic dysfunction in Huntington disease by proton magnetic resonance spectroscopy. Mov Disord 22(14):2052–2056
75. van Oostrom JC et al (2007) 1H magnetic resonance spectroscopy in preclinical Huntington disease. Brain Res 1168:67–71
76. Gomez-Anson B et al (2007) Decreased frontal choline and neuropsychological performance in preclinical Huntington disease. Neurology 68(12):906–910

Chapter 7
MRS in Motor Neuron Diseases

Varan Govind

Abstract Upper motor neurons (UMN) in the motor cortex and/or lower motor neurons (LMN) in the brainstem and spinal cord are affected in a heterogeneous group of related or pathologically similar adult-onset diseases, collectively termed as motor neuron diseases (MND). Amyotrophic lateral sclerosis (ALS) is the most predominant (85 %) among MND in which both UMN and LMN progressively degenerate. The etiology of ALS is not known in approximately 90 % of patients; however, inheritance of genetic mutations underlies in about 10 % of them. To date, neither conventional neuroimaging techniques nor laboratory tests of body fluids or biopsy samples yielded definitive biomarkers for ALS. Furthermore, molecular and cellular mechanisms involved in this disease remain poorly understood. Proton MRS is well suited to quantify metabolite alterations in the brain, and it has been used on patients with ALS in an effort to gain insight into the pathophysiology of ALS. A brief description on MND and their subtypes, limitations of clinical neuroimaging methods to diagnose ALS, a number of brain metabolites that can be quantified by in vivo MRS and their relevance to MND, and MRS techniques and their limitations are provided. Use of a whole-brain MRSI approach to fully characterize changes within the brain due to disease is described with a sample dataset obtained in a patient with ALS. Changes in the concentration or ratio of brain metabolites in patients with ALS and their pathophysiological significances are described. Finally, the use of MRS methods for longitudinal and therapeutic evaluation studies and future perspectives are discussed.

Keywords Motor neuron • MND • ALS • MR spectroscopy • Spectroscopic imaging

---

V. Govind, Ph.D. (✉)
Department of Radiology, University of Miami School of Medicine, Miami, FL, USA
e-mail: VGovind@med.miami.edu

G. Öz (ed.), *Magnetic Resonance Spectroscopy of Degenerative Brain Diseases*,
Contemporary Clinical Neuroscience, DOI 10.1007/978-3-319-33555-1_7

Background

A motor neuron is a nerve cell present in the central nervous system (CNS). Upper motor neurons (UMN) are located in the motor cortex of the brain. Lower motor neurons (LMN) are located in the anterior horn of the spinal cord and the cranial nerve nuclei of the brainstem. The UMN axons coalesce and descend through the spinal cord and synapse with LMNs at spinal nerve root levels, as shown in Fig. 7.1. The LMN axons innervate skeletal muscle fibers.

The motor neurons are affected in a group of adult-onset diseases, collectively termed motor neuron diseases (MND). In these diseases, the common pathological evidence is the progressive degeneration or loss of either UMN or LMN, or both. Based on the type of motor neurons affected, MNDs are broadly categorized into primary lateral sclerosis (PLS) [1] with UMN degeneration, progressive muscular atrophy (PMA) with spinal LMN degeneration, progressive bulbar palsy (PBP) with bulbar LMN degeneration, and amyotrophic lateral sclerosis (ALS) with both UMN and LMN degeneration [2–5]. Among patients with MND, approximately 85 % are diagnosed with ALS [6, 7]. Furthermore, some patients diagnosed with PMA [8] or PLS [9] show evidence of ALS, thereby making distinctions among the MNDs

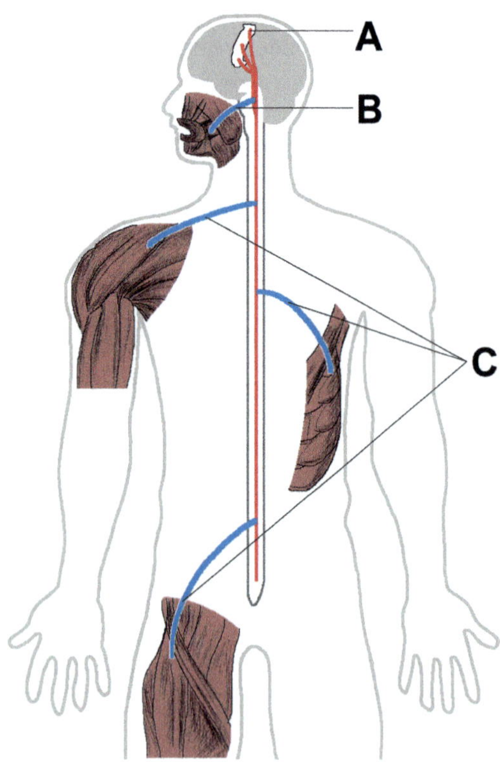

Fig. 7.1 A schematic diagram to show the anatomical locations of upper motor neurons and lower motor neurons. Upper motor neuronal cell bodies are located in the motor cortex (**a**) and their axonal bundles descend through the internal capsule and spinal cord and connect with lower motor neurons at the appropriate brainstem and spinal nerve root levels. The axonal bundles of lower motor neurons with their cell bodies in the brain stem (**b**) that control the facial muscles and tongue, and the axonal bundles of lower motor neurons with their cell bodies in the spinal cord (**c**) that control the limb and respiratory muscles

unclear. Recent studies further indicate that ALS has also been observed in extra-motor brain areas such as frontotemporal regions [10–16]. Approximately 50 % of patients with ALS show cognitive impairments and behavioral abnormalities [17–19] and have clinical, genetic, and pathological overlap with frontotemporal lobar degeneration (FTLD) [20–24]. Hence, ALS is increasingly recognized as a multisystem disorder with variable involvement of UMN, LMN, and extra-motor brain regions in patients.

Clinical examination of patients with ALS reveals spasticity (or stiffness) of limb muscles, presence of the Babinski and Hoffman signs, hyperreflexia, and pseudobulbar affect as signs of UMN involvement and fasciculation (i.e., spontaneous muscle twitching), decreased reflexes, weakening of muscles, and muscle atrophy in affected anatomical regions as signs of LMN involvement of MND [25].

The etiology of ALS is not known in approximately 90 % of patients with ALS, termed as sporadic ALS (sALS), and the remaining 10 % are known to inherit disease-causing genetic mutations and categorized as familial ALS (fALS) [26, 27]. The molecular and cellular mechanisms involved in the pathogenesis of ALS remain poorly understood, but a general consensus evolved among researchers is that it is a syndrome with multiple mechanisms contributing to the disease [28]. Evidence from previous research studies suggest that oxidative damage [29] or stress [30], abnormal accumulation of neurofilaments in axons [31], altered cytoskeletal protein metabolism [32], glutamate excitotoxicity [33], intracellular aggregates or protein misfolding [34], cellular transport abnormalities [35], mutations in DNA-/RNA-binding proteins [36, 37], mitochondrial dysfunction [38], and neuroinflammation [39] are some of the mechanisms implicated to cause the disease. The pathological features of ALS include selective degeneration and subsequent loss of the motor neurons in the motor cortex, brainstem, and spinal cord and degeneration of the white matter fibers (axons that descend from the motor neurons) in the corticospinal tracts (CST). In histopathological studies [40, 41], loss of the pyramidal Betz cells and gliosis in motor cortex, degeneration of CST, appearance of myelin pallor, and shrinkage and loss of anterior horn cells in spinal cord were observed. Neither the anatomical location of the onset of this disease nor its progression is clear—it may begin within the motor neuron cell body (cortex), the axons, or at the neuromuscular junction and progress either away from (anterograde or "dying forward") or toward the motor cortex (retrograde or "dying back"), with evidence for both the possibilities [42–45].

To date, there is no treatment available either to prevent further progression or to reverse or cure ALS completely. The only FDA-approved drug for ALS, riluzole [46], is known to delay mortality for up to three months [47] and its therapeutic mechanism is based on anti-excitotoxic properties.

Potential Clinical Utility of MRS

Diagnosis of ALS in patients is made upon determination of functional deficits in both UMN and LMN of them. Deficits in LMN function of patients can be evaluated by performing electromyography and nerve conduction studies [48–50]. However, despite evidence of UMN degeneration or loss in postmortem brain specimens of patients died with ALS [51], no definitive laboratory or in vivo imaging method is available to assess UMN involvement in the brain. As a result, determination of UMN deficits relies on clinical assessments (in which the clinical symptoms and signs observed in patients are interpreted) and findings of conventional imaging and laboratory body fluid tests to rule out other diseases that mimic MND.

Conventional MR imaging methods show signal intensity changes mainly in the CST and motor cortex [52–57] and atrophy of the motor cortex [58, 59] in patients with ALS. However, these changes are not consistently found in all patients [57]. Similar intensity changes observed from MRIs of some control subjects make a reliable diagnosis of ALS even more difficult [60]. However, clinical MRI is still useful for exclusion of a diagnosis of other pathological conditions that mimic ALS. Among other brain imaging modalities that have been evaluated to assess UMN function in patients with ALS include transcranial magnetic stimulation [61, 62], positron emission tomography [63, 64], and other MR-based techniques such as MR spectroscopy and diffusion tensor imaging [65].

As a noninvasive technique, in vivo proton (i.e., hydrogen or ^{1}H) MR spectroscopy (MRS) offers a unique tool to quantitate a number of metabolites in the brain of patients and healthy normal subjects [66]. The cerebral metabolites readily observed by ^{1}H MRS include *N*-acetyl aspartate (NAA; localized primarily in neurons and axons; described as a neuronal marker), total creatine (Cre; sum of creatine and phosphocreatine; an indicator of cellular energetics), and total choline (Cho; sum of choline, phosphocholine, and glycerophosphocholine; an indicator of cell membrane integrity, synthesis, and degradation). Other, not so easily observable metabolites include glutamate (Glu), glutamine (Gln), and γ-aminobutyric acid (GABA) (excitatory or inhibitory neurotransmitters), myo-inositol (mIns; an osmolyte, present mostly in glial cells; an indicator of glial cell proliferation or gliosis), and glutathione (an antioxidant). These metabolites are involved in various physiological and pathological processes. Since the molecular and cellular processes in the CNS are strongly considered to be the origin and development of MND, it is reasonable to expect that alterations in the brain metabolites will precede the CNS tissue parenchymal changes such as atrophy and thinning of cortex. Thus, quantitation of these brain metabolites will not only provide the earliest indication of disease but also enable to identify specific metabolic pathways that initiate UMN degeneration (e.g., glutamate excitotoxicity), to aid in the assessment of UMN degeneration/deficit (e.g., alterations in NAA, Cho, and mIns), to monitor the progression of UMN degeneration (e.g., spatiotemporal changes in metabolites), and to evaluate the efficacy of potential therapeutics. In Fig. 7.2, shown are the

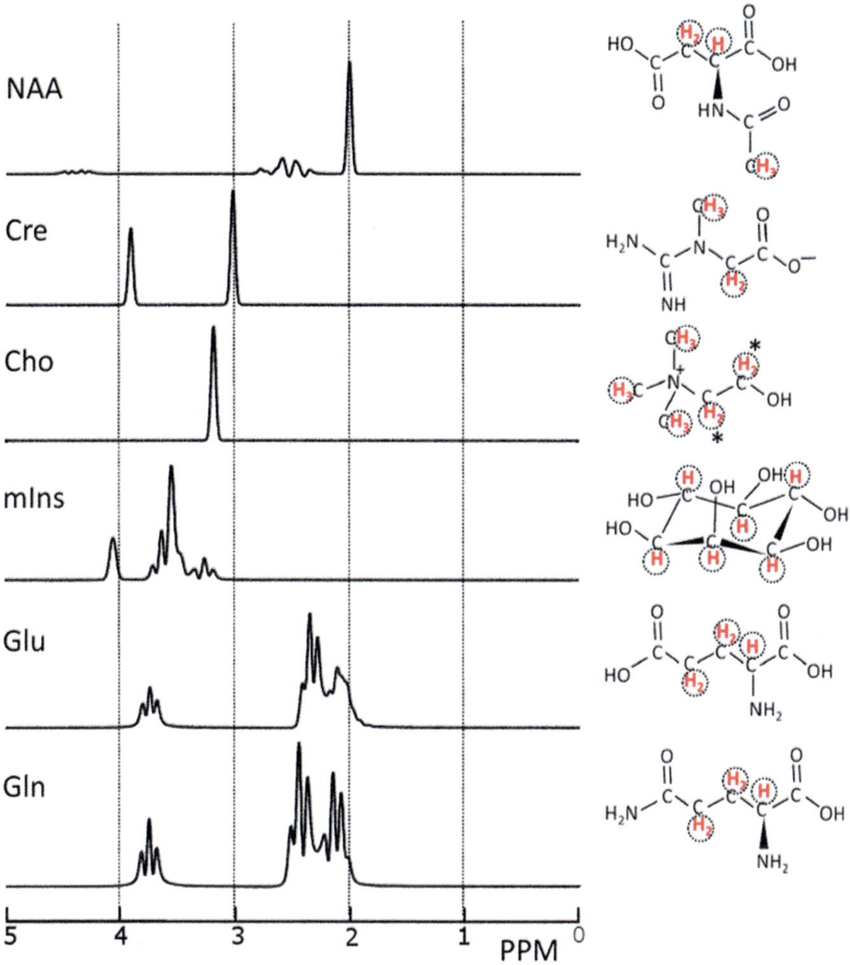

Fig. 7.2 Proton MR spectra of the brain metabolites of primary interest in motor neuron diseases. Spectra shown were simulated for point-resolved spectroscopy (PRESS) sequence at 3 tesla with a TE value of 30 ms (TE1 = TE2 = 15 ms), an identical concentration for all the metabolites and a 6 Hz line broadening. The spectral lines shown originate from the circled hydrogen atom(s). The hydrogen atoms with *asterisks* for choline were not included in the simulation. NAA, *N*-acetyl aspartate; Cre, creatine; Cho, choline; mIns, myo-inositol; Glu, glutamate; Gln, glutamine; PPM, parts per million

theoretically simulated proton MR spectra of the brain metabolites, observed at 3 tesla using a commonly used single-volume selection method for data acquisition, point-resolved spectroscopy (PRESS [67]), with an echo time (TE) of 30 ms. For the simulations, the GAMMA library [68] and published spectral parameters of the brain metabolites [66] were used.

Technical Considerations

ALS is for long thought to affect selectively the upper and lower motor neurons and their axons in the brain and spinal cord. However, results from recent neuropathological and imaging studies [10, 14, 51, 69, 70] provide evidence of involvement of non-motor brain regions that are spread across much of the whole brain. Though detectable brain metabolite alterations might be found at the time of disease onset (or soon after the clinical signs of disease appear) only in few discrete anatomical locations such as the primary motor cortex and corticospinal tracts, the progression of disease is expected to produce concomitant spatiotemporal changes in the metabolites. Thus, a whole-brain MRSI data acquisition approach is necessary to follow the brain metabolite changes longitudinally to assess fully the anatomical spread of the disease and also to evaluate efficacy of potential future therapeutics in patients with ALS.

Spectroscopic data can be obtained using single volume (or voxel) spectroscopy (SVS) or multi-voxel MR spectroscopic imaging (MRSI) techniques. SVS techniques are well suited to pathologies confined to a discrete anatomical location and pathologically homogenous and also to metabolites that are difficult to quantitate using conventional single-shot acquisition methods, for example, GABA in the brain, which requires use of a spectral editing method [71]. Typically, SVS techniques take about 10 min to acquire data from an approximately 8 mL voxel volume. Despite the fact that MRSI is technically challenging on the brain of human subjects and requires longer data acquisition time (>15 min), it is still preferred owing to its wide spatial (or anatomical) coverage. Among the 3D-MRSI techniques, a unique whole-brain MRSI data acquisition technique developed by Maudsley et al. [72] is well suited for cross-sectional and longitudinal studies. This echo-planar spin-echo-based data acquisition technique obtains data from the whole brain including cortical regions and yields reliable data from much of it after processing. In Fig. 7.3 are shown the whole-brain MRSI data obtained from a patient with ALS (45 years old male) at 3T with a TE of 70 ms and an acquisition time of approximately 25 min. The quality of spectra obtained using this technique from discrete anatomical locations in the corticospinal tract of a patient with ALS, as shown in Fig. 7.4, can be gauged from small linewidth values (<10 Hz) and a reasonably high signal-to-noise ratio of the spectral lines. At a TE of 70 ms, only the prominent metabolites such as NAA, Cre, and Cho can be observed. However, a parallel-imaging version of this sequence [73] can be used to acquire whole-brain MRS data in approximately 18 min at a TE value of 17 ms to observe additional metabolites such as myo-inositol, glutamate, and glutamine. For quantitation of GABA, the recently published spectral editing sequence with 3D spatial coverage [74] may be utilized; however, it still leaves out the cortical brain regions that are most relevant to motor neuron diseases and thus its use for MND will be limited.

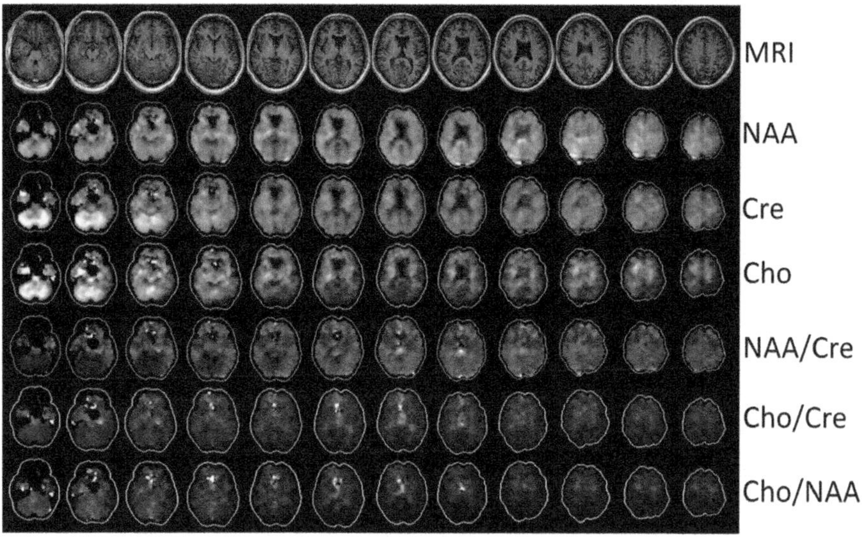

Fig. 7.3 Whole-brain MRSI and T1-MRI data acquired from a subject with ALS (45 years, male) are shown. Metabolite distributions in 12 contiguous slices of 5 mm thickness and T1-MRIs matched to the same slice thickness (i.e., 5 mm) are shown. MRSI data were acquired from 18 slices of 10 mm each that covered the whole brain, and the data were processed to 32 slices of 5 mm each. For brevity, data from only 12 MRSI slices that encompassed most of the brain are shown. Abbreviations used include NAA for *N*-acetyl aspartate, Cre for total creatine, and Cho for total choline. The outlines of the metabolite images correspond to the outer edge of the brain on their respective T1-MRI. The black colored voxels within the brain outline on the metabolite images indicate that the metabolites in these voxels were not quantified either due to the poor quality of data or no metabolite is present such as voxels with CSF. The white colored voxels on some of the metabolite images indicate that the quality of spectral data from those voxels was poor due to high magnetic susceptibility variations in the anatomical region and/or insufficient suppression of water signal

Most published MRS studies in MND reported metabolite quantitative data either as ratios with respect to a co-observed metabolite (for example, Cre) or as values in an institutional unit (i.e., T1 and T2 values of the metabolites are not taken into account for calculations of their absolute concentrations). Furthermore, these studies compared metabolite values or ratios either between patient and healthy control groups in cross-sectional studies or between time points in intra-subject longitudinal studies. Measurements of metabolite T1 and T2 involve experiments with multiple repetition time (TR) or TE values and thereby require long data acquisition times. As a consequence, absolute concentration calculations of the metabolites within the whole brain, comprised of multiple voxels, are not feasible within clinically acceptable scan times. This is one of the major limitations of in vivo MRS studies.

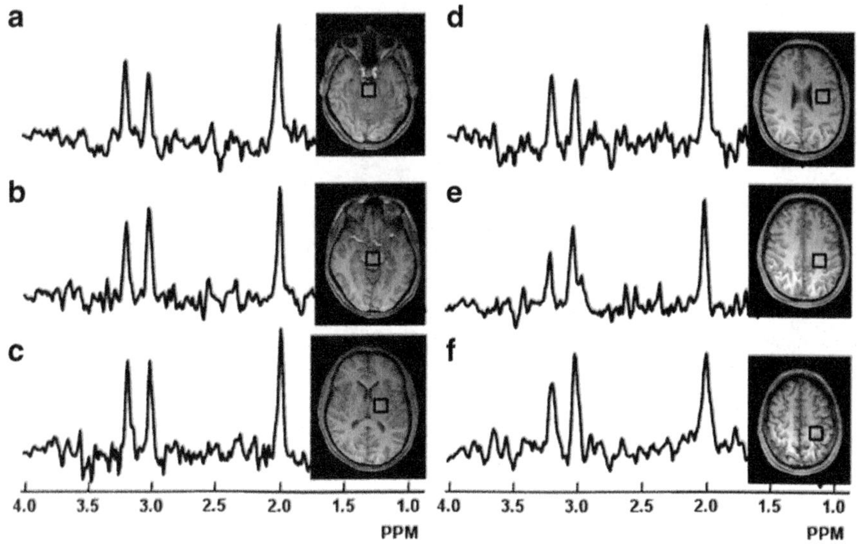

Fig. 7.4 Representative proton MR spectra from discrete anatomical regions in the left corticospinal tract, medulla, and pons/midbrain of a patient with ALS (43 years, male) are shown. The spectrum shown for each region was obtained by averaging spectra from contiguous voxels in that region and the voxel volume for the spectra ranges from 2.7 mL to 6.0 mL. Spectra shown are from medulla (**a**), pons/midbrain (**b**), left internal capsule (**c**), left corona radiata (**d**), left centrum semiovale (**e**), left precentral gyrus/subcortical white matter (**f**). (Adopted from [120])

Overview of MRS in Animal Models of MND

Animal models of ALS provide insights into alterations in cellular and molecular processes that enable researchers to identify possible targets for drug development and to evaluate efficacy of potential experimental therapeutic agents. Etiological models of ALS have been developed in mouse and rat species ([75–82]; see reviews [83, 84]) for four major mutated genes that are known to cause over 60 % of familial ALS and a small fraction of sporadic ALS [85]. These genes are comprised of copper, zinc-superoxide dismutase (SOD1; [29]), transactive response DNA-binding protein (TDP-43; [21, 86]), fused in sarcoma (FUS; [36, 87]), and C9orf72 [23, 24]. Despite some critical differences between these animal models and the human disease [85, 88], only SOD1 rodent models produce phenotypes that recapitulate several core clinical and pathological features of human ALS [83, 85]. Other models either do not mimic the core clinical and pathological features seen in ALS patients or are yet to be evaluated for their recapitulation of ALS pathology (i.e., C9orf72 models).

Prior MRS studies, performed on the wobbler mouse [89] and transgenic mice models with human SOD1 mutations, utilized three different sampling approaches to obtain tissue metabolite profiles that consist of in vivo [89–91], ex vivo intact tissue [92], and tissue extracts [90, 93].

In an animal model of ALS, Pioro et al. [89] have acquired in vivo proton MRS data using PRESS sequence (TR/TE: 2500/135 ms) at 4. 7T in the brain of homozygous mutant wobbler mice ($n = 5$) and age-matched unaffected littermates ($n = 5$) for comparisons. The voxel localized for MRS included neocortex, caudate, putamen, hippocampus, and thalamus. They found significantly decreased NAA/Cre and no difference in Cho/Cre in the brain of mutant wobbler mice. Andreassen et al. [91] have evaluated the concentration of brain metabolites at 4.7T in the sensory cortex region of transgenic mice with a SOD1 mutation and wild-type groups and, furthermore, the effect of long-term creatine supplementation on brain metabolite levels in the transgenic mice group. Their findings included a significantly increased Glx/Cre in the mutant mice group ($n = 9$) as compared to the wild-type group ($n = 6$), and a significantly lower Glx/Cho at 80 day-old but no difference at 110-day-old creatine-fed group ($n = 5$) as compared to the normal diet similar age groups ($n = 5$). Creatine supplementation has significantly increased longevity and motor performance of the transgenic mice group. Their results provided evidence for impaired glutamate transport and marginal beneficial effect of creatine supplementation in mutant SOD1 transgenic mice. In a longitudinal in vivo MRS study, Choi et al. [90] have acquired SVS data at 9.4T in the motor and cingulate cortex regions of mutant transgenic SOD1 and wild-type mice groups. Data were categorized for analyses into three groups ($n = 5$–7) based on the anatomical extent of motor symptoms (i.e., unaffected, paralysis of hind limbs only, and paralysis of hind limbs and partial paralysis of front limbs). The results included significantly decreased NAA and Glx and increased taurine in mice with four limbs affected, compared to a group with no limbs affected.

Ghoddoussi et al. [92] have evaluated the effect of chronic treatment of methionine sulfoximine (MSO), an inhibitor of glutamine synthetase, on the brain metabolites of transgenic mutant SOD1 mice (a mouse model of familial ALS) using high-resolution magic angle spinning (HR-MAS) MRS method with intact ex vivo tissue samples of the motor cortex and anterior striatum. Compared to saline-treated control animals ($n = 6$), significantly decreased concentrations of Gln and Glu in the motor cortex and striatum were found in the MSO treated mice ($n = 6$). Furthermore, the MSO treatment extended the lifespan of mice by 8 %. This study showed that MSO can be used to reduce glutamine synthetase (GS) activity and thereby finding GS as an effective drug target to manipulate the levels of Glu and Gln so as to reduce the extracellular glutamate excitotoxicity-initiated neurodegenerative damages in the brain.

Two proton MRS studies [90, 93] assessed changes in the concentration of metabolites in the tissue extracts of the brain and spinal cord of a mouse model of ALS. Niessen et al. [93] have evaluated alterations in the concentration of metabolites in the tissue extracts of the spinal cord, brainstem, cortex, and cerebellum of transgenic mutant SOD1 ($n = 5$) and non-transgenic ($n = 5$) mice. Data were acquired at 34, 75, 90, and 120 days postpartum and at 14.1T. The first two time points (34 and 75 days old) reflected a stage before the disease onset with no significant motor neuron loss and no evidence of clinical symptoms, and the remaining two time points indicated the average disease onset (90 days old) and terminal (120 days old) stages. Their results included significantly decreased NAA, glutamine, and GABA in the spinal cord and brainstem of the mutant SOD1 mouse group in the presymptomatic phase of the disease ($\leq$75 days old). At 90 days and thereafter, a significantly altered concentration was found for most of the metabolites in the spinal cord and brainstem of the SOD1 mouse group. There were no significant changes in the concentrations of NAA in the cortex and cerebellum of the SOD1 mice at all the four time points, indicating that this animal model does not replicate the cortical motor neuron pathology found in human ALS. Furthermore, myo-inositol concentration in the spinal cord of the mutant SOD1 mice decreased at 120 days, which is in contrast to elevated myo-inositol levels reported in the motor cortex of patients with ALS [55, 94, 95]. This finding indicates further differences between the pathogenesis of ALS in this mouse model and the human ALS. Choi et al. [90] have reported regional variations in the concentration of metabolites in the tissue extracts of the spinal cord, medulla, cerebellum, and sensorimotor cortex of transgenic mutant SOD1 (familial ALS mouse model; FALS) and wild-type (WT) mice at four growth time points (84, 110, 114, 142 days of age) that reflect different stages of the disease. At 84 days of age (close to the disease onset stage), they found a significantly increased glutamate in the sensorimotor cortex of FALS mice ($n = 9$), compared with the WT mice ($n = 9$). Significantly increased glycine, taurine, glutamine, and glutamate at 114 days of age (close to terminal disease stage) and further increase in taurine and glutamine and decreased NAA at 142 days of age (terminal stage with partial paralysis) were reported in the extracts of sensorimotor cortex. They also assessed the effects of 2 % creatine in the diet to animals on the concentration of metabolites in the tissue extracts from the cerebellum, medulla, and cortex regions. Their results included no loss of NAA in medulla, providing evidence of a protective effect of creatine supplementation and significantly increased creatine in medulla and cerebellum.

Though most of the metabolite alterations observed in the animal models of ALS paralleled the changes found in patients with ALS [65], there were few metabolite changes as described above that differed findings from human ALS studies. The possible reasons for the noted discrepancies in animal models include different pathogenic and temporal disease evolution trajectories in these models that do not completely imitate the disease and its temporal evolution in humans. Furthermore, drugs found to alter the disease processes in animal models of ALS have failed to translate in patients with ALS [88, 96–98]. As a result, researchers and clinicians question the usefulness of existing animal models and look for new animal models [96] that mimic the pathophysiology of human ALS at cellular (e.g., motor neurons and astrocytes), molecular (e.g., protein aggregation, and dysfunction, degeneration, apoptosis of motor neurons), and morphological or clinical symptomatic (e.g., muscle weakness, tremor, paresis, and paralysis) levels.

Overview of Clinical MRS

Despite the clinical and pathological evidence of degeneration or loss of UMN and/or LMN in the brain and spinal cord of patients with MND [51], MRS studies have been performed mostly in the brain of patients, owing to technical challenges associated with the acquisition of reliable MRS data in the spinal cord. A brief review of some of the MRS studies performed in the brain and spinal cord of patients with MND is provided here.

Brain

In 1994, Pioro et al. [99] have evaluated for the first time changes in the concentration of the metabolites of the brain of 12 patients with MND and 6 healthy controls using a PRESS-volume selected single-slice MRSI sequence at 1.5T with a TE of 272 ms. The brain regions chosen for this acquisition included the primary motor cortex, premotor, primary sensory cortex, and superior parietal regions. They found a significantly decreased NAA/Cre in the primary motor cortex (PMC) of ten patients (i.e., 5 with signs of both UMN and LMN and 5 with only signs of UMN) as compared to six healthy controls. In contrast, there was no significant difference in NAA/Cre between two patients with progressive spinal muscular atrophy (a - LMN-only disease) and healthy controls. The results from this first published MRS study in MND [99] indicated that NAA/Cre ratio can be considered to be a marker for evaluation of the degree of dysfunction or loss of cortical motor neurons (i.e., UMN). Since then several studies reported altered metabolite concentrations in the brain of patients with MND. These studies are reviewed here by grouping them based on the brain anatomical region evaluated.

Motor Cortex

Several studies reported significantly decreased NAA [59, 100–106], NAA/Cre [99, 104, 105, 107–116], NAA/Cre [94], NAA/Cho [94, 104, 105, 109, 110, 115, 117, 118], NAA/(Cre+Cre) [117], and NAA/(Cre+Cho) [105, 109, 119] in the primary motor cortex (PMC) of patients with ALS compared to healthy control subjects. Furthermore, some of these studies reported increased Cho/Cre [94, 104, 112, 115], Cho/NAA [105, 120] and Cho [55, 102, 105] and decreased Cre [103, 104] in the PMC of ALS patients. Those studies that reported robust results, due to either use of a robust technology/methodology or inclusion of a large number of patients, are briefly reviewed here.

Block et al. [94] have evaluated MRS spectra acquired from the motor cortex of 33 patients with ALS, 20 healthy volunteers, and four patients with multifocal motor neuropathy (MMN; a LMN-only disease). The patient group comprised of

22 patients with clinical signs of UMN involvement and 11 without it. Data were acquired at 1.5T with a 30 cm^3 single volume selected within the motor cortex using the PRESS SVS sequence with TEs of 30 ms and 272 ms. Relaxation times (T1, T2) of NAA, choline, and phosphocreatine were measured in subsets of patients and control subjects. They reported reduced NAA/Cho and NAA/Cre in patients with ALS compared to healthy control subjects. There was a significant decrease in NAA/Cho even in the patients without signs of UMN involvement subgroup. There were no metabolite T1 or T2 differences between the patient and control subject groups. Most importantly, the metabolite values in the MMN patient group matched with the control group values and not with the patient group. The follow-up studies showed a trend toward a lower NAA/Cho as time after the first examination increased. The results of this study [94] suggested that MRS has the potential to provide quantitative markers for motor neuron degeneration, to classify clinically distinct MND, and to track their progression in longitudinal studies.

In 2001, Pohl et al. [103] have reported results of MRS data acquired from 70 patients with ALS and 48 healthy control subjects. The patient group comprised of 15 suspected, 18 possible, 10 probable, and 27 definite ALS, classified according to El Escorial criteria [121]. Data were acquired from a 30 cm^3 voxel localized within the left and right sides of the primary motor cortices and used PRESS SVS sequence with a TE of 272 ms. In addition to metabolite ratios, they also calculated absolute concentrations of NAA, Cre, and Cho using published metabolite T1 and T2 values [94]. They reported significantly reduced concentrations of NAA and Cre and no change in Cho concentration in the PMC of the patient group. NAA/Cho and NAA/Cre were found to be reduced across all the patient subgroups. Furthermore, there was a significant correlation between NAA/Cho in a side of the PMC and the clinical signs of the contralateral affected side. Weak correlations between NAA/Cho and disease duration and severity were also noted. The results of this study [103] confirmed that absolute concentrations of NAA and Cre in the PMC were significantly altered in patients with ALS, and changes in NAA/Cho in the PMC corresponded with the lateralization of clinical signs.

In an effort to identify metabolite markers of UMN involvement in ALS, Kaufmann et al. [62] have acquired MRS data from the brain of a large number of patients ($n = 151$). The study group consists of patients with possible, laboratory-supported probable, and definite ALS ($n = 124$; with clinically identified probable- and definite-UMN signs), and progressive muscular atrophy (a pure LMN disease; $n = 27$). The MRS data were acquired at 1.5T using PRESS SVS sequence with a TE of 272 ms for localization of 8 cm^3 voxels within the left and right sides of the precentral gyrus of patients. NAA/Cre ratio alone identified UMN involvement in 86 % and 63 % of patients with UMN and LMN signs, respectively. They found a significant correlation between the degree of abnormal levels of NAA/Cre in the PMC and UMN signs. Furthermore, they determined that NAA/Cre has the sensitivity and specificity of 0.86 and 0.37, respectively, to detect UMN involvement in patients.

Following Pioro et al. [99], other investigators have applied single-slice MRSI approaches to obtain data from the brain of patients with MND in axial [99, 106,

113, 122] and coronal orientations [70, 115]. Expansion of this approach led to Weiner and colleagues to come up with a 3-slice spin-echo-based MRSI data acquisition method to encompass a wider region of the brain for evaluation in ALS [102, 105, 109, 119]. In one of their studies [109] that included 47 patients with ALS and 17 healthy control subjects, MRSI data were acquired at 1.5T from three axial slices of 15 mm thickness each, positioned in such a way to include the primary motor cortex, centrum semiovale, and internal capsule. Of the 47 patients, 21 were with possible or suspected ALS and 24 with probable/definite ALS. Data acquisition of the 3 slices took 30 min with TR of 1800 ms and TE of 70 ms and nominal in-plane resolution of 7.5×7.5 mm^2. Segmentation of the T1-MRI acquired in the same data acquisition session provided white matter, gray matter, and cerebrospinal fluid partial volume information in each MRSI voxel for calculation of ratios or absolute concentrations of metabolites by tissue types (i.e., gray matter and white matter). Decreased NAA/Cho and NAA/(Cre+Cho) were found in the motor and other regions in both the most and least affected hemispheres of the probable or definite ALS group. A subgroup of 28 patients underwent four follow-up MRI scans with an interval of 3 months in-between. Though a significant decrease in NAA was observed at 3- and 9-month follow-up scans within or outside the motor cortex, no consistent metabolite change over time was found across the subjects scanned. These results led them [109] to conclude that a change in the brain NAA concentration may not be a reliable surrogate marker for therapeutic efficacy trials.

In 2007, Mitsumoto et al. [106] have reported results of MRSI data acquired at 1.5T from the brain of 64 patients with MND (43 sporadic ALS; 6 familial ALS, 9 PMA, and 6 PLS) and 29 healthy control subjects. They found markedly decreased NAA in the primary motor cortex of patients with ALS and all UMN diseases combined (i.e., patients with ALS, PLS, or fALS) groups, compared to the healthy control group. NAA concentration was not significantly different between each of the PMA, PLS, and fALS groups versus the control group. Similar results were noted for NAA/Cre in the PMC of the above two groups compared with the control group, but the group differences were larger (for example, a mean difference in NAA in the ALS group was 11 % versus a mean NAA/Cre difference in the same group was 24 %) and so the significance level was much higher ($p = 0.009$ versus $p < 0.0005$). They found strong correlations ($r = 0.38$–0.45; $p \leq 0.001$) between NAA concentrations in the PMC of all patients and measures of UMN function (i.e., finger tapping speed, average grip strength, and average pinch strength) but less strongly with amyotrophic lateral sclerosis functional rating scale—revised (ALSFRS-R [123]) scores ($r = 0.30$; $p = 0.03$). The results of this study [106] suggested that the brain metabolite measures in the PMC have potential to differentiate ALS, its subtypes, and healthy control subjects.

Govind et al. [120] have published the first whole-brain MRSI study of patients with ALS in which data from the upper limb somatotopy of the PMC homunculus and the corticospinal tracts were analyzed (see Fig. 7.5). MRSI and clinical assessment data from 38 patients with sporadic definite ALS and 70 healthy control subjects were analyzed. In the ALS group compared to the control group, they

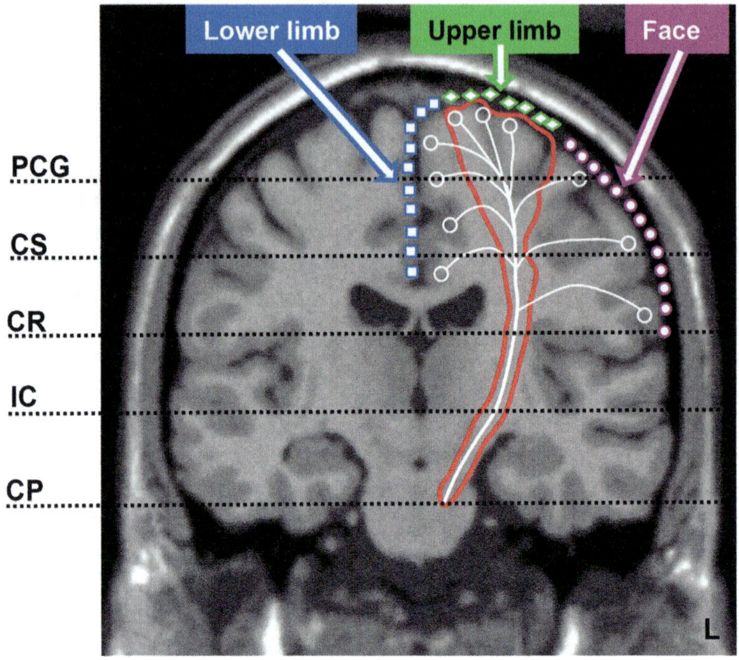

Fig. 7.5 An atlas comprised of the corticospinal tract (CST) and the upper limb/respiratory muscles somatotopy in the left side was overlaid onto a coronal orientation T1-MRI template in Montreal Neurological Institute coordinate space. The somatotopic arrangement within the primary motor cortex and its approximate anatomical regions are shown (the filled circles, diamonds, and squares represent the face, upper limb, and lower limb somatotopies, respectively). The circles and the lines represent the motor nuclei and the axons, respectively; the axons coalesce to form CST, which traverses inferiorly toward the brainstem. The horizontal dashed lines indicate the approximate anatomical levels at which the five segments were made along the length of the CST. PCG, precentral gyrus; CS, centrum semiovale; CR, corona radiata; IC, internal capsule or posterior limb of IC; CP, cerebral peduncle; L, left side of the subject

found significantly decreased NAA and increased Cho and Cho/NAA in the precentral gyrus (PCG) regions of either both the left and right sides or one of the sides (see Fig. 7.6). A strong association between the left-PCG Cho/NAA concentration and the right finger tap rate (a measure of UMN function) was found in this study. Two other studies evaluated brain metabolite alterations in patients with ALS using the same whole-brain MRSI approach [124, 125] following this study.

Despite the fact that glutamate-induced excitotoxicity is one of the possible mechanisms of ALS pathogenesis [33], not many MRS studies evaluated alterations in Glu concentration in the brain of patients with ALS. The MRS-observed Glu spectral peaks in the brain have contributions from both the metabolic (intracellular) and neurotransmitter (extracellular) Glu pools. Quantification of only the neurotransmitter pool of Glu that is responsible to cause excitotoxicity is technically not feasible using MRS methods. The intracellular metabolic pool (order of 10 mmol/L) is relatively much larger than the smaller extracellular neurotransmitter

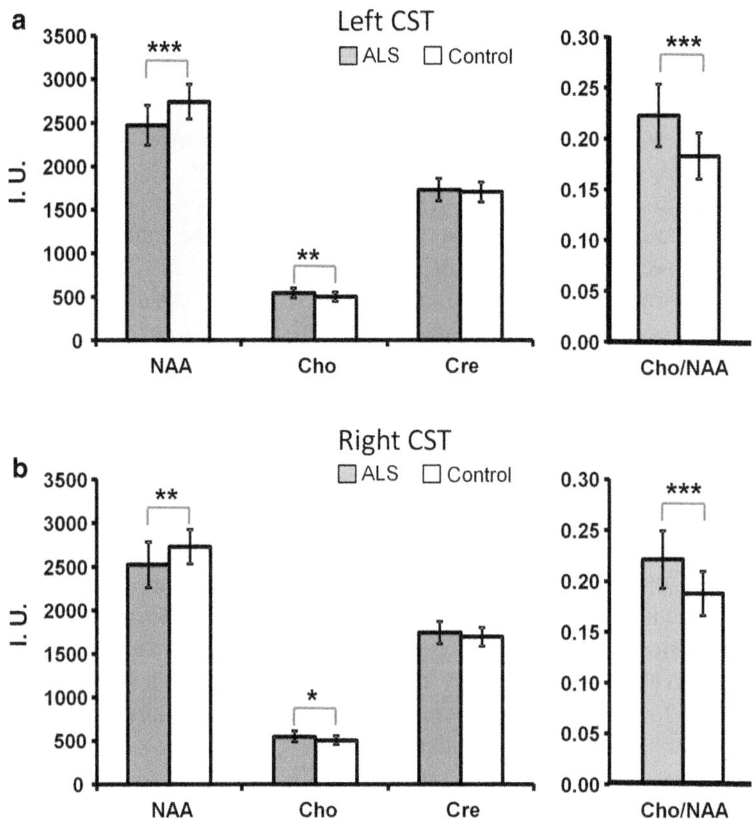

Fig. 7.6 Proton metabolite values and their ratio in the intracranial corticospinal tracts of patients with ALS and matched healthy control groups are shown. The histograms indicate the metabolite values and their ratio (mean± standard deviation) within the entire intracranial corticospinal tracts of the left and right sides, between the precentral gyrus and the cerebral peduncle regions, of the ALS ($n = 38$) and control ($n = 70$) groups. The age of the subjects included ranged from 34 to 65 years. For group comparisons, ANCOVA was used. The *asterisks* indicate significant between-group differences with $*p < 0.05$, $**p < 0.005$, and $***p < 0.0001$. The between-group percentage differences found include −8.9 % (*left*) and −6.2 (*right*) for *N*-acetyl aspartate (NAA), +8.1 (*left*) and +7.5 (*right*) for total choline (Cho), and +20.0 (*left*) and +15.7 (*right*) for Cho/NAA, with the negative and positive signs representing decreased and increased values, respectively, in the ALS group as compared to the control group values. I.U.: institutional unit. (Adopted from [120])

pool (order of few μmol/L) [55], and so the MRS-quantified Glu represents mostly the concentration of its metabolic pool in the brain. Han et al. [116] used TE-averaged PRESS sequence [126] at 3T to quantitate Glu in the brain of patients with ALS. Increased Glu/Cre and Glx/Cre were found in the PMC and posterior limb of internal capsule (PLIC) regions of patients. Another study reported a significant decrease in Glu [55] in the PMC of patients and it used STEAM [127] sequence to acquire data at 1.5T. It is not clear whether the contradictory results of the studies are due to a difference in the acquisition sequences used, the use of different quantitation metrics (ratios versus concentration values), or differences in the patient groups included.

Few studies evaluated changes in myo-inositol concentration in the PMC of patients. All studies reported increased mIns values (mIns [55]) or its ratio with a co-observed metabolite (mIns/Cre [94, 110, 112] or mIns/Cre [94]) in patients. Increased mIns in ALS is compatible with the concept of glial cell proliferation (i.e., gliosis) in the space left behind by loss or degeneration of neurons or changes in the phospholipid composition of myelin [55]. Except one study at 3T [110], all other studies were performed at 1.5T [55, 94, 112].

Despite technical difficulties to quantify GABA in the brain, two recent MRS studies reported lower levels of GABA [128, 129] in the PMC of patients with ALS compared to healthy control subjects. Both the studies utilized MEGA-PRESS spectral editing method at 3T [130] to quantitate GABA because conventional MRS methods are not useful to quantitate it. Foerster et al. [129] have reported lower levels of GABA and higher levels of Glx (i.e., Glu+Gln) in the PMC of patients who were drug naive, compared to the levels in healthy control subjects. Their results suggest that an imbalance between the inhibitory (i.e., GABA) and excitatory (i.e., glutamate) neurotransmitters exists in ALS.

Another pathogenic cause implicated in ALS is oxidative stress-mediated neuronal cell death [30]. Glutathione (GSH) is an important antioxidant in the CNS, and its depletion in the brain is hypothesized to cause motor neuron degeneration [131] or death [132]. A recent MRS study [133] reported lower levels of GSH (36 % with respect to Cre) in patients with ALS compared to healthy control subjects using a spectral editing technique at 3T [134].

Corticospinal Tracts

Despite evidence of significant tissue structural changes in the intracranial corticospinal tracts (CST) of patients with ALS, not many studies [102, 108, 115, 116, 120] evaluated changes in the concentration of metabolites in CST. These studies evaluated MRS data from discrete locations along the length of CST that include centrum semiovale [120], posterior limb of internal capsule [102, 116, 120], periventricular white matter [108], corona radiata [115, 120], cerebral peduncle [120], and whole intracranial CSTs [120]. The metabolite changes reported in these studies include decreased NAA/Cre [116] and NAA [120], increased Cho [120] and Cho/Cre, increased Cho/NAA [120], and increased Glx/Cre [116] and Glu/Cre [116].

In order to assess the degree of metabolite changes at specific anatomical levels along the length of intracranial CST of patients, Govind et al. [120] have divided the intracranial CST into 5 segments at appropriate anatomical levels such as precentral gyrus, centrum semiovale, corona radiata, and posterior limb of internal capsule (see Fig. 7.5). They found significant metabolite changes in these segments in 38 patients with definite ALS compared to 70 healthy control subjects. The changes observed in specific metabolite measures as shown in Fig. 7.7 include increased Cho and Cho/NAA in the segments of left and right CSTs and reduced

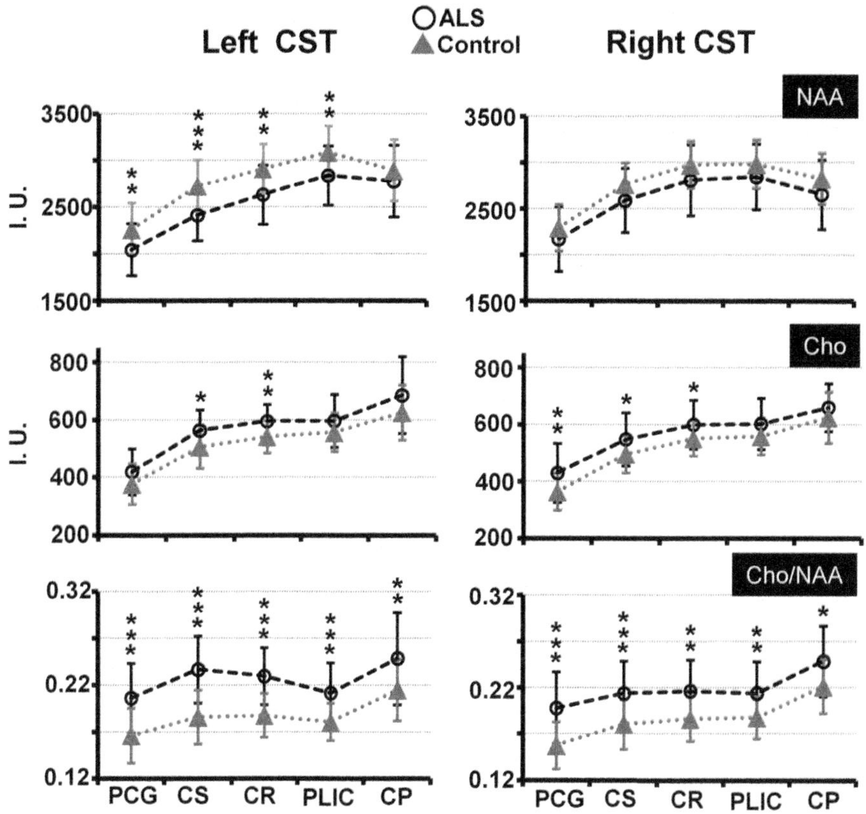

Fig. 7.7 Comparisons of the metabolite values at five discrete regions in the length of the intracranial corticospinal tracts in ALS ($n = 38$) and control ($n = 70$) groups. The age of the subjects included in the two groups ranged from 34 to 65 years. ANCOVA was used for statistical analysis. The *asterisks* indicate significant group differences with $*p < 0.05$, $**p < 0.005$, and $***p < 0.0001$. NAA, *N*-acetyl aspartate; Cho, total choline; PCG, precentral gyrus; CS, centrum semiovale; CR, corona radiata; PLIC, posterior limb of internal capsule; CP, cerebral peduncle; I. U., institutional unit. (Adopted from [120])

NAA in the left CST segments. There were significant correlations between Cho/NAA in the CST) and measures of UMN function (i.e., finger and foot tap rates) and disease severity (i.e., forced vital capacity) in the patient group. Furthermore, the observed correlations showed the expected contralateral anatomical relationships between the regions localizing the motor somatotopies in the PMC and their associated body parts. The observed correlations between the metabolite and clinical measures indicate that the MRS metrics can be used as potential biomarkers for evaluation of UMN dysfunction in the relevant body region of the PMC. The data acquisition and analysis methodology utilized in this study can be used in longitudinal disease progression monitoring and therapeutic efficacy evaluation studies.

Extra-Motor Brain Regions

For a longtime, ALS is considered to affect selectively the motor neurons and their axons in the primary motor cortex and corticospinal tracts of patients; however, accumulating evidence from recent neuropathological and neuroimaging studies [10, 51, 69, 70, 124, 135, 136] indicates that it is a multisystem disorder in which extra-motor brain regions are also affected. The extra-motor brain regions that were found to be affected in ALS by MRS studies include thalamus [135], basal ganglia [135], mesial prefrontal cortex [137], mid cingulate cortex [70] and other regions [109, 124].

Brainstem

The descending corticospinal and corticobulbar tracts (i.e., UMN tracts) and the motor nuclei of cranial nerves (i.e., LMN) are located within the brainstem, and these are known to be affected in ALS [138]. Though it is technically challenging to obtain reproducible MRS data in the brainstem that is comprised of the midbrain, pons, and medulla, few studies reported altered metabolite levels in the brainstem [101, 104, 139–141] of patients with ALS. In brainstem, NAA/Cre was found to be lower in patients with severe bulbar weakness or spasticity, whereas it is near normal in patients with predominantly LMN weakness [139]. Pioro et al. [141] have found 55 % higher levels of Glx (i.e., Glu+Gln) in the medulla of patients with ALS, which supports the concept of Glu-induced excitotoxicity in ALS.

Spinal Cord

The corticospinal tracts (i.e., UMN tracts) and/or the anterior horn cells (i.e., LMN) in the spinal cord of patients are known to be affected in MND. But acquisition of reliable MR spectroscopic data in spinal cord is difficult due to the technical difficulty associated with MRS data acquisition in the anatomical region in which it is located. This explains why very few MRS studies have been published thus far. Three MRS studies reported findings in the spinal cord of patients with ALS [142, 143] and of presymptomatic subjects tested positive for superoxide dismutase 1 (SOD1+) [144]. The findings included decreased NAA/Cre and NAA/mIns in C1–C2 region of the spinal cord of presymptomatic-familial ALS (SOD1+) subjects [144] and patients with ALS [142, 144]. Furthermore, compared to controls, decreased mIns/Cre was found only in SOD1+ subjects and not in patients with ALS. The observed metabolite changes in the SOD1+ group with no signs of disease and normal clinical measures suggest that there are altered metabolic processes preceding the onset of disease. Serial follow-up studies in such a group will establish the relationship between the metabolite changes and first appearance of symptoms.

Relaxation Times of Brain Metabolites

Due to long measurement times needed to acquire additional datasets for the calculation of metabolite relaxation times (i.e., T1 and T2), it is a common practice among in vivo MRS researchers to report metabolite concentrations and between-group metabolite changes (e.g., patients versus control subjects) without accounting for relaxation times of individual metabolites or metabolite relaxation time differences between the groups compared. Most of published MRS studies in MND are no exception to this practice. Two studies reported T1 [94] and/or T2 [94, 140] values of the prominently observed brain metabolites in ALS. Block et al. [94] found no T1 and T2 differences between ALS and control groups for NAA, Cre, and Cho in the brain whereas Hanstock et al. [140] reported decreased T2 for NAA and Cre in the brainstem of patients with ALS compared to healthy control subjects.

MRS in Longitudinal Studies

Few longitudinal brain MRS studies followed patients with ALS for up to 28 months and reported ongoing changes in NAA/Cho [103, 104] and NAA [105, 109, 145] in the primary motor cortex. However, there was no consistent pattern of progressive metabolite changes overtime in patients included in these studies [103, 104, 109]. Several issues may have contributed to this result such as flooring of metabolite levels at first time point as this leads to no further changes in subsequent measurement times, change in the metabolite relaxation times across follow-up studies, inclusion of heterogeneous patient population, and the data acquisition and analysis methods. With regard to correlations between changes in MRS metrics and clinical assessments, Pohl et al. [103] reported a positive correlation between NAA/Cho in the PMC and disease progression rate ($r = 0.55$) of patients.

MRS in Therapeutic Efficacy Evaluation Studies

MRS has been used to evaluate the effects of therapeutics on brain metabolites in patients with ALS [111, 146–150]. Kalra et al. [146] have reported a 6.1 % increase of NAA/Cre in the PMC of patients after 3 weeks of treatment with riluzole, the only FDA-approved drug for ALS. Despite the brain metabolite changes noted in the patients [146], there was no concomitant symptomatic and/or functional changes found in them. The effects of minocycline [147] and creatine [148] on brain metabolites were evaluated after short-term administration of these experimental drugs in patients with ALS.

Future Perspectives

MRI studies performed at the whole-brain level provided evidence of morphological and microstructural changes in the motor and extra-motor brain regions of patients with ALS [65]. Despite the evidence of disease-related tissue structural alterations beyond the motor regions, findings from most of the previous MRS studies in ALS have been limited to data acquired in the primary motor cortex and few other discrete brain anatomical regions, using single-voxel MRS or multi-voxel MRSI methods. The reasons for the limited brain coverage in MRS data acquisitions include poor quality of data obtained from some parts of the brain regions (e.g., the cortical gray matter, and temporal lobe and inferior frontal brain regions) and requirements of long acquisition times to collect multi-voxel data from the whole brain (>30 min). The availability of a true whole-brain MRS data acquisition method [72] together with a comprehensive suite of automated data processing and convenient analysis tools [151] allowed few research groups to acquire whole-brain MRS data from patients with ALS [120, 124, 125]. It is hoped that future cross-sectional and longitudinal studies will use a whole-brain MRS data acquisition method as that would allow comprehensive evaluation of metabolite changes throughout much of the whole brain in patients with MND.

Conventional MRS data acquisition methods (e.g., spin-echo, PRESS, and STEAM sequences) pose difficulties to readily observe and accurately quantitate glutamate and GABA. These metabolites are excitatory (glutamate) and inhibitory (GABA) neurotransmitters and are suggested to be involved in the primary pathophysiological processes in ALS. Spectral editing [74] or other [126] techniques are lately used to have an improved quantitation of these metabolites. However, use of such methods comes with issues such as different editing methods for each metabolite and that leads to long acquisition times even for a single volume of interest, and therefore currently their use for the whole brain does not seem feasible in clinical settings. Perhaps, an MRI-based indirect quantitation method [152] at magnetic field strengths $\geq$ 7T may be explored to map the distributions of these and other metabolites at the whole-brain level within a timeframe suitable for clinical diagnostic scans.

In order to evaluate the true concentration differences of the brain metabolites either between pathological and healthy normal conditions or in longitudinal patient studies, it is necessary to correct for differences in their relaxation times between the conditions or time points. Though it would be ideal to have the metabolite relaxation times measured experimentally, it is impractical in clinical settings, especially for whole-brain multi-voxel MRSI acquisitions and metabolites with multiplet spectral patterns (e.g., glutamate and GABA). This has been the biggest impediment to realize an MRS-based absolute concentration measurement method for metabolites in human subjects. Further development of novel parallel-imaging accelerated sparse-sampling high-speed MRSI data acquisition techniques [153–155] may lead to a relaxation time accounted absolute concentration quantitation method for the metabolites within the whole brain.

ALS is a rare and uniformly fatal disease with rapid progression rates, and death occurs typically between 1 and 5 years after disease symptom onset. Along the course of disease progression, the physical condition of a significant proportion of patients deteriorate to a level at which they can no longer be scanned. Therefore, multi-site studies are necessary to enroll sufficiently large number of patients to conduct statistically meaningful longitudinal pathogenic progression and therapeutic efficacy trial studies. Development and standardization of a common MRI/MRS protocol for use across multiple sites and scanner platforms [156] and availability of a large normative database of metabolite concentrations and tissue type fractional volume compositions (i.e., white matter, gray matter and cerebrospinal fluid) with appropriate clinical variables such as age and gender [72] would greatly facilitate the ease of performing future MRS research studies in human subjects.

Acknowledgments This work was funded by a grant from the National Institutes of Health (R01 NS060874), USA. The author gratefully acknowledges Drs. Maudsley and Sharma for their collaboration in this project. Mr. Sulaiman Sheriff is thanked for his help to make a figure included in this chapter.

References

1. Pringle CE, Hudson AJ, Munoz DG, Kiernan JA, Brown WF, Ebers GC (1992) Primary lateral sclerosis. Clinical features, neuropathology and diagnostic criteria. Brain 115 (Pt 2):495–520
2. Brooks BR, Miller RG, Swash M, Munsat TL (2000) El Escorial revisited: revised criteria for the diagnosis of amyotrophic lateral sclerosis. Amyotroph Lateral Scler Other Motor Neuron Disord 1(5):293–299
3. Traynor BJ, Codd MB, Corr B, Forde C, Frost E, Hardiman OM (2000) Clinical features of amyotrophic lateral sclerosis according to the El Escorial and Airlie House diagnostic criteria: A population-based study. Arch Neurol 57(8):1171–1176
4. Kiernan MC, Vucic S, Cheah BC, Turner MR, Eisen A, Hardiman O, Burrell JR, Zoing MC (2011) Amyotrophic lateral sclerosis. Lancet 377(9769):942–955
5. Tandan R, Bradley WG (1985) Amyotrophic lateral sclerosis: Part 1. Clinical features, pathology, and ethical issues in management. Ann Neurol 18(3):271–280
6. Turner MR, Talbot K (2013) Mimics and chameleons in motor neurone disease. Pract Neurol 13(3):153–164
7. Cui F, Liu M, Chen Y, Huang X, Cui L, Fan D, Pu C, Lu J, Zhou D, Zhang C, Yan C, Li C, Ding X, Liu Y, Li X, Jiang Y, Zhang J, Shang H, Yao X, Ding Y, Niu Q, Wang L (2014) Epidemiological characteristics of motor neuron disease in Chinese patients. Acta Neurol Scand 130(2):111–117
8. Brownell B, Oppenheimer DR, Hughes JT (1970) The central nervous system in motor neurone disease. J Neurol Neurosurg Psychiatry 33(3):338–357
9. Chio A, Calvo A, Moglia C, Mazzini L, Mora G, group Ps (2011) Phenotypic heterogeneity of amyotrophic lateral sclerosis: a population based study. J Neurol Neurosurg Psychiatry 82 (7):740–746
10. Ikemoto A, Hirano A, Akiguchi I (2000) Neuropathology of amyotrophic lateral sclerosis with extra-motor system degeneration: characteristics and differences in the molecular pathology between ALS with dementia and Guamanian ALS. Amyotroph Lateral Scler Other Motor Neuron Disord 1(2):97–104

11. Abrahams S, Goldstein LH, Suckling J, Ng V, Simmons A, Chitnis X, Atkins L, Williams SC, Leigh PN (2005) Frontotemporal white matter changes in amyotrophic lateral sclerosis. J Neurol 252(3):321–331
12. Sage CA, Peeters RR, Gorner A, Robberecht W, Sunaert S (2007) Quantitative diffusion tensor imaging in amyotrophic lateral sclerosis. Neuroimage 34(2):486–499
13. Witgert M, Salamone AR, Strutt AM, Jawaid A, Massman PJ, Bradshaw M, Mosnik D, Appel SH, Schulz PE (2010) Frontal-lobe mediated behavioral dysfunction in amyotrophic lateral sclerosis. Eur J Neurol 17(1):103–110
14. Agosta F, Valsasina P, Riva N, Copetti M, Messina MJ, Prelle A, Comi G, Filippi M (2012) The cortical signature of amyotrophic lateral sclerosis. PLoS One 7(8), e42816
15. Goldstein LH, Abrahams S (2013) Changes in cognition and behaviour in amyotrophic lateral sclerosis: nature of impairment and implications for assessment. Lancet Neurol 12 (4):368–380
16. Kasper E, Schuster C, Machts J, Kaufmann J, Bittner D, Vielhaber S, Benecke R, Teipel S, Prudlo J (2014) Microstructural white matter changes underlying cognitive and behavioural impairment in ALS--an in vivo study using DTI. PLoS One 9(12), e114543
17. Lomen-Hoerth C, Murphy J, Langmore S, Kramer JH, Olney RK, Miller B (2003) Are amyotrophic lateral sclerosis patients cognitively normal? Neurology 60(7):1094–1097
18. Phukan J, Pender NP, Hardiman O (2007) Cognitive impairment in amyotrophic lateral sclerosis. Lancet Neurol 6(11):994–1003
19. Giordana MT, Ferrero P, Grifoni S, Pellerino A, Naldi A, Montuschi A (2011) Dementia and cognitive impairment in amyotrophic lateral sclerosis: a review. Neurol Sci 32(1):9–16
20. Lomen-Hoerth C, Anderson T, Miller B (2002) The overlap of amyotrophic lateral sclerosis and frontotemporal dementia. Neurology 59(7):1077–1079
21. Neumann M, Sampathu DM, Kwong LK, Truax AC, Micsenyi MC, Chou TT, Bruce J, Schuck T, Grossman M, Clark CM, McCluskey LF, Miller BL, Masliah E, Mackenzie IR, Feldman H, Feiden W, Kretzschmar HA, Trojanowski JQ, Lee VM (2006) Ubiquitinated TDP-43 in frontotemporal lobar degeneration and amyotrophic lateral sclerosis. Science 314 (5796):130–133
22. Lillo P, Hodges JR (2009) Frontotemporal dementia and motor neurone disease: overlapping clinic-pathological disorders. J Clin Neurosci 16(9):1131–1135
23. Renton AE, Majounie E, Waite A, Simon-Sanchez J, Rollinson S, Gibbs JR, Schymick JC, Laaksovirta H, van Swieten JC, Myllykangas L, Kalimo H, Paetau A, Abramzon Y, Remes AM, Kaganovich A, Scholz SW, Duckworth J, Ding J, Harmer DW, Hernandez DG, Johnson JO, Mok K, Ryten M, Trabzuni D, Guerreiro RJ, Orrell RW, Neal J, Murray A, Pearson J, Jansen IE, Sondervan D, Seelaar H, Blake D, Young K, Halliwell N, Callister JB, Toulson G, Richardson A, Gerhard A, Snowden J, Mann D, Neary D, Nalls MA, Peuralinna T, Jansson L, Isoviita VM, Kaivorinne AL, Holtta-Vuori M, Ikonen E, Sulkava R, Benatar M, Wuu J, Chio A, Restagno G, Borghero G, Sabatelli M, Consortium I, Heckerman D, Rogaeva E, Zinman L, Rothstein JD, Sendtner M, Drepper C, Eichler EE, Alkan C, Abdullaev Z, Pack SD, Dutra A, Pak E, Hardy J, Singleton A, Williams NM, Heutink P, Pickering-Brown S, Morris HR, Tienari PJ, Traynor BJ (2011) A hexanucleotide repeat expansion in C9ORF72 is the cause of chromosome 9p21-linked ALS-FTD. Neuron 72(2):257–268
24. DeJesus-Hernandez M, Mackenzie IR, Boeve BF, Boxer AL, Baker M, Rutherford NJ, Nicholson AM, Finch NA, Flynn H, Adamson J, Kouri N, Wojtas A, Sengdy P, Hsiung GY, Karydas A, Seeley WW, Josephs KA, Coppola G, Geschwind DH, Wszolek ZK, Feldman H, Knopman DS, Petersen RC, Miller BL, Dickson DW, Boylan KB, Graff-Radford NR, Rademakers R (2011) Expanded GGGGCC hexanucleotide repeat in noncoding region of C9ORF72 causes chromosome 9p-linked FTD and ALS. Neuron 72(2):245–256
25. Swinnen B, Robberecht W (2014) The phenotypic variability of amyotrophic lateral sclerosis. Nat Rev Neurol 10(11):661–670
26. Rowland LP, Shneider NA (2001) Amyotrophic lateral sclerosis. N Engl J Med 344 (22):1688–1700

27. Renton AE, Chio A, Traynor BJ (2014) State of play in amyotrophic lateral sclerosis genetics. Nat Neurosci 17(1):17–23
28. Strong MJ (2001) Progress in clinical neurosciences: the evidence for ALS as a multisystems disorder of limited phenotypic expression. Can J Neurol Sci 28(4):283–298
29. Rosen DR, Siddique T, Patterson D, Figlewicz DA, Sapp P, Hentati A, Donaldson D, Goto J, O'Regan JP, Deng HX et al (1993) Mutations in Cu/Zn superoxide dismutase gene are associated with familial amyotrophic lateral sclerosis. Nature 362(6415):59–62
30. Bains JS, Shaw CA (1997) Neurodegenerative disorders in humans: the role of glutathione in oxidative stress-mediated neuronal death. Brain Res Brain Res Rev 25(3):335–358
31. Hirano A, Donnenfeld H, Sasaki S, Nakano I (1984) Fine structural observations of neurofilamentous changes in amyotrophic lateral sclerosis. J Neuropathol Exp Neurol 43 (5):461–470
32. Strong MJ, Kesavapany S, Pant HC (2005) The pathobiology of amyotrophic lateral sclerosis: a proteinopathy? J Neuropathol Exp Neurol 64(8):649–664
33. Van Den Bosch L, Van Damme P, Bogaert E, Robberecht W (2006) The role of excitotoxicity in the pathogenesis of amyotrophic lateral sclerosis. Biochim Biophys Acta 1762 (11-12):1068–1082
34. Bruijn LI, Houseweart MK, Kato S, Anderson KL, Anderson SD, Ohama E, Reaume AG, Scott RW, Cleveland DW (1998) Aggregation and motor neuron toxicity of an ALS-linked SOD1 mutant independent from wild-type SOD1. Science (New York, NY) 281 (5384):1851–1854
35. Wu CH, Fallini C, Ticozzi N, Keagle PJ, Sapp PC, Piotrowska K, Lowe P, Koppers M, McKenna-Yasek D, Baron DM, Kost JE, Gonzalez-Perez P, Fox AD, Adams J, Taroni F, Tiloca C, Leclerc AL, Chafe SC, Mangroo D, Moore MJ, Zitzewitz JA, Xu ZS, van den Berg LH, Glass JD, Siciliano G, Cirulli ET, Goldstein DB, Salachas F, Meininger V, Rossoll W, Ratti A, Gellera C, Bosco DA, Bassell GJ, Silani V, Drory VE, Brown RH Jr, Landers JE (2012) Mutations in the profilin 1 gene cause familial amyotrophic lateral sclerosis. Nature 488(7412):499–503
36. Vance C, Rogelj B, Hortobagyi T, De Vos KJ, Nishimura AL, Sreedharan J, Hu X, Smith B, Ruddy D, Wright P, Ganesalingam J, Williams KL, Tripathi V, Al-Saraj S, Al-Chalabi A, Leigh PN, Blair IP, Nicholson G, de Belleroche J, Gallo JM, Miller CC, Shaw CE (2009) Mutations in FUS, an RNA processing protein, cause familial amyotrophic lateral sclerosis type 6. Science 323(5918):1208–1211
37. Sreedharan J, Blair IP, Tripathi VB, Hu X, Vance C, Rogelj B, Ackerley S, Durnall JC, Williams KL, Buratti E, Baralle F, de Belleroche J, Mitchell JD, Leigh PN, Al-Chalabi A, Miller CC, Nicholson G, Shaw CE (2008) TDP-43 mutations in familial and sporadic amyotrophic lateral sclerosis. Science 319(5870):1668–1672
38. Sasaki S, Iwata M (2007) Mitochondrial alterations in the spinal cord of patients with sporadic amyotrophic lateral sclerosis. J Neuropathol Exp Neurol 66(1):10–16
39. Bruijn LI, Miller TM, Cleveland DW (2004) Unraveling the mechanisms involved in motor neuron degeneration in ALS. Annu Rev Neurosci 27:723–749
40. Eisen A, Weber M (2001) The motor cortex and amyotrophic lateral sclerosis. Muscle Nerve 24(4):564–573
41. Takahashi T, Yagishita S, Amano N, Yamaoka K, Kamei T (1997) Amyotrophic lateral sclerosis with numerous axonal spheroids in the corticospinal tract and massive degeneration of the cortex. Acta Neuropathol 94(3):294–299
42. Fischer LR, Culver DG, Tennant P, Davis AA, Wang M, Castellano-Sanchez A, Khan J, Polak MA, Glass JD (2004) Amyotrophic lateral sclerosis is a distal axonopathy: evidence in mice and man. Exp Neurol 185(2):232–240
43. Frey D, Schneider C, Xu L, Borg J, Spooren W, Caroni P (2000) Early and selective loss of neuromuscular synapse subtypes with low sprouting competence in motoneuron diseases. J Neurosci 20(7):2534–2542
44. Kennel PF, Finiels F, Revah F, Mallet J (1996) Neuromuscular function impairment is not caused by motor neurone loss in FALS mice: an electromyographic study. Neuroreport 7 (8):1427–1431

45. Chiu AY, Zhai P, Dal Canto MC, Peters TM, Kwon YW, Prattis SM, Gurney ME (1995) Age-dependent penetrance of disease in a transgenic mouse model of familial amyotrophic lateral sclerosis. Mol Cell Neurosci 6(4):349–362
46. Bensimon G, Lacomblez L, Meininger V (1994) A controlled trial of riluzole in amyotrophic lateral sclerosis. ALS/Riluzole Study Group. N Engl J Med 330(9):585–591
47. Lacomblez L, Bensimon G, Leigh PN, Guillet P, Meininger V (1996) Dose-ranging study of riluzole in amyotrophic lateral sclerosis. Amyotrophic Lateral Sclerosis/Riluzole Study Group II. Lancet 347(9013):1425–1431
48. Erminio F, Buchthal F, Rosenfalck P (1959) Motor unit territory and muscle fiber concentration in paresis due to peripheral nerve injury and anterior horn cell involvement. Neurology 9:657–671
49. Carvalho MD, Swash M (2009) Awaji diagnostic algorithm increases sensitivity of El Escorial criteria for ALS diagnosis. Amyotroph Lateral Scler 10(1):53–57
50. Douglass CP, Kandler RH, Shaw PJ, McDermott CJ (2010) An evaluation of neurophysiological criteria used in the diagnosis of motor neuron disease. J Neurol Neurosurg Psychiatry 81(6):646–649
51. Piao YS, Wakabayashi K, Kakita A, Yamada M, Hayashi S, Morita T, Ikuta F, Oyanagi K, Takahashi H (2003) Neuropathology with clinical correlations of sporadic amyotrophic lateral sclerosis: 102 autopsy cases examined between 1962 and 2000. Brain Pathol 13 (1):10–22
52. Goodin DS, Rowley HA, Olney RK (1988) Magnetic resonance imaging in amyotrophic lateral sclerosis. Ann Neurol 23(4):418–420
53. Hofmann E, Ochs G, Pelzl A, Warmuth-Metz M (1998) The corticospinal tract in amyotrophic lateral sclerosis: an MRI study. Neuroradiology 40(2):71–75
54. Peretti-Viton P, Azulay JP, Trefouret S, Brunel H, Daniel C, Viton JM, Flori A, Salazard B, Pouget J, Serratrice G, Salamon G (1999) MRI of the intracranial corticospinal tracts in amyotrophic and primary lateral sclerosis. Neuroradiology 41(10):744–749
55. Bowen BC, Pattany PM, Bradley WG, Murdoch JB, Rotta F, Younis AA, Duncan RC, Quencer RM (2000) MR imaging and localized proton spectroscopy of the precentral gyrus in amyotrophic lateral sclerosis. AJNR Am J Neuroradiol 21(4):647–658
56. Zhang L, Ulug AM, Zimmerman RD, Lin MT, Rubin M, Beal MF (2003) The diagnostic utility of FLAIR imaging in clinically verified amyotrophic lateral sclerosis. J Magn Reson Imaging 17(5):521–527
57. Agosta F, Chio A, Cosottini M, De Stefano N, Falini A, Mascalchi M, Rocca MA, Silani V, Tedeschi G, Filippi M (2010) The present and the future of neuroimaging in amyotrophic lateral sclerosis. AJNR Am J Neuroradiol 31(10):1769–1777
58. Chan S, Shungu DC, Douglas-Akinwande A, Lange DJ, Rowland LP (1999) Motor neuron diseases: comparison of single-voxel proton MR spectroscopy of the motor cortex with MR imaging of the brain. Radiology 212(3):763–769
59. Sarchielli P, Pelliccioli GP, Tarducci R, Chiarini P, Presciutti O, Gobbi G, Gallai V (2001) Magnetic resonance imaging and 1H-magnetic resonance spectroscopy in amyotrophic lateral sclerosis. Neuroradiology 43(3):189–197
60. Abe K, Fujimura H, Kobayashi Y, Fujita N, Yanagihara T (1997) Degeneration of the pyramidal tracts in patients with amyotrophic lateral sclerosis. A premortem and postmortem magnetic resonance imaging study. J Neuroimaging 7(4):208–212
61. Osei-Lah AD, Mills KR (2004) Optimising the detection of upper motor neuron function dysfunction in amyotrophic lateral sclerosis--a transcranial magnetic stimulation study. J Neurol 251(11):1364–1369
62. Kaufmann P, Pullman SL, Shungu DC, Chan S, Hays AP, Del Bene ML, Dover MA, Vukic M, Rowland LP, Mitsumoto H (2004) Objective tests for upper motor neuron involvement in amyotrophic lateral sclerosis (ALS). Neurology 62(10):1753–1757

63. Takahashi H, Snow BJ, Bhatt MH, Peppard R, Eisen A, Calne DB (1993) Evidence for a dopaminergic deficit in sporadic amyotrophic lateral sclerosis on positron emission scanning. Lancet 342(8878):1016–1018
64. Turner MR, Leigh PN (2000) Positron emission tomography (PET)--its potential to provide surrogate markers in ALS. Amyotroph Lateral Scler Other Motor Neuron Disord 1(Suppl 2): S17–S22
65. Turner MR, Agosta F, Bede P, Govind V, Lule D, Verstraete E (2012) Neuroimaging in amyotrophic lateral sclerosis. Biomark Med 6(3):319–337
66. Govindaraju V, Young K, Maudsley AA (2000) Proton NMR chemical shifts and coupling constants for brain metabolites. NMR Biomed 13(3):129–153
67. Bottomley PA (1987) Spatial localization in NMR spectroscopy in vivo. Ann N Y Acad Sci 508:333–348
68. Smith SA, Levante TO, Meier BH, Ernst RR (1994) Computer simulations in magnetic resonance. An object oriented programming approach. J Magn Reson, Series A 106 (1):75–105
69. van der Graaff MM, de Jong JM, Baas F, de Visser M (2009) Upper motor neuron and extra-motor neuron involvement in amyotrophic lateral sclerosis: a clinical and brain imaging review. Neuromuscul Disord 19(1):53–58
70. Sudharshan N, Hanstock C, Hui B, Pyra T, Johnston W, Kalra S (2011) Degeneration of the mid-cingulate cortex in amyotrophic lateral sclerosis detected in vivo with MR spectroscopy. AJNR Am J Neuroradiol 32(2):403–407
71. Harris AD, Puts NA, Barker PB, Edden RA (2014) Spectral-editing measurements of GABA in the human brain with and without macromolecule suppression. Magn Reson Med. doi:10. 1002/mrm.25549
72. Maudsley AA, Domenig C, Govind V, Darkazanli A, Studholme C, Arheart K, Bloomer C (2009) Mapping of brain metabolite distributions by volumetric proton MR spectroscopic imaging (MRSI). Magn Reson Med 61(3):548–559
73. Ding XQ, Maudsley AA, Sabati M, Sheriff S, Dellani PR, Lanfermann H (2014) Reproducibility and reliability of short-TE whole-brain MR spectroscopic imaging of human brain at 3T. Magn Reson Med 73(3):921–928
74. Bogner W, Gagoski B, Hess AT, Bhat H, Tisdall MD, van der Kouwe AJ, Strasser B, Marjanska M, Trattnig S, Grant E, Rosen B, Andronesi OC (2014) 3D GABA imaging with real-time motion correction, shim update and reacquisition of adiabatic spiral MRSI. Neuroimage 103C:290–302
75. Gurney ME (1997) Transgenic animal models of familial amyotrophic lateral sclerosis. J Neurol 244(Suppl 2):S15–S20
76. Howland DS, Liu J, She Y, Goad B, Maragakis NJ, Kim B, Erickson J, Kulik J, DeVito L, Psaltis G, DeGennaro LJ, Cleveland DW, Rothstein JD (2002) Focal loss of the glutamate transporter EAAT2 in a transgenic rat model of SOD1 mutant-mediated amyotrophic lateral sclerosis (ALS). Proc Natl Acad Sci U S A 99(3):1604–1609
77. Yang C, Wang H, Qiao T, Yang B, Aliaga L, Qiu L, Tan W, Salameh J, McKenna-Yasek DM, Smith T, Peng L, Moore MJ, Brown RH Jr, Cai H, Xu Z (2014) Partial loss of TDP-43 function causes phenotypes of amyotrophic lateral sclerosis. Proc Natl Acad Sci U S A 111 (12):E1121–E1129
78. Huang C, Zhou H, Tong J, Chen H, Liu YJ, Wang D, Wei X, Xia XG (2011) FUS transgenic rats develop the phenotypes of amyotrophic lateral sclerosis and frontotemporal lobar degeneration. PLoS Genet 7(3), e1002011
79. Mitchell JC, McGoldrick P, Vance C, Hortobagyi T, Sreedharan J, Rogelj B, Tudor EL, Smith BN, Klasen C, Miller CC, Cooper JD, Greensmith L, Shaw CE (2013) Overexpression of human wild-type FUS causes progressive motor neuron degeneration in an age- and dose-dependent fashion. Acta Neuropathol 125(2):273–288
80. Suzuki N, Maroof AM, Merkle FT, Koszka K, Intoh A, Armstrong I, Moccia R, Davis-Dusenbery BN, Eggan K (2013) The mouse C9ORF72 ortholog is enriched in neurons known to degenerate in ALS and FTD. Nat Neurosci 16(12):1725–1727

81. Duchen LW, Strich SJ (1968) An hereditary motor neurone disease with progressive denervation of muscle in the mouse: the mutant 'wobbler'. J Neurol Neurosurg Psychiatry 31 (6):535–542
82. Panda SK, Wefers B, Ortiz O, Floss T, Schmid B, Haass C, Wurst W, Kuhn R (2013) Highly efficient targeted mutagenesis in mice using TALENs. Genetics 195(3):703–713
83. Laferriere F, Polymenidou M (2015) Advances and challenges in understanding the multifaceted pathogenesis of amyotrophic lateral sclerosis. Swiss Med Wkly 145:w14054
84. Moser JM, Bigini P, Schmitt-John T (2013) The wobbler mouse, an ALS animal model. Mol Genet Genomics 288(5-6):207–229
85. Turner MR, Hardiman O, Benatar M, Brooks BR, Chio A, de Carvalho M, Ince PG, Lin C, Miller RG, Mitsumoto H, Nicholson G, Ravits J, Shaw PJ, Swash M, Talbot K, Traynor BJ, Van den Berg LH, Veldink JH, Vucic S, Kiernan MC (2013) Controversies and priorities in amyotrophic lateral sclerosis. Lancet Neurol 12(3):310–322
86. Arai T, Hasegawa M, Akiyama H, Ikeda K, Nonaka T, Mori H, Mann D, Tsuchiya K, Yoshida M, Hashizume Y, Oda T (2006) TDP-43 is a component of ubiquitin-positive tau-negative inclusions in frontotemporal lobar degeneration and amyotrophic lateral sclerosis. Biochem Biophys Res Commun 351(3):602–611
87. Kwiatkowski TJ Jr, Bosco DA, Leclerc AL, Tamrazian E, Vanderburg CR, Russ C, Davis A, Gilchrist J, Kasarskis EJ, Munsat T, Valdmanis P, Rouleau GA, Hosler BA, Cortelli P, de Jong PJ, Yoshinaga Y, Haines JL, Pericak-Vance MA, Yan J, Ticozzi N, Siddique T, McKenna-Yasek D, Sapp PC, Horvitz HR, Landers JE, Brown RH Jr (2009) Mutations in the FUS/TLS gene on chromosome 16 cause familial amyotrophic lateral sclerosis. Science 323(5918):1205–1208
88. Turner BJ, Talbot K (2008) Transgenics, toxicity and therapeutics in rodent models of mutant SOD1-mediated familial ALS. Prog Neurobiol 85(1):94–134
89. Pioro EP, Wang Y, Moore JK, Ng TC, Trapp BD, Klinkosz B, Mitsumoto H (1998) Neuronal pathology in the wobbler mouse brain revealed by in vivo proton magnetic resonance spectroscopy and immunocytochemistry. Neuroreport 9(13):3041–3046
90. Choi JK, Kustermann E, Dedeoglu A, Jenkins BG (2009) Magnetic resonance spectroscopy of regional brain metabolite markers in FALS mice and the effects of dietary creatine supplementation. Eur J Neurosci 30(11):2143–2150
91. Andreassen OA, Jenkins BG, Dedeoglu A, Ferrante KL, Bogdanov MB, Kaddurah-Daouk R, Beal MF (2001) Increases in cortical glutamate concentrations in transgenic amyotrophic lateral sclerosis mice are attenuated by creatine supplementation. J Neurochem 77 (2):383–390
92. Ghoddoussi F, Galloway MP, Jambekar A, Bame M, Needleman R, Brusilow WS (2010) Methionine sulfoximine, an inhibitor of glutamine synthetase, lowers brain glutamine and glutamate in a mouse model of ALS. J Neurol Sci 290(1-2):41–47
93. Niessen HG, Debska-Vielhaber G, Sander K, Angenstein F, Ludolph AC, Hilfert L, Willker W, Leibfritz D, Heinze HJ, Kunz WS, Vielhaber S (2007) Metabolic progression markers of neurodegeneration in the transgenic G93A-SOD1 mouse model of amyotrophic lateral sclerosis. Eur J Neurosci 25(6):1669–1677
94. Block W, Karitzky J, Traber F, Pohl C, Keller E, Mundegar RR, Lamerichs R, Rink H, Ries F, Schild HH, Jerusalem F (1998) Proton magnetic resonance spectroscopy of the primary motor cortex in patients with motor neuron disease: subgroup analysis and follow-up measurements. Arch Neurol 55(7):931–936
95. Kalra S, Arnold DL (2006) Magnetic resonance spectroscopy for monitoring neuronal integrity in amyotrophic lateral sclerosis. Adv Exp Med Biol 576:275–282, discussion 361-273
96. Ludolph AC (2006) Matrix metalloproteinases--a conceptional alternative for disease-modifying strategies in ALS/MND? Exp Neurol 201(2):277–280
97. Benatar M (2007) Lost in translation: treatment trials in the SOD1 mouse and in human ALS. Neurobiol Dis 26(1):1–13

98. Scott S, Kranz JE, Cole J, Lincecum JM, Thompson K, Kelly N, Bostrom A, Theodoss J, Al-Nakhala BM, Vieira FG, Ramasubbu J, Heywood JA (2008) Design, power, and interpretation of studies in the standard murine model of ALS. Amyotroph Lateral Scler 9(1):4–15
99. Pioro EP, Antel JP, Cashman NR, Arnold DL (1994) Detection of cortical neuron loss in motor neuron disease by proton magnetic resonance spectroscopic imaging in vivo. Neurology 44(10):1933–1938
100. Gredal O, Rosenbaum S, Topp S, Karlsborg M, Strange P, Werdelin L (1997) Quantification of brain metabolites in amyotrophic lateral sclerosis by localized proton magnetic resonance spectroscopy. Neurology 48(4):878–881
101. Bradley WG, Bowen BC, Pattany PM, Rotta F (1999) ^{1}H-magnetic resonance spectroscopy in amyotrophic lateral sclerosis. J Neurol Sci 169(1-2):84–86
102. Schuff N, Rooney WD, Miller R, Gelinas DF, Amend DL, Maudsley AA, Weiner MW (2001) Reanalysis of multislice ^{1}H MRSI in amyotrophic lateral sclerosis. Magn Reson Med 45 (3):513–516
103. Pohl C, Block W, Karitzky J, Traber F, Schmidt S, Grothe C, Lamerichs R, Schild H, Klockgether T (2001) Proton magnetic resonance spectroscopy of the motor cortex in 70 patients with amyotrophic lateral sclerosis. Arch Neurol 58(5):729–735
104. Block W, Traber F, Flacke S, Jessen F, Pohl C, Schild H (2002) In-vivo proton MR-spectroscopy of the human brain: assessment of N-acetylaspartate (NAA) reduction as a marker for neurodegeneration. Amino Acids 23(1-3):317–323
105. Suhy J, Miller RG, Rule R, Schuff N, Licht J, Dronsky V, Gelinas D, Maudsley AA, Weiner MW (2002) Early detection and longitudinal changes in amyotrophic lateral sclerosis by ^{1}H MRSI. Neurology 58(5):773–779
106. Mitsumoto H, Ulug AM, Pullman SL, Gooch CL, Chan S, Tang MX, Mao X, Hays AP, Floyd AG, Battista V, Montes J, Hayes S, Dashnaw S, Kaufmann P, Gordon PH, Hirsch J, Levin B, Rowland LP, Shungu DC (2007) Quantitative objective markers for upper and lower motor neuron dysfunction in ALS. Neurology 68(17):1402–1410
107. Abe K, Takanashi M, Watanabe Y, Tanaka H, Fujita N, Hirabuki N, Yanagihara T (2001) Decrease in N-acetylaspartate/creatine ratio in the motor area and the frontal lobe in amyotrophic lateral sclerosis. Neuroradiology 43(7):537–541
108. Yin H, Lim CC, Ma L, Gao Y, Cai Y, Li D, Liang Y, Guo X (2004) Combined MR spectroscopic imaging and diffusion tensor MRI visualizes corticospinal tract degeneration in amyotrophic lateral sclerosis. J Neurol 251(10):1249–1254
109. Rule RR, Suhy J, Schuff N, Gelinas DF, Miller RG, Weiner MW (2004) Reduced NAA in motor and non-motor brain regions in amyotrophic lateral sclerosis: a cross-sectional and longitudinal study. Amyotroph Lateral Scler Other Motor Neuron Disord 5(3):141–149
110. Kalra S, Hanstock CC, Martin WR, Allen PS, Johnston WS (2006) Detection of cerebral degeneration in amyotrophic lateral sclerosis using high-field magnetic resonance spectroscopy. Arch Neurol 63(8):1144–1148
111. Kalra S, Tai P, Genge A, Arnold DL (2006) Rapid improvement in cortical neuronal integrity in amyotrophic lateral sclerosis detected by proton magnetic resonance spectroscopic imaging. J Neurol 253(8):1060–1063
112. Lombardo F, Frijia F, Bongioanni P, Canapicchi R, Minichilli F, Bianchi F, Hlavata H, Rossi B, Montanaro D (2009) Diffusion tensor MRI and MR spectroscopy in long lasting upper motor neuron involvement in amyotrophic lateral sclerosis. Arch Ital Biol 147 (3):69–82
113. Charil A, Corbo M, Filippi M, Kesavadas C, Agosta F, Munerati E, Gambini A, Comi G, Scotti G, Falini A (2009) Structural and metabolic changes in the brain of patients with upper motor neuron disorders: a multiparametric MRI study. Amyotroph Lateral Scler 10 (5-6):269–279
114. Sivak S, Bittsansky M, Kurca E, Turcanova-Koprusakova M, Grofik M, Nosal V, Polacek H, Dobrota D (2010) Proton magnetic resonance spectroscopy in patients with early stages of amyotrophic lateral sclerosis. Neuroradiology 52(12):1079–1085

115. Pyra T, Hui B, Hanstock C, Concha L, Wong JC, Beaulieu C, Johnston W, Kalra S (2010) Combined structural and neurochemical evaluation of the corticospinal tract in amyotrophic lateral sclerosis. Amyotroph Lateral Scler 11(1-2):157–165
116. Han J, Ma L (2010) Study of the features of proton MR spectroscopy (^{1}H-MRS) on amyotrophic lateral sclerosis. J Magn Reson Imaging 31(2):305–308
117. Jones AP, Gunawardena WJ, Coutinho CM, Gatt JA, Shaw IC, Mitchell JD (1995) Preliminary results of proton magnetic resonance spectroscopy in motor neurone disease (amyotrophic lateral sclerosis). J Neurol Sci 129(Suppl):85–89
118. Giroud M, Walker P, Bernard D, Lemesle M, Martin D, Baudouin N, Brunotte F, Dumas R (1996) Reduced brain N-acetyl-aspartate in frontal lobes suggests neuronal loss in patients with amyotrophic lateral sclerosis. Neurol Res 18(3):241–243
119. Rooney WD, Miller RG, Gelinas D, Schuff N, Maudsley AA, Weiner MW (1998) Decreased N-acetylaspartate in motor cortex and corticospinal tract in ALS. Neurology 50 (6):1800–1805
120. Govind V, Sharma KR, Maudsley AA, Arheart KL, Saigal G, Sheriff S (2012) Comprehensive evaluation of corticospinal tract metabolites in amyotrophic lateral sclerosis using whole-brain ^{1}H MR spectroscopy. PLoS One 7(4), e35607
121. Brooks BR (1994) El Escorial World Federation of Neurology criteria for the diagnosis of amyotrophic lateral sclerosis. Subcommittee on Motor Neuron Diseases/Amyotrophic Lateral Sclerosis of the World Federation of Neurology Research Group on Neuromuscular Diseases and the El Escorial "Clinical limits of amyotrophic lateral sclerosis" workshop contributors. J Neurol Sci 124:96–107
122. Wang S, Poptani H, Woo JH, Desiderio LM, Elman LB, McCluskey LF, Krejza J, Melhem ER (2006) Amyotrophic lateral sclerosis: diffusion-tensor and chemical shift MR imaging at 3.0 T. Radiology 239(3):831–838
123. Cedarbaum JM, Stambler N, Malta E, Fuller C, Hilt D, Thurmond B, Nakanishi A (1999) The ALSFRS-R: a revised ALS functional rating scale that incorporates assessments of respiratory function. BDNF ALS Study Group (Phase III). J Neurol Sci 169(1-2):13–21
124. Verma G, Woo JH, Chawla S, Wang S, Sheriff S, Elman LB, McCluskey LF, Grossman M, Melhem ER, Maudsley AA, Poptani H (2013) Whole-brain analysis of amyotrophic lateral sclerosis by using echo-planar spectroscopic imaging. Radiology 267(3):851–857
125. Stagg CJ, Knight S, Talbot K, Jenkinson M, Maudsley AA, Turner MR (2013) Whole-brain magnetic resonance spectroscopic imaging measures are related to disability in ALS. Neurology 80(7):610–615
126. Hurd R, Sailasuta N, Srinivasan R, Vigneron DB, Pelletier D, Nelson SJ (2004) Measurement of brain glutamate using TE-averaged PRESS at 3T. Magn Reson Med 51(3):435–440
127. Frahm J, Bruhn H, Gyngell ML, Merboldt KD, Hanicke W, Sauter R (1989) Localized high-resolution proton NMR spectroscopy using stimulated echoes: initial applications to human brain in vivo. Magn Reson Med 9(1):79–93
128. Foerster BR, Callaghan BC, Petrou M, Edden RA, Chenevert TL, Feldman EL (2012) Decreased motor cortex gamma-aminobutyric acid in amyotrophic lateral sclerosis. Neurology 78(20):1596–1600
129. Foerster BR, Pomper MG, Callaghan BC, Petrou M, Edden RA, Mohamed MA, Welsh RC, Carlos RC, Barker PB, Feldman EL (2013) An imbalance between excitatory and inhibitory neurotransmitters in amyotrophic lateral sclerosis revealed by use of 3-T proton magnetic resonance spectroscopy. JAMA Neurol 70(8):1009–1016
130. Mescher M, Merkle H, Kirsch J, Garwood M, Gruetter R (1998) Simultaneous in vivo spectral editing and water suppression. NMR Biomed 11(6):266–272
131. Chi L, Ke Y, Luo C, Gozal D, Liu R (2007) Depletion of reduced glutathione enhances motor neuron degeneration in vitro and in vivo. Neuroscience 144(3):991–1003
132. Li Y, Maher P, Schubert D (1997) A role for 12-lipoxygenase in nerve cell death caused by glutathione depletion. Neuron 19(2):453–463

133. Weiduschat N, Mao X, Hupf J, Armstrong N, Kang G, Lange DJ, Mitsumoto H, Shungu DC (2014) Motor cortex glutathione deficit in ALS measured in vivo with the J-editing technique. Neurosci Lett 570:102–107
134. Terpstra M, Henry PG, Gruetter R (2003) Measurement of reduced glutathione (GSH) in human brain using LCModel analysis of difference-edited spectra. Magn Reson Med 50(1):19–23
135. Sharma KR, Saigal G, Maudsley AA, Govind V (2011) ^{1}H MRS of basal ganglia and thalamus in amyotrophic lateral sclerosis. NMR Biomed 24(10):1270–1276
136. Smith MC (1960) Nerve Fibre Degeneration in the Brain in Amyotrophic Lateral Sclerosis. J Neurol Neurosurg Psychiatry 23(4):269–282
137. Usman U, Choi C, Camicioli R, Seres P, Lynch M, Sekhon R, Johnston W, Kalra S (2011) Mesial prefrontal cortex degeneration in amyotrophic lateral sclerosis: a high-field proton MR spectroscopy study. AJNR Am J Neuroradiol 32(9):1677–1680
138. Lawyer T Jr, Netsky MG (1953) Amyotrophic lateral sclerosis. AMA Arch Neurol Psych 69(2):171–192
139. Cwik VA, Hanstock CC, Allen PS, Martin WR (1998) Estimation of brainstem neuronal loss in amyotrophic lateral sclerosis with in vivo proton magnetic resonance spectroscopy. Neurology 50(1):72–77
140. Hanstock CC, Cwik VA, Martin WR (2002) Reduction in metabolite transverse relaxation times in amyotrophic lateral sclerosis. J Neurol Sci 198(1-2):37–41
141. Pioro EP, Majors AW, Mitsumoto H, Nelson DR, Ng TC (1999) ^{1}H-MRS evidence of neurodegeneration and excess glutamate + glutamine in ALS medulla. Neurology 53(1):71–79
142. Carew JD, Nair G, Pineda-Alonso N, Usher S, Hu X, Benatar M (2011) Magnetic resonance spectroscopy of the cervical cord in amyotrophic lateral sclerosis. Amyotroph Lateral Scler 12(3):185–191
143. Ikeda K, Murata K, Kawase Y, Kawabe K, Kano O, Yoshii Y, Takazawa T, Hirayama T, Iwasaki Y (2013) Relationship between cervical cord 1H-magnetic resonance spectroscopy and clinoco-electromyographic profile in amyotrophic lateral sclerosis. Muscle Nerve 47(1):61–67
144. Carew JD, Nair G, Andersen PM, Wuu J, Gronka S, Hu X, Benatar M (2011) Presymptomatic spinal cord neurometabolic findings in SOD1-positive people at risk for familial ALS. Neurology 77(14):1370–1375
145. Unrath A, Ludolph AC, Kassubek J (2007) Brain metabolites in definite amyotrophic lateral sclerosis. A longitudinal proton magnetic resonance spectroscopy study. J Neurol 254(8):1099–1106
146. Kalra S, Cashman NR, Genge A, Arnold DL (1998) Recovery of N-acetylaspartate in corticomotor neurons of patients with ALS after riluzole therapy. Neuroreport 9(8):1757–1761
147. Khiat A, D'Amour M, Souchon F, Boulanger Y (2010) MRS study of the effects of minocycline on markers of neuronal and microglial integrity in ALS. Magn Reson Imaging 28(10):1456–1460
148. Atassi N, Ratai EM, Greenblatt DJ, Pulley D, Zhao Y, Bombardier J, Wallace S, Eckenrode J, Cudkowicz M, Dibernardo A (2010) A phase I, pharmacokinetic, dosage escalation study of creatine monohydrate in subjects with amyotrophic lateral sclerosis. Amyotroph Lateral Scler 11(6):508–513
149. Kalra S, Cashman NR, Caramanos Z, Genge A, Arnold DL (2003) Gabapentin therapy for amyotrophic lateral sclerosis: lack of improvement in neuronal integrity shown by MR spectroscopy. AJNR Am J Neuroradiol 24(3):476–480
150. Kalra S, Genge A, Arnold DL (2003) A prospective, randomized, placebo-controlled evaluation of corticoneuronal response to intrathecal BDNF therapy in ALS using magnetic resonance spectroscopy: feasibility and results. Amyotroph Lateral Scler Other Motor Neuron Disord 4(1):22–26

151. Maudsley AA, Darkazanli A, Alger JR, Hall LO, Schuff N, Studholme C, Yu Y, Ebel A, Frew A, Goldgof D, Gu Y, Pagare R, Rousseau F, Sivasankaran K, Soher BJ, Weber P, Young K, Zhu X (2006) Comprehensive processing, display and analysis for in vivo MR spectroscopic imaging. NMR Biomed 19(4):492–503
152. Cai K, Haris M, Singh A, Kogan F, Greenberg JH, Hariharan H, Detre JA, Reddy R (2012) Magnetic resonance imaging of glutamate. Nat Med 18(2):302–306
153. Chatnuntawech I, Gagoski B, Bilgic B, Cauley SF, Setsompop K, Adalsteinsson E (2014) Accelerated H MRSI using randomly undersampled spiral-based k-space trajectories. Magn Reson Med. doi:10.1002/mrm.25394
154. Cao P, Wu EX (2015) Accelerating phase-encoded proton MR spectroscopic imaging by compressed sensing. J Magn Reson Imaging 41(2):487–495
155. Wilson NE, Iqbal Z, Burns BL, Keller M, Thomas MA (2015) Accelerated five-dimensional echo planar J-resolved spectroscopic imaging: Implementation and pilot validation in human brain. Magn Reson Med. doi:10.1002/mrm.25605
156. Sabati M, Sheriff S, Gu M, Wei J, Zhu H, Barker PB, Spielman DM, Alger JR, Maudsley AA (2014) Multivendor implementation and comparison of volumetric whole-brain echo-planar MR spectroscopic imaging. Magn Reson Med. doi:10.1002/mrm.25510

Chapter 8
MR Spectroscopy in Multiple Sclerosis

Niamh Cawley and Olga Ciccarelli

Abstract Multiple Sclerosis is a chronic inflammatory demyelinating disorder of the central nervous system. ^{1}H-MRS provides us with additional information on the chemical pathology within the brain and spinal cord in MS patients, when compared to conventional MRI. This has enhanced our understanding of the pathogenesis and natural history in multiple sclerosis. It has proved a useful biomarker of neurodegeneration, as reflected by a decrease in the levels of the neuronal marker, *N*-acetylaspartate. Changes in choline and myo-inositol have also informed us of the importance of monitoring changes in myelin damage and repair and of the extent of gliosis. This chapter will provide an overview of ^{1}H-MRS in the brain and spinal cord in clinical and preclinical studies in multiple sclerosis. We will also discuss the future potential of ^{1}H-MRS in multiple sclerosis and its promising applications.

Keywords Multiple sclerosis • MR spectroscopy • Brain • Spinal cord • White matter • Grey matter • Disability • Neurodegeneration • Inflammation • Progression

Introduction

Prevalence, Incidence and MS Sub-Types

Multiple sclerosis (MS) is a chronic inflammatory disease of the brain and spinal cord with both demyelinating and neuroaxonal degenerative components. It is the most common cause of neurological disability in young adults worldwide.

Jean-Martin Charcot was the first to describe the clinicopathological features of MS in 1868, although, there had been earlier case presentations in the 1820s.

N. Cawley, M.B. Bch. B.A.O. (✉)
NMR Research Unit, Queen Square MS Centre, UCL Institute of Neurology, London, UK
e-mail: n.cawley@ucl.ac.uk

O. Ciccarelli, Ph.D., F.R.C.P.
NMR Research Unit, Queen Square MS Centre, UCL Institute of Neurology, London, UK

NIHR, UCL/UCLH Biomedical Research Centre, London, UK
e-mail: o.ciccarelli@ucl.ac.uk

G. Öz (ed.), *Magnetic Resonance Spectroscopy of Degenerative Brain Diseases*, Contemporary Clinical Neuroscience, DOI 10.1007/978-3-319-33555-1_8

However, the aetiology and pathophysiology of the disease course has only recently started to be understood.

MS is a global disease with an estimated 2.3 million sufferers worldwide. MS has a prevalence of 97–184 per 100,000 [1]. The prevalence of MS varies greatly, with the highest prevalence in North America and Europe (140 and 108 per 100,000, respectively). The prevalence is lowest in Sub-Saharan Africa and East Asia at 2.1 and 2.2 per 100,000 [2]. The incidence of MS worldwide tends to increase with increasing latitude, with the highest disease prevalence seen in Europe between 45° and 65° north latitude, North America, Southern Canada, and Northern Australia [3]. Geographical variations are seen over relatively small distances. For example, MS is more common in Scotland with an incidence almost double that of Wales [4], while in South America the prevalence of MS in Argentina is 18 per 100,000, which is six times higher than Ecuador with a prevalence of 3.2 per 100,000 [5].

The incidence of MS has been rising around the world over the last few decades. This is likely due to a rising incidence in women, which may support a sex-linked environmental factor [6]. Potential explanations for this include a rise in female smoking, later age of first pregnancy, declining birth rates, and obesity. The exact aetiology of MS is unknown, but it is widely recognised that it has a complex causation that involves both genetic and environmental factors [6].

MS can be divided into different sub-types, characterised by a different clinical course. The classification of disease sub-types is very important due to the differences in prognosis. To date, disease-modifying therapy has been proven to be effective in the relapsing–remitting form of MS [2]. The initial phase of relapsing–remitting MS (RRMS) is characterised by episodes of active disease with neurological dysfunction (relapse) during which demyelinating lesions form in the central nervous system (CNS), interspersed with periods of clinical inactivity [7, 8]. There is a variable degree of recovery between relapses. The disease course for RRMS is highly unpredictable; both the frequency of relapses and the accumulation of disability are linked to the sex of the patient as well as age at disease onset. About 85 % of all MS patients present with RRMS [9]. About 40 % of these patients with RRMS progress into a secondary progressive form after a period of 10 years [10]. After 20 years, 80 % of patients with RRMS have developed secondary progressive MS [11, 12].

Secondary Progressive MS (SPMS) is diagnosed retrospectively by a history of gradual worsening after an initial relapsing disease course, with or without acute exacerbations during the progressive course [13]. To date, there are no clear clinical, imaging, immunological, or pathological criteria to determine the transition point when RRMS converts to SPMS; the transition is usually gradual [13]. This has limited our ability to study the imaging and biomarker characteristics that may distinguish this course. The median time to SPMS is 21.4 years from the time of initial diagnosis [14]. Male gender, age at onset, and motor symptoms at first presentation are associated with the onset of secondary progression in untreated MS patients [14].

A minority of patients with MS (10 %) have a slow accumulation of disability from disease onset and are described as having primary progressive MS (PPMS) [15]. About 10 % of patients with PPMS have relapses [16]. PPMS differs from other types of MS as the age of symptom onset is typically older (fourth or fifth decade) when compared to patients presenting with RRMS (second or third decade) [2, 6]. Studies to date show that PPMS has an equal sex distribution. This is in contrast to RRMS which has a female preponderance at 2–3:1 [2, 6]. It has been suggested that PPMS may represent a distinct, non-, or at least less inflammatory, pathological form of MS [17], but clinical, imaging, and genetic data suggest that PPMS is a part of the spectrum of progressive MS phenotypes and that any differences are relative rather than absolute [18]. Analyses of natural history cohorts demonstrate that worsening proceeds at a similar rate in SPMS and PPMS [8, 13, 19].

A subgroup of patients with RRMS seems to suffer minimal disease-related deficits, as reflected by the expanded disability status scale (EDSS) score. Benign MS (BMS) is defined in various ways in relation to the length of disease duration (10–15 years) and the severity of the disability (EDSS $\leq$2 or 3) [9]. BMS is seen in 20 % of patients and they tend to have little or no disability after 15 years [9]. The terms "benign" and "malignant" are not MS phenotypes, but they are intended to provide an indication of disease severity over time and were described "by consensus" [9]. These terms can, in theory, apply to any MS phenotype depending on the degree of activity over time or impairment/disability at any given point in time. In a long-term disease like MS, the severity and activity of the disease can change significantly and unpredictably, and therefore the applicability of these terms should be re-evaluated [13]. As clinical features alone seem to be insufficient when defining BMS, imaging parameters are looked at for additional metrics that may better characterise this disease course [20].

Pathology of MS

Traditionally MS was considered to be an autoimmune inflammatory disorder, which was mediated by a T cell attack against CNS elements, in particular myelin. Patients with RRMS display a compelling autoimmune and inflammatory phenotype on laboratory and radiological tests [21]. Although the same pathological abnormalities are present in all forms of MS, they vary both quantitatively and qualitatively between the three forms (RR, SP, and PP) of MS [17], with the degree of inflammation higher during the early relapsing stage of the disease and lower in the later phase of the disease [22].

Pathological examination of biopsied or post-mortem brains of patients with MS shows characteristic perivascular inflammatory infiltrates, which consist mainly of lymphocytes and macrophages. Inflammatory infiltrates mainly contain CD8+ T lymphocytes, which are clonally expanded in active lesions [23]. CD4+ T cells and B-lymphocytes are less numerous than CD8+ T-cells and mainly accumulate in the perivascular space and meninges [24, 25]. Microglial activation in the initial lesions is associated with demyelination and neurodegeneration with subsequent recruitment of macrophage as a result of myelin breakdown [22, 26].

White Matter

The understanding of MS has evolved significantly over the last number of years. Traditionally it was thought that MS was solely a disease of the white matter, but improvements in radiology and pathology in more recent years indicate that grey matter structures (i.e. neurons and synapses) of the brain are also affected [27].

Early RRMS is predominantly a disease of the white matter, characterised by confluent plaques of demyelination. The active demyelination is associated with inflammation and blood–brain barrier leakage [28]. The progressive stages of the disease are characterised by the gradual expansion of white matter lesions in the absence of blood–brain barrier leakage, with widespread cortical demyelination and widespread injury of the normal appearing white and grey matter. This ultimately results in extensive brain tissue loss and injury [28].

In the past, the majority of the evidence suggested that the disease began due to an immune dysregulation. This is based on the "outside-in" model of MS [21]. The resulting inflammatory reaction, which typically follows a RR course in the initial stages, causes further demyelination and tissue injury [21].

In more recent years, doubt has been cast on this model due to some inconsistencies and an "inside-out" model of MS has been proposed. This model proposes that the initial malfunction occurs within the CNS, like other neurodegenerative conditions, for example Parkinson's and Alzheimer's disease. Primary neurodegeneration, possibly involving the oligodendrocyte–myelin complex, may be involved in the initial event. This then releases highly antigenic constituents, which promote secondary autoimmune and inflammatory responses in the predisposed individual [29, 30].

Progressive stages of MS remain largely refractory to treatment. It is widely suggested that accumulating axonal loss is responsible for clinical progression. Pathological changes associated with clinical progression include inflammatory changes becoming increasingly compartmentalised in the perivascular and subarachnoid spaces behind a relatively intact blood–brain barrier [31].

Grey Matter

Extensive pathology is seen in the grey matter in the CNS. Grey matter pathology is characterised by demyelination with a relative absence of immune cell infiltrates. The number of cortical lesions increases at a significantly greater rate in patients with actively progressive disease and correlates with increasing disability [32]. Therefore, there is increasing evidence that accumulating cortical grey matter pathology plays an important role in the severity of both physical and cognitive disability [33]. The degree of cortical atrophy is not constant throughout the brain, with marked regional variations in cortical involvement. There is more extensive involvement seen in the hippocampus, frontal, and temporal cortices as well as the cingulate gyrus, and less extensive involvement in the occipital lobe and primary motor cortex [34]. The predilection for the frontal lobe and the hippocampus is likely to account for the predominance of cognitive over motor disability related to cortical pathology [34].

Little is known about the mechanism of lesion formation in the cortical grey matter and how the accumulation of cortical pathology may contribute to clinical progression [35]. Post-mortem studies have shown that inflammatory infiltrates are commonly found in the cerebral leptomeninges in SPMS [22]. In a proportion of SPMS cases, large B cell aggregates have been identified in the subarachnoid space with some of the characteristics of ectopic B cell follicles [22, 36]. The ectopic B cell follicle-like structures can be found distributed throughout the cerebral meninges and they are variable in number and size [36]. They are mainly present in the cerebral sulci which suggests that reduced CSF flow and the microenvironment of these locations favour the homing and retention of inflammatory cells which gives rise to an inflammatory milieus in the CSF [36]. MS cases with these meningeal lymphoid-like structures are associated with a younger age at disease onset, a shorter time to disease progression and wheelchair use, and shorter disease duration [37].

Potential Clinical Utility of Magnetic Resonance Spectroscopy (MRS)

In MS, conventional MRI is sensitive to grey and white matter atrophy, and it can distinguish active from chronic lesions. However, it lacks specificity to gliosis, inflammation, demyelination, and neuronal loss [38]. Additionally, lesions in the grey matter are difficult to identify with conventional MRI (Fig. 8.1). MRS can be seen as a complementary technique to MRI because it helps to enhance sensitivity to pathology; it also provides chemical and biophysical information about the microenvironment of tissues.

Resonances in MR spectra are mainly identified by their frequency, that is to say their position in the spectrum and expressed as the shift in frequency in parts-per-million (ppm) relative to a standard [40]. MRS has the unique ability to characterise the chemical pathology of brain and spinal cord lesions in MS, as well as regions of the brain or spinal cord that are not associated with evident structural abnormalities on conventional MRI [41]. This results in improved diagnosis, it helps to better define the natural history of the disease process and, occasionally, it can help in monitoring metabolic responses to therapy [42, 43].

Quantitative MRI techniques, such as MRS, have been used to measure disease burden within focal lesions, both acute and chronic, as well as in normal appearing

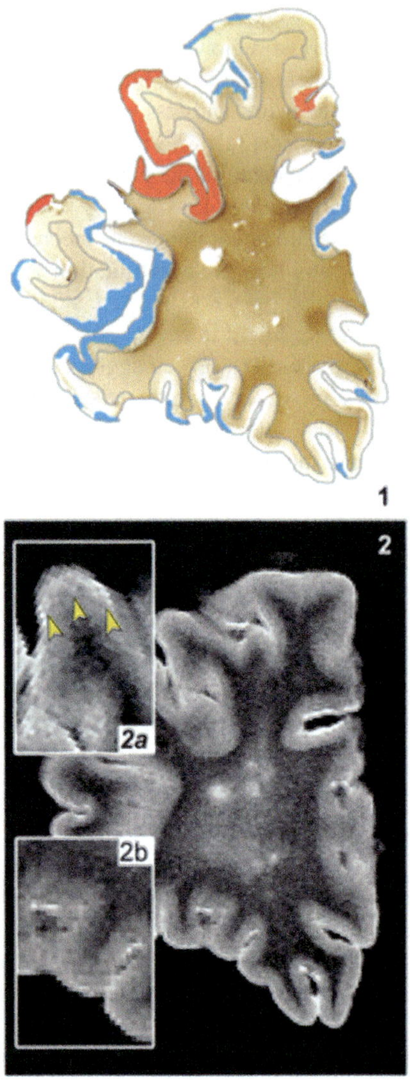

Fig. 8.1 Comparison between magnetic resonance imaging (MRI)-visible and invisible cortical lesions. *1*: Cortical lesions were assessed on proteolipid protein-stained tissue sections (original magnification _0.7), and after comparison with the corresponding MRI images, marked as visible (*red*) and invisible (*blue*). *2*: Corresponding proton density-weighted MR image of the same brain slice. *2a*: MRI-visible lesion. *2b*: MRI invisible lesion (Seewann et al. [39]); (with permission)

white and grey matter of MS patients at different stages of the disease. ^{1}H-MRS has demonstrated that although grey and white matter appear normal on conventional MRI, they may be affected in the early phases of the disease [44]. ^{1}H-MRS has the specificity to detect and differentiate between neuronal and glial abnormalities in the absence of distinct structural injury. This has greatly improved our understanding of the factors involved in the evolution/natural history of MS [38].

Studies carried out to date in MS have demonstrated changes in various metabolites, including *N*-acetylaspartate (NAA), choline (Cho), myo-inositol (mIns), glutamate (Glu), lactate, and creatine (Cr) [40, 45], which suggest the presence of specific pathological abnormalities (see Box 1). By quantifying NAA levels, a marker of neuroaxonal integrity, which reflects both axonal loss and metabolic dysfunction [46], ^{1}H-MRS studies have highlighted the role of neuroaxonal damage in MS [47]. Cho is associated with cell membrane phospholipids and is a marker of cell membrane turnover and myelin breakdown and repair [47, 48]. Cr and mIns are both markers of gliosis [47, 48]. The excitatory neurotransmitter, Glu is associated with Glu excitotoxicity and neuroaxonal loss [49, 50], while lactate can be increased due to the metabolism of inflammatory cells.

In the following sections of this chapter, we will briefly describe the pathological correlates of spectroscopic abnormalities. We will first review MRS changes in the acute and chronic lesions and in the normal appearing white matter and grey matter in patients with MS when compared with healthy controls. We will then discuss the use of MRS to help with the diagnosis and prognosis of MS. Finally, we will review the potential use of MRS for monitoring patients' response to treatments.

Box 1. Commonly Quantified Metabolites in MS and the Corresponding Underlying Pathological Abnormalities

- *N-acetylaspartate (NAA)* (usually this refers to total NAA (tNAA), which is equal to the sum of NAA and NAA-glutamate (NAAG)), is a marker of neuroaxonal integrity and/or metabolic mitochondrial function [46, 51].
- *Myo-inositol (mIns)* is a glial cell marker and increases are associated with astrocytic activation and proliferation [52].
- *Choline (Cho)* reflects changes in steady-state levels of membrane phospholipids released during myelin breakdown [53]. Increases are associated with demyelination or inflammation with increased membrane turnover.
- *Creatine (Cr)*, if increased, is a marker of gliosis.
- *Glutamate (Glu)* is the major excitatory neurotransmitter in the CNS. It is a marker of the neurotransmitter pool. While in the brain it is possible to quantify Glu using special techniques, in the spinal cord, Glu is typically quantified with glutamine (Gln) (Glu + Gln=Glx) [54].
- *Lactate*, if increased, is a marker of inflammation, and indicates the presence of macrophage infiltrate in large acute MS lesions.

Acute Lesions

In acute demyelinating lesions in the brain of patients with MS, decreases in NAA are seen, while Cr, Cho, mIns, lactate, and Glu are increased [45, 47, 50]. The abnormal concentrations of metabolites in acute demyelinating plaques can change over time, with normalisation of NAA, Cr, and lactate [47]. The recovery of NAA to (almost) normal levels is likely due to the resolution of oedema associated with acute inflammation, in addition to remyelination [47]. Partial normalisation of NAA levels after acute lesions has also been observed in the spinal cord, in association with development of cord atrophy, suggesting that this may also indicate increased neuronal mitochondrial metabolism [55]. Additionally, modelling NAA, which reflects axonal structural integrity and mitochondrial metabolism, with imaging measures of axonal integrity (axial diffusivity and cord cross-sectional area), enables one to extract the mitochondrial metabolic component of NAA [56]. Lower residual variance in NAA, reflecting reduced mitochondrial metabolism, was associated with greater clinical disability in MS, independent of structural damage [56], suggesting that modelling metabolites with structural imaging measures has potential and important clinical applications.

Chronic Lesions

Typically, in chronic lesions in the brain, NAA is significantly reduced, with mIns remaining increased [45]. The ongoing decrease in NAA in chronic plaques is likely due to ongoing neuroaxonal loss [47, 50]. One study did not demonstrate a change in glutamate levels in chronic T1-hypointense areas when compared to control white matter areas [50].

Normal Appearing White Matter

A number of studies have shown a reduction in NAA in normal appearing white matter in patients with MS (and these decreases seem to vary with distance from the focal white matter lesions), while Glu, mIns, and Cr are increased in normal appearing white matter [47, 50, 57]. Not all studies have demonstrated a decrease in NAA in normal appearing white matter, as results from a few studies have shown an unchanged concentration of NAA [50, 57]. The differences in results may be related to differences in the acquisition parameters, differences in quantification methods used, the location of the voxel, the volume of tissue examined and the sample size [50, 57]. Several studies have looked at the metabolic changes in early RRMS, and they have demonstrated that the concentration of NAA levels at baseline in MS patients was reduced in the normal appearing white matter compared to controls [58, 59]. The normal appearing white matter NAA tended to recover from baseline. This demonstrates that neuroaxonal damage is seen in the early stages of the disease, but part of this damage maybe reversible given the increase in NAA from baseline [58]. Another study in early RRMS [60] demonstrated that glial abnormalities (due to myelin breakdown and astrogliosis) were more widespread and were larger in magnitude compared to axonal abnormalities. This was represented by patients' white matter Cr, Cho, and mIns being higher than controls. White matter NAA was lower in patients than controls with some recovery of NAA levels, indicating partial recovery of the axonal abnormalities [60]. This is likely due to the effects of treatment [60]. This demonstrates that MRS may be a good indicator of early myelin disorders and acute inflammation [47, 61].

In individuals presenting with a clinically isolated syndrome, a number of studies have shown a decrease in NAA with normal or elevated mIns and Cho levels [62, 63]. This indicates neuroaxonal damage occurring during the first demyelinating event, but absent or very little increase in the activity of glial cells.

The concentrations of Cr in lesional tissue and normal appearing white matter in patients with MS have been inconsistent among studies. Cr values in normal appearing white matter of patients with MS decreased [64], increased [65–67], or were unaffected [64]. Similarly, Cr values in lesional white matter of patients with MS decreased [64], increased [68], or were unchanged [66]. These conflicting results are likely due to differences in patient selection, and in acquisition and quantification methods used [69].

A number of MS studies have measured metabolites as a ratio to Cr, rather than individual metabolite measures. The use of ratios is potentially less sensitive to disease effects in MS than the use of absolute values [63]. The use of the ratio to Cr is based on the assumption that Cr is normal in the patient's brain. However, MS studies have demonstrated abnormalities of Cr in demyelinating lesions and normal appearing white matter in MS [57, 63, 70, 71]. A meta-analysis carried out on the variability of NAA and Cr suggests that if there is a large discrepancy between the individual variability of these two metabolites, one needs to be cautious with the ratios obtained [69].

Grey Matter

MR spectroscopy has demonstrated a reduction in NAA, Cho, and Glu in the grey matter of the brain of MS patients [49, 72]. These reductions tend to be more pronounced in patients with progressive disease [72]. A reduction in whole brain NAA in MS patients and patients presenting with a clinically isolated syndrome has also been found [44]. This raises the possibility that MRS may be a useful marker of global neurodegeneration [44].

A ^{1}H-MRS study by Geurts et al. [73], demonstrated alterations in metabolite concentrations in the thalamus and hippocampus in MS patients, which were related to T2 lesion load, but no alteration was seen in the cortical grey matter. In the thalamus there was a decrease in NAA, which was most prominent in the PPMS group, with an increase in mIns in the thalamus, which was greatest in the SPMS cohort [73]. The increase in mIns is likely to reflect gliosis. Brain pathology in PPMS patients tends to show less of an inflammatory component, displays more widespread damage to neuroaxonal elements [74, 75], with less focal pathology in the brain than the spinal cord [76], which is in keeping with spectroscopic findings. The mIns concentration in the hippocampus increased, while there was no change in any of the other metabolites [73]. The increase in mIns in both the thalamus and hippocampus correlates with increased T2 lesion load, which suggests that gliosis increases as the lesion volume increases in the brain [73].

A recent 3T study [49] looked first at Glu levels in the hippocampus, thalamus, cingulate, and parietal cortices in patients with RRMS compared to controls. Figure 8.2 demonstrates the voxel placement in these four grey matter regions.

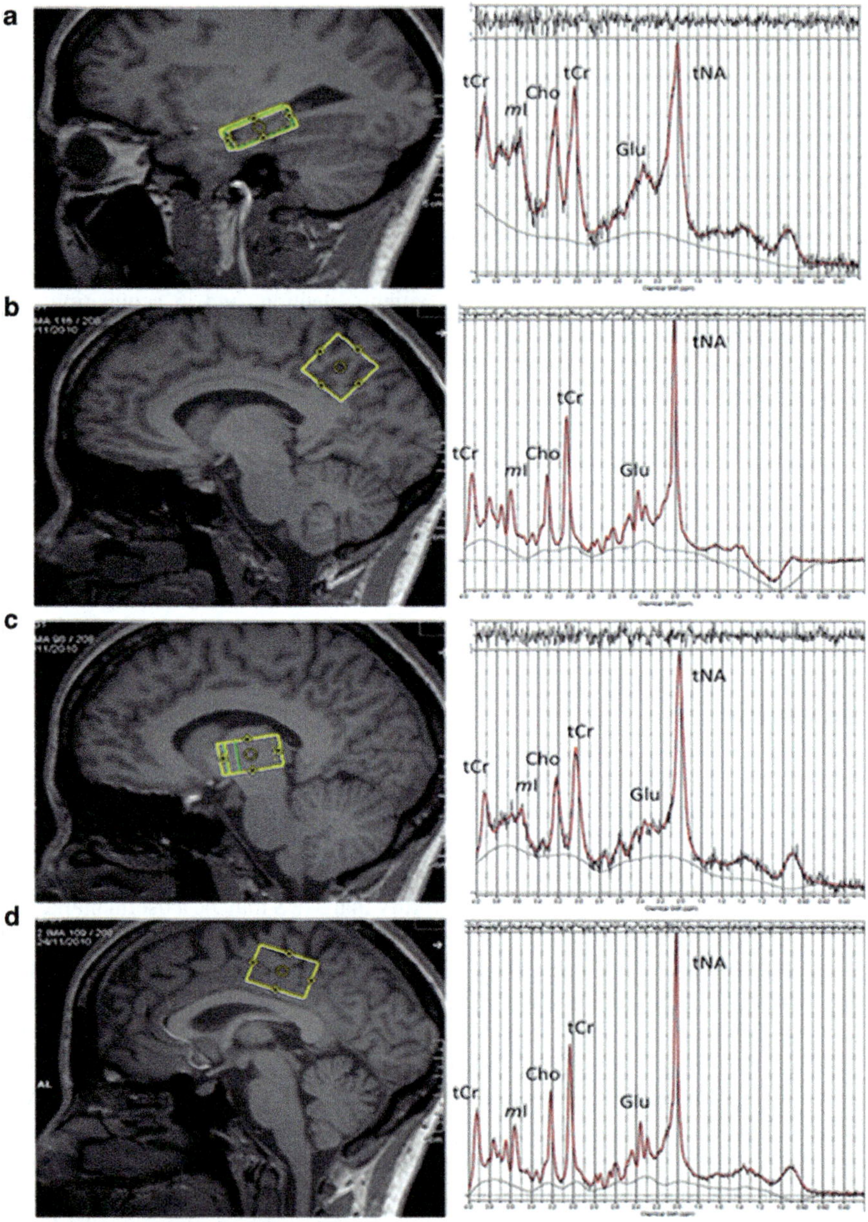

Fig. 8.2 Examples of placement of the MR spectroscopy voxels (*left*) with their corresponding MRS spectra (*right*) in the right hippocampus (**a**), parietal cortex (**b**), right thalamus (**c**), and cingulate (**d**) (Muhlert et al. [49]; with permission)

Second, it investigated if there was a relationship between Glu concentration and memory performance [49]. There was significantly lower Glu in the cingulate and parietal cortices in patients compared to controls, with a trend towards lower Glu in the right hippocampus in patients. As Glu is much more abundant inside cells than in the extracellular space, a possible explanation for these results is reduced synapses and neuroaxonal degeneration, which has been described in post-mortem studies [77]. A positive correlation between Glu concentrations and cognitive function was demonstrated, with worse visuospatial memory linked to lower Glu in the hippocampal, thalamic, and cingulate regions in patients with RRMS [49]. These findings using ^{1}H-MRS may lead to new treatment approaches, which specifically target this neurotransmitter in MS patients with cognitive impairment [49].

Differential Diagnosis

^{1}H-MRS can be useful in differentiating between different pathologies. This was demonstrated in a recent study of MS patients and neuromyelitis optica (NMO)-related cases, with each case having a lesion in the upper cervical cord [78]. The key finding from this study demonstrated that the mIns/Cr ratio was reduced in the upper cervical cord which contained the longitudinally extensive NMO lesions when compared to healthy controls and MS patients, who showed lesions in the same region. Increased concentration of mIns is thought to be a marker of astrocytic activation and proliferation [79]. Therefore, the reduction in mIns is likely to reflect astrocytic necrosis, which is pathological for NMO lesions and is thought to mediate oligodendrocyte injury [80]. This raises the possibility of developing novel imaging spinal cord markers to distinguish NMO from MS and it may also have value in future clinical trials in NMO with remyelinating or neuroprotective therapies [78]. Idiopathic inflammatory demyelinating diseases can present as single focal brain lesions, which can be clinically and radiologically indistinguishable from tumours [81]. Some conventional MR imaging features (Balo-like pattern) can help differentiate tumours from inflammatory lesions, but in many situations this information is not sufficient. ^{1}H-MRS can provide additional diagnostic information by assessing the concentrations of NAA, Cho, mIns, and Glx [82]. This can help in distinguishing between tumour and solitary inflammatory lesions, which may avoid invasive diagnostic and therapeutic procedures [81].

Prognosis

Radiologically Isolated Syndrome (RIS)

A radiologically isolated syndrome (RIS) is defined as the unanticipated MRI finding of brain spatial dissemination of focal white matter lesions highly suggestive of MS in subjects without symptoms typical for MS and with normal neurological examination [83]. There is an urgent clinical need to be able to more accurately differentiate individuals with just an incidental radiological finding from those individuals who are likely to go on to develop MS [84]. ^{1}H-MRS can provide valuable information on the disease processes as well as vital insights into the underlying pathological changes. One study demonstrated a decrease in the levels of NAA/Cr levels in a group of RIS subjects [84] when compared to healthy controls. These findings indicate the presence of neuroaxonal damage or loss, even in this preclinical condition [83]. These findings may be helpful in stratifying RIS individuals with a high risk of progression to MS [84].

Clinically Isolated Syndrome (CIS)

Quantifying metabolic damage at the earliest clinical stage of disease has been investigated in order to better determine predictors of disease evolution in subjects presenting with a Clinically Isolated Syndrome (CIS). A study demonstrated reduced NAA/Cr, when combined with brain atrophy and measures of focal demyelinating lesions, helped predict short-term disease evolution in patients with CIS [85]. Another study showed a significantly lower NAA in the normal appearing white matter in patients presenting with a CIS, who converted to MS, when compared to non-converters [86]. This indicates that neuroaxonal damage at baseline in patients with CIS was more prominent in those who went on to develop clinically definite MS, suggesting that NAA may be a useful prognostic marker for patients with a higher risk of conversion to clinically definite MS [62, 86]. There was no evidence of increased glial cell activity, as reflected by mIns levels, in the normal appearing white matter in patients with CIS [62].

A study in patients who had presented with a CIS 14 years previously demonstrated elevated mIns in patients relative to controls in both lesions and normal appearing white matter [87]. The mIns concentration positively correlated with EDSS score at 14 years. After 14 years, 78 % of the CIS patients converted to clinically definite MS, while 22 % remained as a clinically isolated syndrome. Interestingly, it was the change in mIns concentration in the normal appearing white and grey matter which separated these two groups, with those developing clinically definite MS having increased levels of mIns [87]. The increase in mIns early in MS may have a potential role in the future to help determine diagnosis, future relapses, and disability [88].

RRMS

A recent study by Achtnichts et al. [89] looked at whole brain NAA in two RRMS cohorts of $\geq$15 year's duration. One group had an EDSS score of $\leq$3 (benign MS) and the other group had an EDSS score of $\geq$3 (non-benign MS). However, no differences were seen in the whole brain NAA between the patient groups. The preservation of neuronal integrity in both normal appearing white matter and focal lesions does not seem to be a discriminating factor in distinguishing BMS from non-BMS patients of long disease duration. Another study that estimated whole brain NAA in RRMS [90] found that atrophy increased and NAA declined linearly with increasing MS disease duration, and that the rate of NAA decline was 3.6 times faster than atrophy. Further longitudinal studies are needed to understand whether NAA concentration is a useful marker of disease progression [90].

Progressive MS

It is not possible to predict the long-term course of relapse-onset MS, specifically related to progressive disability [67]. It is thought the progressive phase of MS is most likely due to widespread neurodegeneration, mediated through the innate immune responses within the CNS, with microglial activation, astrocyte proliferation, meningeal inflammation, mitochondrial dysfunction, ion channel abnormalities, and oxidative injury [91].

A recently published study [92] investigated the ratio of mIns to NAA longitudinally in MS. This combined measure reflects both the damaged CNS innate immune process (increase in mIns) and the neurodegeneration, which causes long-term disability (reduced NAA) [67]. The mIns/NAA ratio in normal appearing white matter of a cohort of MS patients was significantly associated with future disease progression [92]. These findings were then confirmed in another cohort of MS patients. This clearly demonstrates the robustness of this measure as a biological predictor [67].

Monitoring Treatment Response

To date several longitudinal, multicentre studies in MS using ^{1}H-MRS, have looked at the effects of drug therapies on reversing or stopping the progression of neuroaxonal injury [47]. These studies have demonstrated reproducibility between sites when factors such as method of data acquisition, position and size of the volume of interest, post-processing, and quantification procedures are standardised [93].

^{1}H-MRS longitudinal studies carried out to date in patients on immunomodulatory treatment have demonstrated recovering or stable white matter NAA [60]. This is likely due to treatment effect on axons but not on glial markers (Cr, Cho, and mIns), as these markers continued to increase, indicating patients' glial abnormalities progressed while on immunomodulatory therapy [77].

^{1}H-MRS has a greater pathological specificity for axonal integrity and other pathological processes compared to conventional MRI in MS; however, it is only rarely used in phase III clinical trials [47]. This is primarily due to the technical difficulties related to the protocol implementation and standardisation of data acquisition, position and size of the volume of interest (VOI), post-processing, and quantification procedures between centres [47]. In the future more consensus guidelines need to be available to help research groups overcome these challenges [47].

Clinical Correlations and Considerations for Clinical Applications of MRS

Over the last 10 years, the information acquired from MRS has provided a great contribution to better define the pathogenesis and natural history of MS [94]. Levels of NAA have provided evidence of neurodegeneration from the earliest stages of the disease, which has led us to reconsider the role of neuroaxonal damage in MS [95, 96].

In this section of the chapter, we will review the clinical correlations between MRS in the brain and spinal cord.

Disability

Many studies have demonstrated correlations in ^{1}H-MRS findings with clinical disability, with NAA consistently correlating with the Expanded Disability Status Scale (EDSS) [97–99]. This clearly indicates neuroaxonal damage as a mechanism of disability, as patients with long disease duration, SPMS, and patients with RRMS and advanced clinical disability have the greatest NAA loss [81]. In addition to NAA, a number of studies have shown associations between mIns and Glx in normal appearing white matter and disability [87, 100].

Fatigue

In addition to disability, NAA has also been correlated with fatigue in MS patients. A significant decrease in NAA/Cr in the lentiform nucleus in MS patients was associated with fatigue, supporting the involvement of the basal ganglia as a contributing factor to the development of this MS-related symptom [101].

Cognition

Cognitive impairment is common in MS, occurring at all stages of the disease [102]. Memory impairment in MS has been associated with both grey matter cortical thinning [103] and atrophy in specific grey matter regions, like the hippocampus [104]. ^{1}H-MRS has shown correlations between specific cognitive deficits and NAA levels in certain brain structures [81]. One study found a decrease in NAA in the right locus ceruleus in the pons correlated with selective attention deficits [105]. A previous study by Muhlert et al. [49], mentioned earlier, demonstrated worse visuospatial memory was associated with lower Glu levels in the hippocampus, thalamus, and cingulate regions in patients with RRMS. These studies clearly indicate that changes in metabolites in specific brain regions correlate with specific cognitive deficits.

NAAG is the most abundant neuropeptide in the mammalian brain, and is a selective agonist for metabotropic glutamate receptor 3 (mGluR3) [106]. An MRS study quantified NAAG in the hippocampi of MS patients and found that NAAG correlated with cognitive function [107]. MS patients with cognitive impairment had lower levels of NAAG, while individuals with normal cognitive function had higher levels of NAAG [107]. This raises the possibility that NAAG may be a potential biomarker for cognitive function in MS, although quantification of NAAG is challenging.

Spinal Cord

Due to the technical challenges of ^{1}H-MRS in the spinal cord, there are significantly fewer studies performed to date compared with ^{1}H-MRS of the brain. Most studies published have demonstrated similar findings, with reduced NAA concentrations in the cervical cord in patients with MS compared to healthy controls [108–110].

One study looked at changes in NAA using ^{1}H-MRS following a cervical spinal cord relapse (with a lesion between C1 and C3), in MS patients to investigate mechanisms involved in spinal cord repair and associated clinical recovery [55]. Results demonstrated a reduction in NAA concentration at 1 month and then a persistent increase in NAA from 1 month to 6 months in those individuals who demonstrated a greater clinical recovery. The partial recovery in NAA after the relapse was associated with a decline in spinal cord cross sectional area. One possible theory for the increase in NAA concentration is that this may be driven by increased axonal mitochondrial metabolism [55]. A subsequent study investigated the indirect estimation of mitochondrial metabolism in vivo by statistically modelling the NAA concentration measured in the spinal cord, with other structural imaging measures, such as atrophy and axial diffusivity, which is derived from diffusion tensor imaging (DTI) [56]. These findings are similar to brain studies, which reported temporal changes in NAA after acute lesions [111].

Higher field strength (3T) MRI has enabled the detection of additional metabolite concentrations with spinal cord MRS, such as Glu/Gln [54, 112, 113], with an example of the voxel placement in the cervical spinal cord shown in Fig. 8.3. This has mainly been achieved through the optimisation of the acquisition protocol, the use of a cervical collar, which reduced the macroscopic movement of the neck, triggered iterative shimming and optimised water suppression [114]. Figure 8.4 shows a baseline and follow-up spectra of a healthy control and an MS patient in the cervical spinal cord, with a decrease in NAA and an increase in mIns in the patient, compared to the control [115].

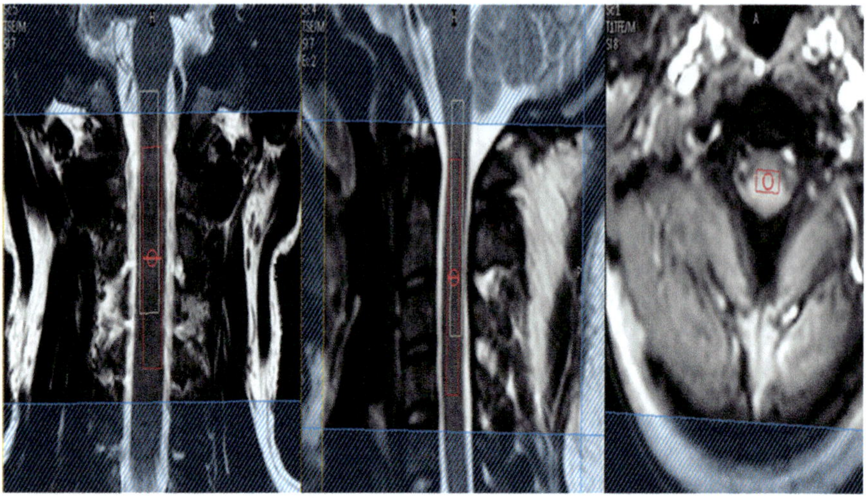

Fig. 8.3 Sagittal, coronal, and axial views of the voxel placement (*red box*) in the cervical cord of one subject. Of particular importance was the avoidance of CSF inclusion in the voxel in all planes. The *white box* is the shifted water voxel and the outer *blue boxes* are the outer volume suppression slabs in the anterior–posterior and foot–head directions (Solanky et al. [54]; with permission)

Overview of MRS in Animal Models of MS

The non-invasiveness of ^{1}H-MRS, as well as its clinical translatability, makes it an ideal tool for parallel experimental paradigms using animal and human studies [116]. Animal models of MS have led to a better understanding of the disease. The three most commonly studied animal models of MS are as follows:

1. The purely autoimmune experimental autoimmune encephalomyelitis (EAE) induces autoimmune-mediated CNS inflammation, demyelination, and axonal damage.
2. The virally induced chronic demyelinating disease models, with the main model of Theiler's murine encephalomyelitis virus (TMEV) infection.
3. Toxin-induced models of demyelination, including the cuprizone model and focal demyelination induced by lyso-phosphatidyl choline [117].

Pre-clinical MR imaging has contributed significantly to enhancing our understanding of the disease pathology in MS. Prior to the development of ^{1}H-MRS, standard EAE studies required histological examination of the tissue which was a terminal experiment at a fixed time point [118]. ^{1}H-MRS allows serial examination of the animal at specific time points, allowing longitudinal studies. This can then be supported with histological examination of the tissue. Scanning an animal at multiple time points also allows for before and after treatment analyses [118].

Studies using ^{1}H-MRS in EAE have consistently found reduced NAA levels [119, 120] in EAE, in keeping with findings in MS in normal appearing white

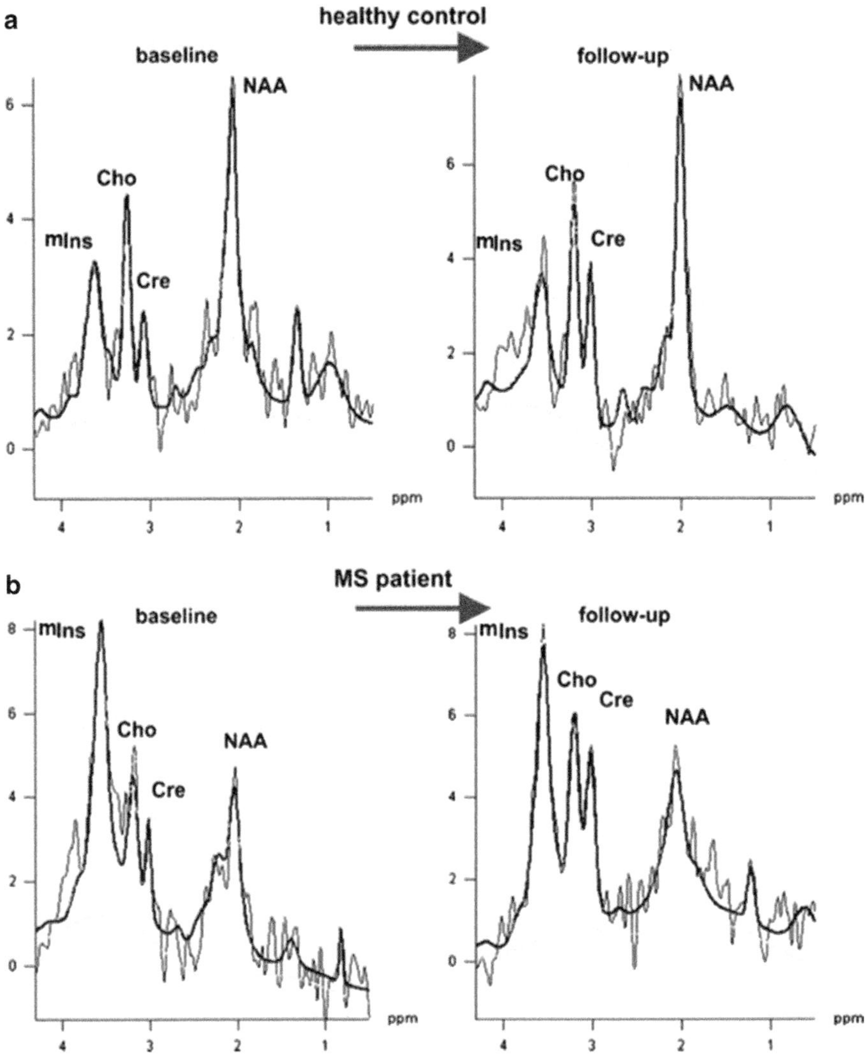

Fig. 8.4 Baseline and follow-up spectra of a healthy control and a MS patient in the cervical cord, depicting a reduction in NAA and an increase in mIns in the patient spectra (Bellenberg et al. [115]; with permission)

matter [60, 99] and at the level of the whole brain [121]. Increased levels of Cho and Cr have also been seen in EAE [120], in agreement with findings from MS studies [57, 65]

More recently, a study group [117] from the US studied pathological differences between two strains of mice, which were identical except for the deletion of β2-microglobulin in one strain. The complete deletion of β2-microglobulin results in the absence of CD8+ T cells among other cell types [122]. They investigated the influence of CD8+ T cells on spinal cord demyelination, remyelination, axonal loss, and brainstem NAA levels in a longitudinal study during acute and chronic phases of TMEV infection [122]. Early in the disease brainstem NAA levels dropped in both strains. However, NAA recovered in β2-microglobulin-negative mice with remyelination and axonal preservation. B2-microglobulin-positive mice lost axons of all calibres, while β2-microglobulin-negative mice had a modest loss of only medium- and large-calibre axons [122]. This study clearly supports the hypothesis of the role of CD8+ T cells in axonal injury and subsequent neurological deficits in TMEV, as well as potentially influencing the degree of demyelination and/or remyelination [118]. This suggests the therapeutic targeting of cytotoxic CD8+ T cells in chronic demyelination in human MS [118]. This provides further evidence of the importance of MRS in animal models of disease.

Future Perspectives of ^{1}H-MRS in Multiple Sclerosis

To date, ^{1}H MRS studies have given us valuable insight into the pathogenesis and natural history of MS. In the future, more multi-centred, longitudinal studies using ^{1} H-MRS are needed in order to clearly determine the role of ^{1}H-MRS in monitoring disease evolution in addition to treatment efficacy in MS.

MR scanners of higher field strength are becoming increasingly popular. One benefit of higher field strength scanners is that the signal-to-noise ratio (SNR) increases linearly with the field strength, in addition to higher spatial resolution, resulting in improved resolution of two adjacent spectral peaks. This allows one to investigate other metabolites such as gamma-aminobutyric acid (GABA), which is overlapped by NAA and Cr at 3T, but resolved at 7T given the larger chemical shift dispersion.

The use of 7T gives additional information from Glu and Gln, which are challenging to resolve at 3T. Higher field strength has mainly been applied to brain imaging in MS, while spinal cord imaging in MS has mainly been conducted at 1.5T and 3T. Higher field strength in the spinal cord needs to overcome problems related to chemical shift displacement and positioning errors, which would make MRS volume of interest placement in the spinal cord more challenging.

Diffusion Tensor Spectroscopy (DTS) combines features of diffusion tensor imaging (DTI) and MRS, resulting in the measurement of diffusion properties of intracellular metabolites [123]. This enables the measurement of the diffusion properties of intracellular, cell-type-specific metabolites [124]. To date one study using a 7T scanner demonstrated reduced NAA parallel diffusivity in MS subjects and this inversely correlated with water parallel diffusivity and clinical severity [124]. These preliminary findings are thought to be able to distinguish axonopathy from other processes such as inflammation and demyelination [124]. Based on this, DTS changes may help interpret findings from more specific techniques, such as DTI, and further clarify their pathophysiological significance [124]. Future studies will be required to further analyse the natural history of DTS measurements in MS [124].

The combination of ^{1}H and ^{31}P MRS provides a more specific approach to metabolite changes. ^{31}P MRS specifically quantifies the high energy metabolite phosphocreatine (PCr), enabling the calculation of a ratio between PCr and Cr, when used in combination with ^{1}H-MRS [125]. This has the potential to provide additional information on the biochemical background related to ^{1}H-detectable metabolite changes in MS by differentiating the components of the ^{1}H metabolite signals [125].

The detection of glutathione (GSH) is technically challenging at field strengths of 1.5T and 3T, mainly due to the low concentration of this metabolite within the brain, and the standard single-echo acquisitions, which cannot be used to resolve GSH from other resonances that it overlaps [126]. The use of spectral editing techniques enables the unobstructed detection of GSH. GSH is the major antioxidant in the human brain and plays a key role in the detoxification of reactive oxidants [127]. Oxidative stress plays an important role in the pathogenesis of MS as macrophages, activated as a result of inflammation, generate reactive oxygen species, with resulting damage to myelin and oligodendrocytes [126]. A study in MS patients demonstrated a reduction in GSH in the grey matter of MS patients when compared to controls [126]. There was no significant difference in the concentration of GSH in the white matter of patients relative to controls [126]. These findings demonstrate the presence of oxidative stress in the normal appearing grey matter. This supports the pathogenic role of oxidative stress contributing to the disease process in MS. GSH has the potential to be a very useful biomarker to predict disease progression, in addition to a potential therapeutic target to combat oxidative stress both in the initial inflammatory period of the disease but also in preventing the neurodegenerative processes associated with progressive MS [127]. MR spectroscopy studies at 7T will allow for improved quantification of metabolite abnormalities, therefore providing valuable new insights into MS pathogenesis, once technical limitations posed by the very high field are overcome.

With further investigation of novel metabolites in MS, we expect that great effort will be put in future on GABA. GABA is the main inhibitory neurotransmitter in the CNS and is thought to be involved in brain plasticity, cortical adaptation, and reorganisation in neurodegenerative disease processes [128]. A number of studies in healthy controls demonstrated a decrease in GABA inhibition facilitated long-term potentiation-like activity in the motor cortex. Very few studies have been carried out to date looking at GABA in MS patients using ^{1}H-MRS. One study demonstrated an inverse correlation between the concentration of GABA in the sensorimotor cortex of patients with RRMS and the 9-hole peg test [128]. These results indicate a worsening of performance with increased GABA levels. These findings suggest that cortical GABA may be a potential marker of function and reorganisation of cortical grey matter in patients with MS [128]. To date one group has looked at quantifying GABA in the spinal cord [129] and demonstrated that it is feasible. In the future, the alterations of this metabolite in MS may have a role in the development of novel and individual treatment concepts but this needs to be investigated [129].

Conclusions

^{1}H MRS complemented by conventional MR techniques provides advanced insights into the pathogenesis of MS by obtaining additional information on the microenvironment within the brain and spinal cord. This enhances our understanding of the mechanisms involved in disease progression, as well as treatment response in patients with MS.

Technical limitations of ^{1}H MRS in MS, such as reproducibility of the examination conditions, can be difficult for any voxel size and position, making its use in phase III clinical trials challenging. In addition, standardised post-processing and quantification procedures between centres remain a challenge. Other challenges include motion artefacts, and image distortions from the magnetic susceptibility at tissue/bone/air interfaces, limiting ^{1}H MRS use in the clinical setting.

The increasing availability of high field scanners may enable the use of ^{1}H MRS to aid the diagnostic criteria of MS, in addition to monitoring disease progression and treatment response. In the future this will enable investigators to measure more novel metabolites, which will allow them to study mechanisms related to relapses and neurological progression.

References

1. Pugliatti M, Rosati G, Carton H, Riise T, Drulovic J, Vecsei L, Milanov I (2006) The epidemiology of multiple sclerosis in Europe. Eur J Neurol 13:700–722
2. Miller DH, Leary SM (2007) Primary-progressive multiple sclerosis. Lancet Neurol 6:903–912
3. Ebers GC, Daumer M (2008) Natural history of MS. Eur J Neurol 15:881–882
4. Mackenzie IS, Morant SV, Bloomfield GA, MacDonald TM, O'Riordan JI (2013) Changing face of multiple sclerosis in the United kingdom 1990-2010. An incidence and prevalence study. J Neurol Neurosurg Psychiatry 84:e2
5. Evans C, Beland SG, Kulaga S, Wolfson C, Kingwell E, Marriott J, Koch M, Makhani N, Morrow S, Fisk J, Dykeman J, Jette N, Pringsheim T, Marrie RA (2013) Incidence and prevalence of multiple sclerosis in the Americas: a systematic review. Neuroepidemiology 40:195–210
6. Compston A, Coles A (2008) Multiple sclerosis. Lancet 372:1502–1517
7. Confavreux C, Compston DA, Hommes OR, McDonald WI, Thompson AJ (1992) EDMUS, a European database for multiple sclerosis. J Neurol Neurosurg Psychiatry 55:671–676
8. Confavreux C, Vukusic S (2006) Natural history of multiple sclerosis: a unifying concept. Brain 129:606–616
9. Lublin FD, Reingold SC (1996) Defining the clinical course of multiple sclerosis: results of an international survey. National Multiple Sclerosis Society (USA) Advisory Committee on Clinical Trials of New Agents in Multiple Sclerosis. Neurology 46:907–911

10. Weinshenker BG, Bass B, Rice GP, Noseworthy J, Carriere W, Baskerville J, Ebers GC (1989) The natural history of multiple sclerosis: a geographically based study. 2. Predictive value of the early clinical course. Brain 112(Pt 6):1419–1428
11. Vukusic S, Confavreux C (2003) Prognostic factors for progression of disability in the secondary progressive phase of multiple sclerosis. J Neurol Sci 206:135–137
12. Myhr KM, Riise T, Vedeler C, Nortvedt MW, Gronning R, Midgard R, Nyland HI (2001) Disability and prognosis in multiple sclerosis: demographic and clinical variables important for the ability to walk and awarding of disability pension. Mult Scler 7:59–65
13. Lublin FD, Reingold SC, Cohen JA, Cutter GR, Sorensen PS, Thompson AJ, Wolinsky JS, Balcer LJ, Banwell B, Barkhof F, Bebo B, Calabresi PA Jr, Clanet M, Comi G, Fox RJ, Freedman MS, Goodman AD, Inglese M, Kappos L, Kieseier BC, Lincoln JA, Lubetzki C, Miller AE, Montalban X, O'Connor PW, Petkau J, Pozzilli C, Rudick RA, Sormani MP, Stuve O, Waubant E, Polman CH (2014) Defining the clinical course of multiple sclerosis: The 2013 revisions. Neurology 83:278–286
14. Koch M, Kingwell E, Rieckmann P, Tremlett H (2010) The natural history of secondary progressive multiple sclerosis. J Neurol Neurosurg Psychiatry 81:1039–1043
15. Koch M, Kingwell E, Rieckmann P, Tremlett H (2009) The natural history of primary progressive multiple sclerosis. Neurology 73:1996–2002
16. Andersson PB, Waubant E, Gee L, Goodkin DE (1999) Multiple sclerosis that is progressive from the time of onset: clinical characteristics and progression of disability. Arch Neurol 56:1138–1142
17. Lassmann H, VAN Horssen J, Mahad D (2012) Progressive multiple sclerosis: pathology and pathogenesis. Nat Rev Neurol 8:647–656
18. Lassmann H, Bruck W, Lucchinetti CF (2007) The immunopathology of multiple sclerosis: an overview. Brain Pathol 17:210–218
19. Ebers GC (2004) Natural history of primary progressive multiple sclerosis. Mult Scler 10 (Suppl 1):8–13, discussion S13-5
20. Benedict RH, Fazekas F (2009) Benign or not benign MS: a role for routine neuropsychological assessment? Neurology 73:494–495
21. Stys PK, Zamponi GW, VAN Minnen J, Geurts JJ (2012) Will the real multiple sclerosis please stand up? Nat Rev Neurosci 13:507–514
22. Frischer JM, Bramow S, Dal-Bianco A, Lucchinetti CF, Rauschka H, Schmidbauer M, Laursen H, Sorensen PS, Lassmann H (2009) The relation between inflammation and neurodegeneration in multiple sclerosis brains. Brain 132:1175–1189
23. Babbe H, Roers A, Waisman A, Lassmann H, Goebels N, Hohlfeld R, Friese M, Schroder R, Deckert M, Schmidt S, Ravid R, Rajewsky K (2000) Clonal expansions of CD8(+) T cells dominate the T cell infiltrate in active multiple sclerosis lesions as shown by micromanipulation and single cell polymerase chain reaction. J Exp Med 192:393–404
24. Ozawa K, Suchanek G, Breitschopf H, Bruck W, Budka H, Jellinger K, Lassmann H (1994) Patterns of oligodendroglia pathology in multiple sclerosis. Brain 117(Pt 6):1311–1322
25. Serafini B, Rosicarelli B, Magliozzi R, Stigliano E, Aloisi F (2004) Detection of ectopic B-cell follicles with germinal centers in the meninges of patients with secondary progressive multiple sclerosis. Brain Pathol 14:164–174
26. Lassmann H (2011) Review: the architecture of inflammatory demyelinating lesions: implications for studies on pathogenesis. Neuropathol Appl Neurobiol 37:698–710
27. Geurts JJ, Stys PK, Minagar A, Amor S, Zivadinov R (2009) Gray matter pathology in (chronic) MS: modern views on an early observation. J Neurol Sci 282:12–20
28. Lassmann H (2014) Mechanisms of white matter damage in multiple sclerosis. Glia 62 (11):1816–1830
29. Hauser SL, Oksenberg JR (2006) The neurobiology of multiple sclerosis: genes, inflammation, and neurodegeneration. Neuron 52:61–76
30. Trapp BD, Nave KA (2008) Multiple sclerosis: an immune or neurodegenerative disorder? Annu Rev Neurosci 31:247–269

31. Meinl E, Krumbholz M, Derfuss T, Junker A, Hohlfeld R (2008) Compartmentalization of inflammation in the CNS: a major mechanism driving progressive multiple sclerosis. J Neurol Sci 274:42–44
32. Calabrese M, Rocca MA, Atzori M, Mattisi I, Favaretto A, Perini P, Gallo P, Filippi M (2010) A 3-year magnetic resonance imaging study of cortical lesions in relapse-onset multiple sclerosis. Ann Neurol 67:376–383
33. Calabrese M, Filippi M, Gallo P (2010) Cortical lesions in multiple sclerosis. Nat Rev Neurol 6:438–444
34. Gilmore CP, Donaldson I, Bo L, Owens T, Lowe J, Evangelou N (2009) Regional variations in the extent and pattern of grey matter demyelination in multiple sclerosis: a comparison between the cerebral cortex, cerebellar cortex, deep grey matter nuclei and the spinal cord. J Neurol Neurosurg Psychiatry 80:182–187
35. Reynolds R, Roncaroli F, Nicholas R, Radotra B, Gveric D, Howell O (2011) The neuropathological basis of clinical progression in multiple sclerosis. Acta Neuropathol 122:155–170
36. Magliozzi R, Howell O, Vora A, Serafini B, Nicholas R, Puopolo M, Reynolds R, Aloisi F (2007) Meningeal B-cell follicles in secondary progressive multiple sclerosis associate with early onset of disease and severe cortical pathology. Brain 130:1089–1104
37. Magliozzi R, Howell OW, Reeves C, Roncaroli F, Nicholas R, Serafini B, Aloisi F, Reynolds R (2010) A Gradient of neuronal loss and meningeal inflammation in multiple sclerosis. Ann Neurol 68:477–493
38. Filippi M, Rocca MA, DE Stefano N, Enzinger C, Fisher E, Horsfield MA, Inglese M, Pelletier D, Comi G (2011) Magnetic resonance techniques in multiple sclerosis: the present and the future. Arch Neurol 68:1514–1520
39. Seewann A, Vrenken H, Kooi EJ, VAN DER Valk P, Knol DL, Polman CH, Pouwels PJ, Barkhof F, Geurts JJ (2011) Imaging the tip of the iceberg: visualization of cortical lesions in multiple sclerosis. Mult Scler 17:1202–1210
40. DE Stefano N, Filippi M (2007) MR spectroscopy in multiple sclerosis. J Neuroimaging 17 (Suppl 1):31S–35S
41. Gass A, Richards TL (2013) Serial proton magnetic resonance spectroscopy of normal-appearing gray and white matter in MS. Neurology 80:17–18
42. Lin A, Ross BD, Harris K, Wong W (2005) Efficacy of proton magnetic resonance spectroscopy in neurological diagnosis and neurotherapeutic decision making. NeuroRx 2:197–214
43. Arnold DL, Matthews PM, Francis G, Antel J (1990) Proton magnetic resonance spectroscopy of human brain in vivo in the evaluation of multiple sclerosis: assessment of the load of disease. Magn Reson Med 14:154–159
44. Miller DH, Thompson AJ, Filippi M (2003) Magnetic resonance studies of abnormalities in the normal appearing white matter and grey matter in multiple sclerosis. J Neurol 250:1407–1419
45. Bakshi R, Thompson AJ, Rocca MA, Pelletier D, Dousset V, Barkhof F, Inglese M, Guttmann CR, Horsfield MA, Filippi M (2008) MRI in multiple sclerosis: current status and future prospects. Lancet Neurol 7:615–625
46. Moffett JR, Ross B, Arun P, Madhavarao CN, Namboodiri AM (2007) N-Acetylaspartate in the CNS: from neurodiagnostics to neurobiology. Prog Neurobiol 81:89–131
47. DE Stefano N, Filippi M, Miller D, Pouwels PJ, Rovira A, Gass A, Enzinger C, Matthews PM, Arnold DL (2007) Guidelines for using proton MR spectroscopy in multicenter clinical MS studies. Neurology 69:1942–1952
48. Moore GR, Laule C (2012) Neuropathologic correlates of magnetic resonance imaging in multiple sclerosis. J Neuropathol Exp Neurol 71:762–778
49. Muhlert N, Atzori M, DE Vita E, Thomas DL, Samson RS, Wheeler-Kingshott CA, Geurts JJ, Miller DH, Thompson AJ, Ciccarelli O (2014) 2014. Memory in multiple sclerosis is linked to glutamate concentration in grey matter regions. J Neurol Neurosurg Psychiatry 85 (8):833–839

50. Srinivasan R, Sailasuta N, Hurd R, Nelson S, Pelletier D (2005) Evidence of elevated glutamate in multiple sclerosis using magnetic resonance spectroscopy at 3T. Brain 128:1016–1025
51. VAN Horssen J, Witte ME, Ciccarelli O (2012) The role of mitochondria in axonal degeneration and tissue repair in MS. Mult Scler 18:1058–1067
52. Brand A, Richter-Landsberg C, Leibfritz D (1993) Multinuclear NMR studies on the energy metabolism of glial and neuronal cells. Dev Neurosci 15:289–298
53. Davie CA, Hawkins CP, Barker GJ, Brennan A, Tofts PS, Miller DH, McDonald WI (1993) Detection of myelin breakdown products by proton magnetic resonance spectroscopy. Lancet 341:630–631
54. Solanky BS, Abdel-Aziz K, Yiannakas MC, Berry AM, Ciccarelli O, Wheeler-Kingshott CA (2013) In vivo magnetic resonance spectroscopy detection of combined glutamate-glutamine in healthy upper cervical cord at 3T. NMR Biomed 26:357–366
55. Ciccarelli O, Altmann DR, Mclean MA, Wheeler-Kingshott CA, Wimpey K, Miller DH, Thompson AJ (2010) Spinal cord repair in MS: does mitochondrial metabolism play a role? Neurology 74:721–727
56. Ciccarelli O, Toosy AT, DE Stefano N, Wheeler-Kingshott CA, Miller DH, Thompson AJ (2010) Assessing neuronal metabolism in vivo by modeling imaging measures. J Neurosci 30:15030–15033
57. Vrenken H, Barkhof F, Uitdehaag BM, Castelijns JA, Polman CH, Pouwels PJ (2005) MR spectroscopic evidence for glial increase but not for neuro-axonal damage in MS normal-appearing white matter. Magn Reson Med 53:256–266
58. Tiberio M, Chard DT, Altmann DR, Davies G, Griffin CM, Mclean MA, Rashid W, Sastre-Garriga J, Thompson AJ, Miller DH (2006) Metabolite changes in early relapsing-remitting multiple sclerosis. A two year follow-up study. J Neurol 253:224–230
59. Chard DT, Griffin CM, Mclean MA, Kapeller P, Kapoor R, Thompson AJ, Miller DH (2002) Brain metabolite changes in cortical grey and normal-appearing white matter in clinically early relapsing-remitting multiple sclerosis. Brain 125:2342–2352
60. Kirov II, Tal A, Babb JS, Herbert J, Gonen O (2013) Serial proton MR spectroscopy of gray and white matter in relapsing-remitting MS. Neurology 80:39–46
61. Miller TR, Mohan S, Choudhri AF, Gandhi D, Jindal G (2014) Advances in multiple sclerosis and its variants: conventional and newer imaging techniques. Radiol Clin North Am 52:321–336
62. Wattjes MP, Harzheim M, Lutterbey GG, Klotz L, Schild HH, Traber F (2007) Axonal damage but no increased glial cell activity in the normal-appearing white matter of patients with clinically isolated syndromes suggestive of multiple sclerosis using high-field magnetic resonance spectroscopy. AJNR Am J Neuroradiol 28:1517–1522
63. Fernando KT, Mclean MA, Chard DT, Macmanus DG, Dalton CM, Miszkiel KA, Gordon RM, Plant GT, Thompson AJ, Miller DH (2004) Elevated white matter myo-inositol in clinically isolated syndromes suggestive of multiple sclerosis. Brain 127:1361–1369
64. Davies SE, Newcombe J, Williams SR, McDonald WI, Clark JB (1995) High resolution proton NMR spectroscopy of multiple sclerosis lesions. J Neurochem 64:742–748
65. Inglese M, Li BS, Rusinek H, Babb JS, Grossman RI, Gonen O (2003) Diffusely elevated cerebral choline and creatine in relapsing-remitting multiple sclerosis. Magn Reson Med 50:190–195
66. Bitsch A, Bruhn H, Vougioukas V, Stringaris A, Lassmann H, Frahm J, Bruck W (1999) Inflammatory CNS demyelination: histopathologic correlation with in vivo quantitative proton MR spectroscopy. AJNR Am J Neuroradiol 20:1619–1627
67. Miller DH (2014) 2014. Magnetic resonance spectroscopy: a possible in vivo marker of disease progression for multiple sclerosis? JAMA Neurol 71(7):828–830
68. Pan JW, Hetherington HP, Vaughan JT, Mitchell G, Pohost GM, Whitaker JN (1996) Evaluation of multiple sclerosis by ^{1}H spectroscopic imaging at 4.1T. Magn Reson Med 36:72–77
69. Caramanos Z, Narayanan S, Arnold DL (2005) ^{1}H-MRS quantification of tNA and tCr in patients with multiple sclerosis: a meta-analytic review. Brain 128:2483–2506

70. Rooney WD, Goodkin DE, Schuff N, Meyerhoff DJ, Norman D, Weiner MW (1997) ^{1}H MRSI of normal appearing white matter in multiple sclerosis. Mult Scler 3:231–237
71. Suhy J, Rooney WD, Goodkin DE, Capizzano AA, Soher BJ, Maudsley AA, Waubant E, Andersson PB, Weiner MW (2000) ^{1}H MRSI comparison of white matter and lesions in primary progressive and relapsing-remitting MS. Mult Scler 6:148–155
72. Filippi M, Rocca MA (2005) MRI evidence for multiple sclerosis as a diffuse disease of the central nervous system. J Neurol 252(Suppl 5):v16–v24
73. Geurts JJ, Reuling IE, Vrenken H, Uitdehaag BM, Polman CH, Castelijns JA, Barkhof F, Pouwels PJ (2006) MR spectroscopic evidence for thalamic and hippocampal, but not cortical, damage in multiple sclerosis. Magn Reson Med 55:478–483
74. Lucchinetti C, Bruck W (2004) The pathology of primary progressive multiple sclerosis. Mult Scler 10(Suppl 1):S23–S30
75. Filippi M, Rovaris M, Rocca MA (2004) Imaging primary progressive multiple sclerosis: the contribution of structural, metabolic, and functional MRI techniques. Mult Scler 10(Suppl 1): S36–S44, discussion S44-5
76. Rovaris M, Bozzali M, Santuccio G, Ghezzi A, Caputo D, Montanari E, Bertolotto A, Bergamaschi R, Capra R, Mancardi G, Martinelli V, Comi G, Filippi M (2001) In vivo assessment of the brain and cervical cord pathology of patients with primary progressive multiple sclerosis. Brain 124:2540–2549
77. Narayanan S, DE Stefano N, Francis GS, Arnaoutelis R, Caramanos Z, Collins DL, Pelletier D, Arnason BGW, Antel JP, Arnold DL (2001) Axonal metabolic recovery in multiple sclerosis patients treated with interferon beta-1b. J Neurol 248:979–986
78. Ciccarelli O, Thomas DL, DE Vita E, Wheeler-Kingshott CA, Kachramanoglou C, Kapoor R, Leary S, Matthews L, Palace J, Chard D, Miller DH, Toosy AT, Thompson AJ (2013) 2013. Low myo-inositol indicating astrocytic damage in a case series of neuromyelitis optica. Ann Neurol 74(2):301–305
79. Sajja BR, Wolinsky JS, Narayana PA (2009) Proton magnetic resonance spectroscopy in multiple sclerosis. Neuroimaging Clin N Am 19:45–58
80. Marignier R, Nicolle A, Watrin C, Touret M, Cavagna S, Varrin-Doyer M, Cavillon G, Rogemond V, Confavreux C, Honnorat J, Giraudon P (2010) Oligodendrocytes are damaged by neuromyelitis optica immunoglobulin G via astrocyte injury. Brain 133:2578–2591
81. Rovira A, Alonso J (2013) ^{1}H magnetic resonance spectroscopy in multiple sclerosis and related disorders. Neuroimaging Clin N Am 23:459–474
82. Majos C, Aguilera C, Alonso J, Julia-Sape M, Castaner S, Sanchez JJ, Samitier A, Leon A, Rovira A, Arus C (2009) Proton MR spectroscopy improves discrimination between tumor and pseudotumoral lesion in solid brain masses. AJNR Am J Neuroradiol 30:544–551
83. Moore F, Okuda DT (2009) Incidental MRI anomalies suggestive of multiple sclerosis: the radiologically isolated syndrome. Neurology 73:1714
84. Stromillo ML, Giorgio A, Rossi F, Battaglini M, Hakiki B, Malentacchi G, Santangelo M, Gasperini C, Bartolozzi ML, Portaccio E, Amato MP, DE Stefano N (2013) Brain metabolic changes suggestive of axonal damage in radiologically isolated syndrome. Neurology 80:2090–2094
85. Sbardella E, Tomassini V, Stromillo ML, Filippini N, Battaglini M, Ruggieri S, Ausili Cefaro L, Raz E, Gasperini C, Sormani MP, Pantano P, Pozzilli C, De Stefano N (2011) Pronounced focal and diffuse brain damage predicts short-term disease evolution in patients with clinically isolated syndrome suggestive of multiple sclerosis. Mult Scler 17:1432–1440
86. Wattjes MP, Harzheim M, Lutterbey GG, Bogdanow M, Schmidt S, Schild HH, Traber F (2008) Prognostic value of high-field proton magnetic resonance spectroscopy in patients presenting with clinically isolated syndromes suggestive of multiple sclerosis. Neuroradiology 50:123–129
87. Kapeller P, Brex PA, Chard D, Dalton C, Griffin CM, Mclean MA, Parker GJ, Thompson AJ, Miller DH (2002) Quantitative ^{1}H MRS imaging 14 years after presenting with a clinically isolated syndrome suggestive of multiple sclerosis. Mult Scler 8:207–210

88. Narayana PA (2005) Magnetic resonance spectroscopy in the monitoring of multiple sclerosis. J Neuroimaging 15:46S–57S
89. Achtnichts L, Gonen O, Rigotti DJ, Babb JS, Naegelin Y, Penner IK, Bendfeldt K, Hirsch J, Amann M, Kappos L, Gass A (2013) Global N-acetylaspartate concentration in benign and non-benign multiple sclerosis patients of long disease duration. Eur J Radiol 82:e848–e852
90. Ge Y, Gonen O, Inglese M, Babb JS, Markowitz CE, Grossman RI (2004) Neuronal cell injury precedes brain atrophy in multiple sclerosis. Neurology 62:624–627
91. Lassmann H (2013) Pathology and disease mechanisms in different stages of multiple sclerosis. J Neurol Sci 333:1–4
92. Llufriu S, Kornak J, Ratiney H, Oh J, Brenneman D, Cree BA, Sampat M, Hauser SL, Nelson SJ, Pelletier D (2014) 2014. Magnetic resonance spectroscopy markers of disease progression in multiple sclerosis. JAMA Neurol 71(7):840–847
93. Sajja BR, Narayana PA, Wolinsky JS, Ahn CW (2008) Longitudinal magnetic resonance spectroscopic imaging of primary progressive multiple sclerosis patients treated with glatiramer acetate: multicenter study. Mult Scler 14:73–80
94. DE Stefano N, Bartolozzi ML, Guidi L, Stromillo ML, Federico A (2005) Magnetic resonance spectroscopy as a measure of brain damage in multiple sclerosis. J Neurol Sci 233:203–208
95. Matthews PM, Arnold DL (2001) Magnetic resonance imaging of multiple sclerosis: new insights linking pathology to clinical evolution. Curr Opin Neurol 14:279–287
96. Trapp BD, Ransohoff R, Rudick R (1999) Axonal pathology in multiple sclerosis: relationship to neurologic disability. Curr Opin Neurol 12:295–302
97. DE Stefano N, Matthews PM, Antel JP, Preul M, Francis G, Arnold DL (1995) Chemical pathology of acute demyelinating lesions and its correlation with disability. Ann Neurol 38:901–909
98. DE Stefano N, Matthews PM, Narayanan S, Francis GS, Antel JP, Arnold DL (1997) Axonal dysfunction and disability in a relapse of multiple sclerosis: longitudinal study of a patient. Neurology 49:1138–1141
99. Aboul-Enein F, Krssak M, Hoftberger R, Prayer D, Kristoferitsch W (2010) Reduced NAA-levels in the NAWM of patients with MS is a feature of progression. A study with quantitative magnetic resonance spectroscopy at 3 Tesla. PLoS One 5:e11625
100. Sastre-Garriga J, Ingle GT, Chard DT, Ramio-Torrenta L, Mclean MA, Miller DH, Thompson AJ (2005) Metabolite changes in normal-appearing gray and white matter are linked with disability in early primary progressive multiple sclerosis. Arch Neurol 62:569–573
101. Tellez N, Alonso J, Rio J, Tintore M, Nos C, Montalban X, Rovira A (2008) The basal ganglia: a substrate for fatigue in multiple sclerosis. Neuroradiology 50:17–23
102. Langdon DW (2011) Cognition in multiple sclerosis. Curr Opin Neurol 24:244–249
103. Calabrese M, Rinaldi F, Mattisi I, Grossi P, Favaretto A, Atzori M, Bernardi V, Barachino L, Romualdi C, Rinaldi L, Perini P, Gallo P (2010) Widespread cortical thinning characterizes patients with MS with mild cognitive impairment. Neurology 74:321–328
104. Sicotte NL, Kern KC, Giesser BS, Arshanapalli A, Schultz A, Montag M, Wang H, Bookheimer SY (2008) Regional hippocampal atrophy in multiple sclerosis. Brain 131:1134–1141
105. Gadea M, Martinez-Bisbal MC, Marti-Bonmati L, Espert R, Casanova B, Coret F, Celda B (2004) Spectroscopic axonal damage of the right locus coeruleus relates to selective attention impairment in early stage relapsing-remitting multiple sclerosis. Brain 127:89–98
106. Neale JH, Olszewski RT, Zuo D, Janczura KJ, Profaci CP, Lavin KM, Madore JC, Bzdega T (2011) Advances in understanding the peptide neurotransmitter NAAG and appearance of a new member of the NAAG neuropeptide family. J Neurochem 118:490–498
107. Rahn KA, Watkins CC, Alt J, Rais R, Stathis M, Grishkan I, Crainiceau CM, Pomper MG, Rojas C, Pletnikov MV, Calabresi PA, Brandt J, Barker PB, Slusher BS, Kaplin AI (2012) Inhibition of glutamate carboxypeptidase II (GCPII) activity as a treatment for cognitive impairment in multiple sclerosis. Proc Natl Acad Sci U S A 109:20101–20106

108. Marliani AF, Clementi V, Albini Riccioli L, Agati R, Carpenzano M, Salvi F, Leonardi M (2010) Quantitative cervical spinal cord 3T proton MR spectroscopy in multiple sclerosis. AJNR Am J Neuroradiol 31:180–184
109. Ciccarelli O, Wheeler-Kingshott CA, Mclean MA, Cercignani M, Wimpey K, Miller DH, Thompson AJ (2007) Spinal cord spectroscopy and diffusion-based tractography to assess acute disability in multiple sclerosis. Brain 130:2220–2231
110. Blamire AM, Cader S, Lee M, Palace J, Matthews PM (2007) Axonal damage in the spinal cord of multiple sclerosis patients detected by magnetic resonance spectroscopy. Magn Reson Med 58:880–885
111. De Stefano N, Matthews PM, Fu L, Narayanan S, Stanley J, Francis GS, Antel JP, Arnold DL (1998) Axonal damage correlates with disability in patients with relapsing-remitting multiple sclerosis. Results of a longitudinal magnetic resonance spectroscopy study. Brain 121 (Pt 8):1469–1477
112. Wheeler-Kingshott CA, Stroman PW, Schwab JM, Bacon M, Bosma R, Brooks J, Cadotte DW, Carlstedt T, Ciccarelli O, Cohen-Adad J, Curt A, Evangelou N, Fehlings MG, Filippi M, Kelley BJ, Kollias S, MacKay A, Porro CA, Smith S, Strittmatter SM, Summers P, Thompson AJ, Tracey I (2014) The current state-of-the-art of spinal cord imaging: applications. Neuroimage 84:1082–1093
113. Abdel Aziz K, Solanky B, Wheeler-Kingshott C, Cawley N, Yiannakas M, Thompson A, Ciccarelli O. (2013) Evidence for early neuronal damage in the cervical cord of patients with primary progressive MS. ECTRIMS. Multiple Sclerosis J 2013. Abstract No: 197
114. Stroman PW, Wheeler-Kingshott C, Bacon M, Schwab JM, Bosma R, Brooks J, Cadotte D, Carlstedt T, Ciccarelli O, Cohen-Adad J, Curt A, Evangelou N, Fehlings MG, Filippi M, Kelley BJ, Kollias S, MacKay A, Porro CA, Smith S, Strittmatter SM, Summers P, Tracey I (2014) The current state-of-the-art of spinal cord imaging: methods. Neuroimage 84:1070–1081
115. Bellenberg B, Busch M, Trampe N, Gold R, Chan A, Lukas C (2013) 1H-magnetic resonance spectroscopy in diffuse and focal cervical cord lesions in multiple sclerosis. Eur Radiol 23:3379–3392
116. Hermann D, Weber-Fahr W, Sartorius A, Hoerst M, Frischknecht U, Tunc-Skarka N, Perreau-Lenz S, Hansson AC, Krumm B, Kiefer F, Spanagel R, Mann K, Ende G, Sommer WH (2012) Translational magnetic resonance spectroscopy reveals excessive central glutamate levels during alcohol withdrawal in humans and rats. Biol Psychiatry 71:1015–1021
117. Denic A, Johnson AJ, Bieber AJ, Warrington AE, Rodriguez M, Pirko I (2011) The relevance of animal models in multiple sclerosis research. Pathophysiology 18:21–29
118. Lee MR, Denic A, Hinton DJ, Mishra PK, Choi DS, Pirko I, Rodriguez M, Macura SI (2012) Preclinical ^{1}H-MRS neurochemical profiling in neurological and psychiatric disorders. Bioanalysis 4:1787–1804
119. Chen CC, Zechariah A, Hsu YH, Chen HW, Yang LC, Chang C (2008) Neuroaxonal ion dyshomeostasis of the normal-appearing corpus callosum in experimental autoimmune encephalomyelitis. Exp Neurol 210:322–330
120. Brenner RE, Munro PM, Williams SC, Bell JD, Barker GJ, Hawkins CP, Landon DN, McDonald WI (1993) The proton NMR spectrum in acute EAE: the significance of the change in the Cho:Cr ratio. Magn Reson Med 29:737–745
121. Rigotti DJ, Inglese M, Kirov II, Gorynski E, Perry NN, Babb JS, Herbert J, Grossman RI, Gonen O (2012) Two-year serial whole-brain N-acetyl-L-aspartate in patients with relapsing-remitting multiple sclerosis. Neurology 78:1383–1389
122. Denic A, Pirko I, Wootla B, Bieber A, Macura S, Rodriguez M (2012) Deletion of beta-2-microglobulin ameliorates spinal cord lesion load and promotes recovery of brainstem NAA levels in a murine model of multiple sclerosis. Brain Pathol 22:698–708
123. Kan HE, Techawiboonwong A, VAN Osch MJ, Versluis MJ, Deelchand DK, Henry PG, Marjanska M, VAN Buchem MA, Webb AG, Ronen I (2012) Differences in apparent

diffusion coefficients of brain metabolites between grey and white matter in the human brain measured at 7T. Magn Reson Med 67:1203–1209
124. Wood ET, Ronen I, Techawiboonwong A, Jones CK, Barker PB, Calabresi P, Harrison D, Reich DS (2012) Investigating axonal damage in multiple sclerosis by diffusion tensor spectroscopy. J Neurosci 32:6665–6669
125. Hattingen E, Magerkurth J, Pilatus U, Hubers A, Wahl M, Ziemann U (2011) Combined ^{1}H and ^{31}P spectroscopy provides new insights into the pathobiochemistry of brain damage in multiple sclerosis. NMR Biomed 24:536–546
126. Srinivasan R, Ratiney H, Hammond-Rosenbluth KE, Pelletier D, Nelson SJ (2010) MR spectroscopic imaging of glutathione in the white and gray matter at 7T with an application to multiple sclerosis. Magn Reson Imaging 28:163–170
127. Carvalho AN, Lim JL, Nijland PG, Witte ME, Van Horssen J (2014) Glutathione in multiple sclerosis: More than just an antioxidant? Mult Scler. 20:1425–1431
128. Bhattacharyya PK, Phillips MD, Stone LA, Bermel RA, Lowe MJ (2013) Sensorimotor cortex gamma-aminobutyric acid concentration correlates with impaired performance in patients with MS. AJNR Am J Neuroradiol 34:1733–1739
129. Hock A, Wilm B, Zandomeneghi G, Ampanozi G, Franckenberg S, De Zanche N, Nordmeyer-Maßner J, Spyros S. Kollias, Kramer T, Thali M, Ernst M, and Henning A. (2014) Detection of GABA, Aspartate and Glutathione in the Human Spinal Cord. ISMRM. Abstract No: 1712

Chapter 9
Magnetic Resonance Spectroscopy in Ataxias

Pierre-Gilles Henry, H. Brent Clark, and Gülin Öz

Abstract Ataxia refers to deficits in coordination of movement and balance. This chapter focuses on recessively and dominantly inherited as well as sporadic degenerative ataxias. These diseases are characterized by neurodegeneration in the cerebellar system, including the cerebellum and its afferent and efferent connections, and frequently also in other brain regions such as the pontine nuclei. They display great pathological diversity, as well as phenotypic variability; thus there is a great need for imaging biomarkers that reflect the underlying pathology and that can be used for diagnostic, prognostic, and treatment monitoring purposes. Despite technical challenges of magnetic resonance spectroscopy (MRS) in the cerebellum and brainstem, MRS has been shown to be sensitive to neurochemical alterations in various degenerative ataxias. Namely, early neurochemical abnormalities have been detected by MRS in ataxias prior to the structural atrophy detectable by conventional MRI and prior to symptoms. Correlations with clinical status and pathological severity were demonstrated in clinical and animal model studies, respectively. MRS was also shown to distinguish different ataxia subtypes, with potential utility in differential diagnosis, especially valuable for sporadic ataxias in the absence of genetic testing. Finally, a few studies have utilized MRS for treatment monitoring in clinical trials of recessive ataxias, and a great need exists in this area for all degenerative ataxias. More longitudinal investigations and standardization of advanced MRS methodology for multi-site trials will be critical in this respect.

Keywords Spinocerebellar ataxia • Friedreich's ataxia • Multiple system atrophy • Ataxia-telangiectasia • Ataxia with oculomotor apraxia • Autosomal recessive spastic ataxia of Charlevoix-Saguenay

---

P.-G. Henry, Ph.D. • G. Öz, Ph.D. (✉)
Department of Radiology, Center for Magnetic Resonance Research, University of Minnesota, Minneapolis, MN 55455, USA
e-mail: henry@cmrr.umn.edu; gulin@cmrr.umn.edu

H.B. Clark, M.D., Ph.D.
Department of Laboratory Medicine and Pathology, University of Minnesota, Minneapolis, MN 55455, USA
e-mail: clark002@umn.edu

G. Öz (ed.), *Magnetic Resonance Spectroscopy of Degenerative Brain Diseases*, Contemporary Clinical Neuroscience, DOI 10.1007/978-3-319-33555-1_9

Symptoms of Ataxias

The term ataxia is used to describe deficits in coordination of movement and postural instability that occur independently of motor weakness. Typical features of ataxia include abnormalities of posture and gait, loss of coordination of limb movements, poor articulation of speech, and disturbances of eye movements. Postural changes may include instability of the trunk while standing or, when severe, even sitting. Ataxic patients compensate with a wide-based stance, but they still have difficulty maintaining balance. Walking is irregular, wide based, and staggering. The patients cannot walk placing one foot in front of the other. Limb ataxia is characterized by the inability to perform movements with proper timing and trajectory, known as dysmetria. Patients also may have intention tremor, jerky coarse movements, and a reduced ability to perform rapid alternating movements. Their speech disturbances, called dysarthria, consist of disruption of the rhythm, fluency, and articulation of speech. Oculomotor changes are variable and include nystagmus, and problems with visual pursuit, ocular fixation, and alignment, as well as saccadic eye movements. Although ataxias can have multiple causes, most are neurodegenerative in nature. Neurodegenerative ataxias can be hereditary or sporadic and will be the focus of this chapter.

To evaluate the severity of clinical involvement of ataxic patients, clinical assessment scales have been developed. These include the International Cooperative Ataxia Rating Scale (ICARS) [1], the Scale for the Rating and Assessment of Ataxia (SARA) [2], and the Friedreich's Ataxia Rating Scale (FARS) [3, 4].

Pathological Classification of Degenerative Ataxias

The cerebellar system is the principal pathological target of neurodegenerative processes that result in ataxia. This system includes the cerebellum itself and also its afferent and efferent connections. In some forms of ataxias the cerebellar system is the sole target, but many others have a more complex involvement of other brain regions.

Classification of ataxias has historically been problematic because of clinical and pathological variability within individual types of disease. Recent advances in genetic classification of many of the hereditary ataxias have helped somewhat in better categorization, but the phenotypic variability seen in a number of forms remains a confounding factor in characterizing these diseases.

Pathological classifications have emphasized three major patterns of neurodegeneration although they are not always so clearly distinct and do not tightly correlate with clinical entities.

1. The most common pattern is cerebellar cortical degeneration, which is characterized by loss of Purkinje cells. There also may be variable but usually milder loss of granular neurons and cortical interneurons. There often is neuronal loss in the inferior olivary nuclei that is thought to result from retrograde trans-synaptic

degeneration due to loss of climbing-fiber targets. Recent studies have suggested that some of the ataxias thought to be simple cerebellar degenerations may have pathology in brainstem areas that affect eye movements [5]. Examples of this type of pathology include spinocerebellar ataxia 5 (SCA5) and SCA6 and a number of the other dominantly inherited ataxias.

2. Olivopontocerebellar atrophy (OPCA) results from more complex involvement of the cerebellar system and other areas of the nervous system. In addition to cerebellar cortical degeneration there is loss of afferent mossy fibers from the basal pontine nuclei with respective loss of volume in the middle cerebellar peduncles and the cerebellar white matter. Mossy fibers from vestibular and spinocerebellar pathways also frequently are affected. The inferior olivary nuclei likewise have neuronal loss that usually is as great or greater than the loss of Purkinje neurons, suggesting that it is part of the primary process rather than secondary to retrograde trans-synaptic degeneration. If severe, olivary atrophy may be accompanied by atrophy of the inferior cerebellar peduncles. There also is variable involvement of the deep cerebellar nuclei, which are spared in the cerebellar form of multiple system atrophy (MSA-C) and to some extent in SCA2, but affected in SCA1. Loss of neurons in the deep cerebellar nuclei results in atrophy of the superior cerebellar peduncles. Other variably affected areas include the spinal cord, cranial nerve nuclei, substantia nigra, red nuclei, periaquaductal gray nuclei, and the basal ganglia.
3. Spinocerebellar degeneration is a third category and is characterized by loss of cerebellar afferent projections, with milder or no detectable involvement of the cerebellar cortex. The deep cerebellar nuclei also are affected in some instances, such as in Friedreich's ataxia (FRDA) [6] and SCA3. Despite the name, most of the autosomal dominant hereditary ataxias, termed spinocerebellar ataxias, do not necessarily have this pattern of degeneration.

Although the combinations of the affected sites can be quite variable depending on the type of ataxia, there are consistent histological features in the affected areas. Shrinkage of the cerebellar folia with sulcal expansion is characteristic of cerebellar cortical degeneration, principally due to loss of the Purkinje neurons. The Purkinje neuronal loss often is preceded by dendritic and somatic atrophy, which may contribute to the atrophy of the molecular layer of the cortex. Atrophy and death of Purkinje neurons are accompanied by hypertrophy and hyperplasia of Bergmann astrocytes, which reside in the Purkinje cell layer. In later stages there often is at least partial loss of interneurons, as well as the granular neurons. Loss of Purkinje neurons in the cortex results in degeneration of axon terminals in the deep cerebellar nuclei that is accompanied by gliosis with variable loss of deep nuclear neurons.

In ataxias with brainstem and spinal cord involvement, other features often are present. Neuronal loss in the basal pontine nuclei is reflected by gross atrophy of the basis pontis, loss or shrinkage of transverse fibers, and a loss of volume in the middle cerebellar peduncles and cerebellar white matter. These gross changes are accompanied by loss of neurons and their fiber projections with reactive astrocytosis. Similar histopathological changes are seen in other affected brainstem and spinal cord nuclei and tracts.

Recessively Inherited Ataxias

The most common recessive ataxias are FRDA and ataxia-telangiectasia (A-T). FRDA primarily affects sensory afferent projections and deep nuclear efferents of the cerebellum, while largely sparing the cerebellar cortex, olives, and pons. Grossly, the spinal cord may seem small with histological evidence of loss of axons in the posterior columns and the dorsal roots as well as in the spinocerebellar tracts. There also frequently is some degree of pyramidal tract degeneration. The dentate nuclei are severely atrophic, as are their efferent pathways in the superior cerebellar peduncles [7].

In A-T the cerebellar vermis and hemispheres have gross atrophy. Microscopically, there is severe loss of Purkinje cells with surviving cells having bizarre dendritic morphology, and heterotopic somata within the molecular layer. Cortical interneurons and granular neurons are less severely affected with prominent preservation/proliferation of basket fibers around the sites of degenerated Purkinje cells (empty baskets). There is retrograde trans-synaptic degeneration in the inferior olivary nuclei and gliosis in the dentate nuclei without neuronal loss. The basal pons is not affected but the spinal cord can have atrophy of the posterior columns [8].

Ataxia with oculomotor apraxia type 2 (AOA2) also has been termed spinocerebellar ataxia recessive type 1 (SCAR1) because oculomotor apraxia is not always present. Onset of disease occurs in the second decade of life and patients frequently have amyotrophy as well as ataxia, oculomotor apraxia, and neuropathy. This disorder has not been well characterized neuropathologically [9].

Autosomal recessive spastic ataxia of Charlevoix-Saguenay (ARSACS) is an early-onset spastic ataxia. An autopsy study from one patient found atrophy of the anterior cerebellar vermis with loss of Purkinje neurons. The corticospinal tracts were atrophic in the pons and medulla and spinal cord, but there was no involvement of the basal pontine nuclei, the deep cerebellar nuclei, or the inferior olivary nuclei [10].

Dominantly Inherited Ataxias

There are over 30 forms of dominantly inherited ataxias, collectively known as the SCAs, but pathologically very diverse. A number of the more common and better characterized forms of SCA have a polyglutamine-trinucleotide-repeat-expansion mutation. Variability in the length of the expanded repeat has a major influence on the phenotype of these patients such that members of the same kindred can have significant differences in their pathological and clinical features. The polyglutamine-related SCA include SCA1, SCA2, SCA3, SCA6, SCA7, and SCA17. The pathology of SCA1, SCA2, and SCA7 tends to be that of the OPCA pattern in most cases [5, 11, 12]. The deep cerebellar nuclei are more affected in

SCA1 and less so in SCA2 and SCA7 and the pons may be less affected in SCA7. SCA3 has a more unique pattern with spinal and pontine involvement with relative sparing of the cerebellar cortex and olivary nuclei and severe involvement of the dentate nuclei. It should be emphasized, however, that within each disease there is a considerable spectrum of pathology, such that not all cases have the patterns described above as the dominant features. SCA3 not infrequently has extrapyramidal, pyramidal, and even lower motor neuronal involvement as major features superseding or masking ataxia. All of the above with the possible exception of SCA6 can have involvement of other brainstem regions as well as supratentorial regions, including the deep gray nuclei and in some instances cerebral cortex.

SCA6 is mostly a simple cerebellar cortical degeneration although recent studies have shown mild involvement of brainstem and pyramidal motor neurons [5].

Sporadic Ataxias

The sporadic ataxias usually have the OPCA or simple cerebellar atrophic patterns of degeneration. A common sporadic form of OPCA is the cerebellar form of multiple system atrophy (MSA-C) in which there is severe pontine atrophy with corresponding shrinkage of the middle cerebellar peduncles and the cerebellar white matter. The cerebellar cortex is affected but less severely than in many SCA and the deep cerebellar nuclei are spared. Therefore, the outflow areas of the dentate nucleus in the cerebellar white matter and the superior cerebellar peduncles are spared. MSA-C frequently has mild changes that are present to a greater extent in the parkinsonian form, MSA-P, i.e., putaminal and nigral atrophy. There are other forms of sporadic OPCA that are not related to MSA-C and have a better prognosis, but the pathological pattern may be similar [13]. There are a number of apparently sporadic simple cerebellar atrophies mostly of late onset [14] although some of these cases may represent undiagnosed hereditary ataxias. The neuropathology of these cases resembles what is seen in SCA6 with the degeneration affecting Purkinje cells.

Potential Clinical Utility of Magnetic Resonance Spectroscopy

In vivo proton magnetic resonance spectroscopy (^{1}H MRS) enables noninvasive quantification of endogenous metabolites that exist at millimolar concentrations in selected tissue volumes and thereby provides biochemical information that is not available from conventional structural MRI. Specifically, ^{1}H MRS may provide insight into neuronal viability (*N*-acetylaspartate, NAA), cellular proliferation or membrane turnover (choline-containing compounds, tCho), glial activation (*myo-*

inositol, mIns), neurotransmitter activity (glutamate, glutamine, γ-aminobutyric acid (GABA)), oxidative stress (glutathione, vitamin C), and energy metabolism (creatine, phosphocreatine, glucose, lactate) [15–17]. Since biochemical abnormalities precede the ultimate demise of neuronal populations in neurodegenerative diseases such as hereditary and sporadic degenerative ataxias, a primary potential clinical utility of MRS in ataxias is the early detection of neurochemical abnormalities prior to the cerebellar and brainstem atrophy detectable by conventional MRI. This may enable detection of disease onset in mutation carriers of hereditary ataxias and allow the timely administration of potential neuroprotective therapies.

Along the same lines, MRS may help with monitoring of the effects of potential therapies in degenerative ataxias. While ataxias are currently untreatable, neuronal and motor dysfunction has been shown to be reversible in mouse models [18, 19]. Furthermore, advances in understanding of molecular mechanisms of common ataxias have raised a realistic hope for the development of effective therapeutic interventions [20–22]. However, assessing whether therapies can abate progression of neurodegenerative diseases is challenging because of their slow progression and phenotypic variability, which in turn necessitate long clinical trials with large sample sizes, a particular challenge for rare diseases like ataxias. Clinical outcome measures that are typically used in clinical trials have limitations: they do not distinguish between disease-modifying vs. purely symptomatic drug effects, they cannot detect effectiveness of therapies in the presymptomatic stage (when neuroprotective agents are likely to be most effective), they are unable to provide prompt feedback regarding drug effectiveness and usually have poor test-retest reliability [23]. MRS has the potential to directly assess disease-modifying effects of therapeutic interventions in the brain and to gauge their effectiveness quickly and objectively.

Another area where MRS may have clinical impact in ataxias is differential diagnosis. While the definitive diagnosis in hereditary ataxias will always rely on genetic testing, distinguishing ataxias by noninvasive imaging has been a long-term interest in the ataxia field [24–29] because such biomarkers could guide genetic testing in the absence of family history and help with diagnosing sporadic ataxias. Degenerative ataxias are challenging to distinguish due to the substantial overlap in the affected brain regions, pathology, and clinical presentation [30–32]. Therefore there is a need for other neuroimaging modalities such as MRS to provide potentially ataxia subtype-specific information not available from conventional atrophy assessments.

Technical Considerations

The precision and accuracy of in vivo MRS strongly depend on the quality of methodology used to acquire and process the data. Critical factors include good radiofrequency (RF) coils with high transmit efficiency and good receive sensitivity, a well-optimized pulse sequence with efficient outer volume suppression, excellent B_0 shimming, and shot-to-shot data acquisition for subsequent shot-to-shot frequency and phase correction. Single-voxel MRS generally provides the highest spectral quality, and therefore better precision and detection of higher number of metabolites. Up to three regions of interest can be measured sequentially with single-voxel MRS in a 1-h scanning session. Magnetic resonance spectroscopic imaging (MRSI), on the other hand, offers the advantage of larger spatial coverage, often at the expense of lower overall spectral quality.

A short-echo time sequence is advantageous to minimize signal loss due to T_2 relaxation and to retain signals from metabolites with J-coupled resonances, such as glutamate and glutamine. STEAM allows very short echo times and has been used in many studies, but suffers from an intrinsic 50 % loss in signal-to-noise ratio (SNR) compared to full-intensity sequences. PRESS provides full SNR, but suffers from high chemical-shift displacement errors, especially at higher magnetic fields. Newer sequences such as LASER [33] and semi-LASER [34, 35] or SPECIAL [36] offer a number of advantages: insensitivity to B_1 inhomogeneity, low chemical-shift displacement error, and reduction of T_2 loss and J-modulation through the Carr-Purcell-Meiboom-Gill (CPMG) effect.

Various challenges of acquiring high-quality spectra from the cerebellum and brainstem were recently summarized [17] and primarily stem from the caudal location of these structures in the brain and their proximity to bone/air/tissue interfaces. Hence broader intrinsic linewidths are observed relative to other cerebral regions. Despite these challenges, high-quality MR spectra can be obtained from the cerebellum and pons on clinical 3 T platforms both in healthy volunteers and patients with ataxias [37, 38] (Fig. 9.1).

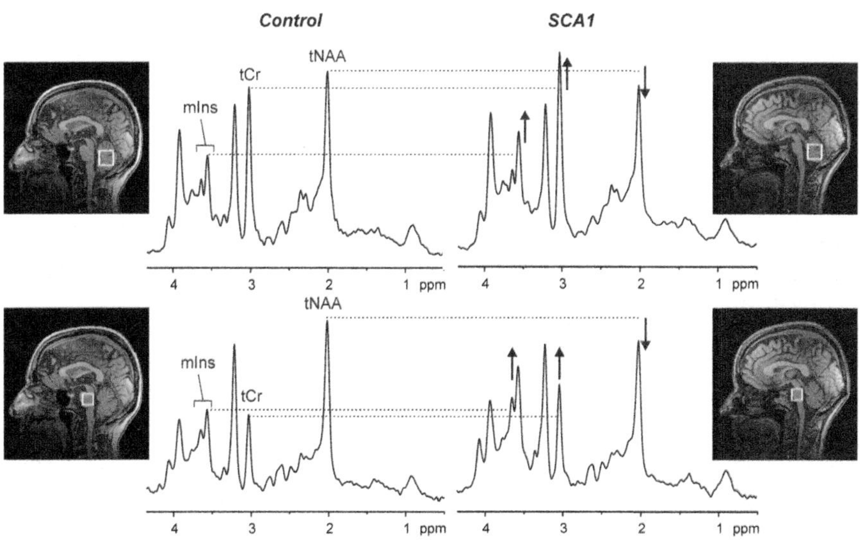

Fig. 9.1 Proton MR spectra obtained from the cerebellar vermis and pons of a healthy control (*left*) and a patient with SCA1 (*right*) at 3T (semi-LASER [34], TR/TE = 5000/28 ms). Voxel positions are shown in T_1-weighted mid-sagittal images. Differences in the spectra from the patient vs. control in total NAA (tNAA), *myo*-inositol (mIns), and total creatine (tCr) are clearly visible

On the processing side, quantification using water as an internal concentration reference has become the standard for quantification of ^{1}H MRS spectra, and is preferable to reporting ratios of concentrations, which can be difficult to interpret. Correction for cerebrospinal fluid (CSF) content in the voxel is also critical to correctly account for loss of tissue during neurodegeneration [39]. This can be achieved either by acquiring the water signal at multiple echo times and fitting the T_2 decay with a bi-exponential curve [40] or by image segmentation [41].

Higher magnetic fields have been shown to be beneficial for MRS, providing higher precision and increasing the number of metabolites that can be measured reliably. From 3T to 7T, gains in sensitivity are particularly pronounced for glutamate, glutamine, and GABA [42]. However, 7T systems are still not widely available in clinical settings.

Overview of MRS in Animal Models of Ataxias

Recessively Inherited Ataxias

To the best of our knowledge, no animal model studies have been reported using in vivo MRS in recessive ataxias. This may in part be due to the fact that the underlying genetic defects in recessive ataxias were identified relatively recently. In addition, the neuropathology of animal models of FRDA, the most common recessive ataxia, has primarily been investigated at the level of the dorsal root ganglia and spinal cord, which are likely too small for in vivo MRS.

Dominantly Inherited Ataxias

MRS studies on animal models of dominantly inherited ataxias are few. Nonetheless, they provided insights into disease and helped validate the biomarker and surrogate marker potential of the metabolites quantified by MRS.

The first study that attempted to define the metabolic phenotype in a dominantly inherited ataxia by MR was not an in vivo investigation, but utilized ^{1}H nuclear magnetic resonance (NMR) spectroscopy of tissue extracts and magic angle spinning ^{1}H NMR of intact tissue [43]. Using these methods, the investigators demonstrated a number of neurochemical alterations in a transgenic mouse model of SCA3, a polyglutamine disease and the commonest dominantly inherited ataxia. They concluded that the differences observed in the cerebellum and cerebrum of SCA3 mice relative to controls, including higher glutamine and lower tCho, mIns, GABA, and lactate levels, were likely caused by neuronal metabolic dysfunction, rather than large amounts of cell death.

Later studies focused on transgenic and knock-in models of another polyglutamine ataxia, SCA1. A longitudinal investigation of a transgenic SCA1 model, which overexpresses the mutant human ataxin-1 in Purkinje cells and reproduces the Purkinje cell pathology and the phenotype of the human disease, demonstrated progressive neurochemical changes, starting at the presymptomatic stage of disease [44]. Specifically, neuronal markers NAA and glutamate were lower and the putative glial marker mIns was higher in SCA1 mice than controls. These abnormalities worsened over time and were correlated with the molecular layer thickness and a pathological severity score, demonstrating that they reflect the progressive pathology of the cerebellar disease (Fig. 9.2). Importantly, these were the same metabolites that correlated with the clinical deficit in patients [45], confirming the validity of the neurochemical abnormalities observed in the mouse model. The same study uncovered another set of early biochemical abnormalities, namely increased total creatine (creatine + phosphocreatine, tCr) and glutamine and decreased taurine, that anticipated disease progression.

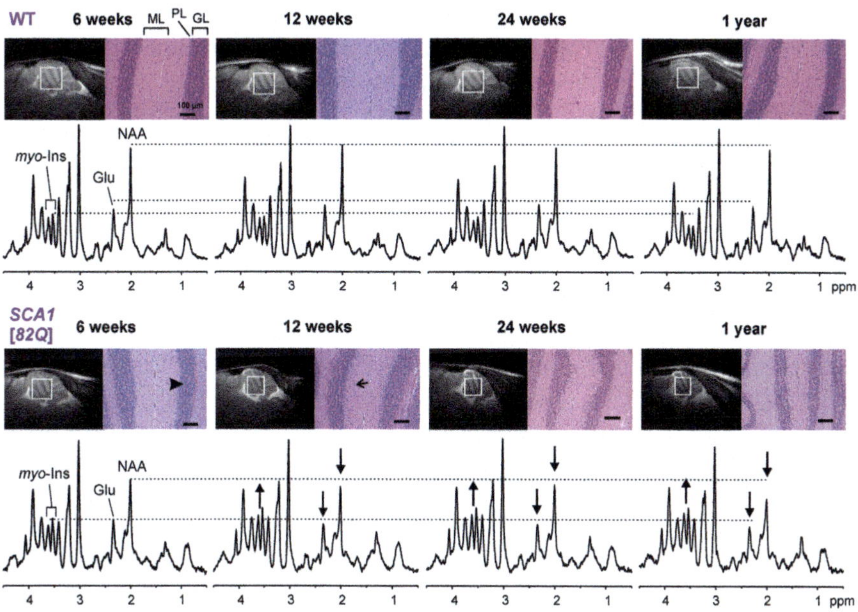

Fig. 9.2 Proton MR spectra (LASER [33], TR/TE = 5000/15 ms), and mid-sagittal T_2-weighted images obtained from one WT (*upper row*) and one SCA1 mouse (*lower row*) across their life-span [44]. Also shown are representative H&E-stained histological slides from different mice at each age. The voxel position for the spectra is shown on each image. The longitudinal changes in NAA, glutamate, and *myo*-inositol in the spectra of the SCA1 mouse are shown with *arrows*. The thinning of the molecular layer in SCA1 mice is visible. Other prominent pathological alterations in this model are early formation of vacuoles (shown with *arrowhead*) and heterotopic Purkinje cells (shown with *arrow*). ML: molecular layer, PL: Purkinje layer, GL: granular layer. Adapted from [44] with kind permission from Society for Neuroscience

Next, a study on a conditional transgenic mouse model of SCA1 demonstrated the sensitivity of metabolite levels to disease reversal [46]. The expression of the ataxin-1 transgene in this conditional transgenic model was put under doxycycline regulation such that administration of doxycycline suppresses transgene expression and thereby rescues the neuronal pathology and motor phenotype [18]. The ultra-high-field MRS study conducted with this mouse line before and after doxycycline administration demonstrated partial reversal of abnormal NAA and mIns concentrations to wild-type (WT) levels, together with complete normalization of taurine and tCr. Therefore the earliest biochemical abnormalities (taurine, tCr) were completely reversible, while the markers of progressive neurodegeneration (NAA, mIns) accurately reflected the partial reversal of pathology, as confirmed by histological measurements. A later study using the same conditional model showed that ultra-high-field MRS accurately reflects transgene expression levels based on a significant correlation between NAA/mIns ratio and quantitative PCR [47]. In addition, receiver operating characteristic (ROC) analyses demonstrated that MRS is as sensitive as the invasive measures histology and qPCR and more sensitive than standard motor behavioral testing (Rotarod) in detecting treatment effects in SCA1.

A knock-in model of SCA1, which displays milder cerebellar pathology than the transgenic model [48], was monitored longitudinally using ultra-high-field MRS for the earliest neurochemical changes related to neurodegeneration in SCA1 [49]. While the cerebellar pathology was found to be very mild even at the latest disease stage investigated, the knock-in mice were distinguishable from WT mice starting at the earliest stage using the MRS-measured neurochemical levels, demonstrating the sensitivity of neurochemical levels to changes in the absence of overt pathology. Consistent with observations in the transgenic model described above, the earliest changes included glutamine, tCr, and taurine. Therefore, MRS studies in SCA1 models revealed neurochemical alterations that were consistent between different models of the disease (at the same disease stages), as well as between the models and the human disease.

Finally, high-field MRS was recently utilized to characterize a novel SCA5 mouse model and demonstrated a higher decline rate of NAA concentrations compared to control mice [50].

Overview of Clinical MRS in Ataxias

Prior to the identification of the causative genes for many hereditary ataxias, studies classified ataxias using either clinically defined criteria [51] or imaging-based criteria, such as olivopontocerebellar atrophy (OPCA) and cerebellar cortical atrophy (CCA) [52, 53]. Considering the pathological heterogeneity between different ataxia subtypes grouped under these classifications, below we will primarily focus on reports of genetically defined ataxias, as well as sporadic ataxias.

Recessively Inherited Ataxias

A number of MRS studies in autosomal recessive cerebellar ataxias have been reported. Most of these studies have been performed in patients with FRDA, the most common form of recessive ataxia. A few additional studies have been reported in A-T, AOA2, and ARSACS. Many of these studies show significant neurochemical differences between patients and controls, with a decrease in total NAA (NAA + *N*-acetylaspartylglutamate, tNAA) and an increase in mIns being the most common findings. However, much remains to be done to assess regional differences, and, more importantly, how early these changes occur in the course of the disease, how they evolve with time, and their correlation with clinical status.

Friedreich's Ataxia: ^{1}H MRS in the Cerebellum

Several studies have shown reduced tNAA-to-tCr ratio (tNAA/tCr) in the cerebellum (both in vermis and cerebellar hemispheres) in patients with FRDA relative to controls [54–57]. This decrease was relatively small, on the order of 10 %, and did not always reach statistical significance in studies with small number of patients [55, 57].

Although most studies reported metabolite ratios, one study quantified metabolites independently using an external concentration reference, which allows quantification of individual metabolites, as opposed to ratios. The study reported a decrease in tNAA and an increase in tCr in vermis [39], suggesting that the decrease in tNAA/tCr ratio observed in the abovementioned studies likely reflects a change in both metabolites, and not only a decrease in tNAA. That study also corrected for CSF contamination in the voxel, a potential source of bias in the presence of atrophy.

Finally, a study at higher magnetic field (4 Tesla), using water as an internal concentration reference, reported decreased tNAA, increased tCr, and increased mIns in the vermis and cerebellar hemispheres of patients with FRDA relative to controls [58] (Fig. 9.3). An increase in glutamine was also observed in the vermis. Although no obvious atrophy of the cerebellum can be seen on MRI images in FRDA in the first years of the disease [59], the MRS study also showed detectable atrophy as evidenced by the increase in CSF fraction in the spectroscopic voxel in the vermis (from 11 % in controls to 18 % in patients).

Overall, most MRS studies in FRDA show relatively small but consistent changes in metabolite concentrations in the cerebellum (both in vermis and in cerebellar hemispheres). The most commonly observed changes are a decrease in tNAA, presumably reflecting neuronal damage, and an increase in tCr and mIns, presumably reflecting gliosis.

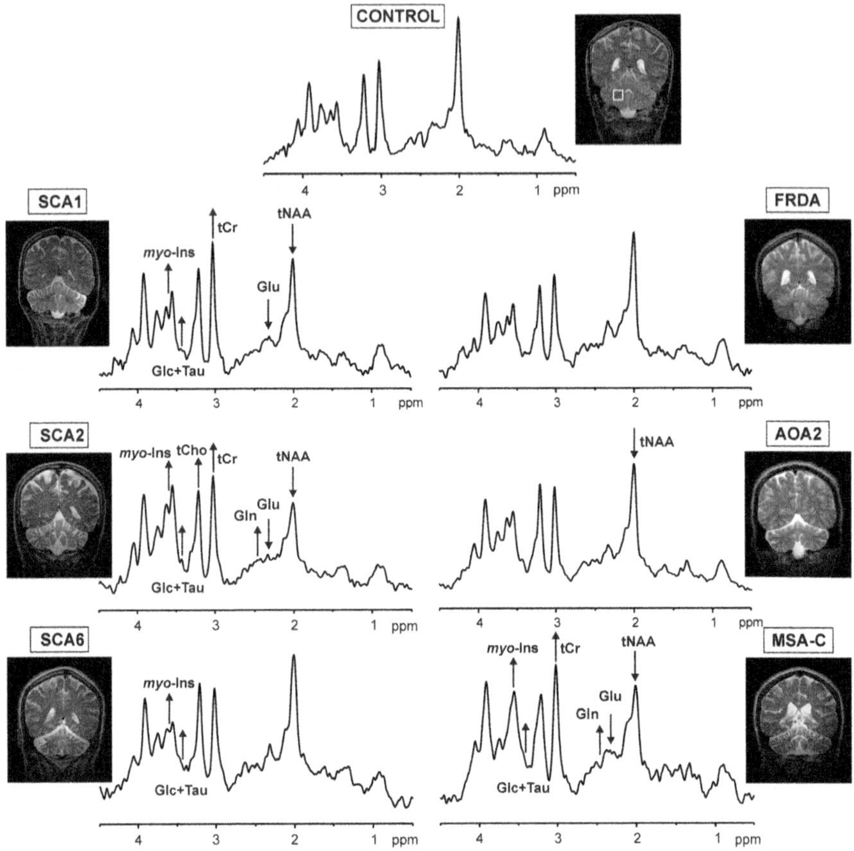

Fig. 9.3 Proton MR spectra obtained from the cerebellar hemisphere of a healthy control (*top*) and patients with dominant (SCA1, SCA2, SCA6), recessive (FRDA, AOA2), and sporadic (MSA-C) ataxias at 4T (STEAM, TR/TE = 4500/5 ms). Voxel position is shown in the T_2-weighted coronal image of the control subject. Reported metabolite alterations in each disease are shown with *arrows* that indicate the direction of change. Adapted from [29, 45] with kind permission from Springer Science+Business Media B.V and John Wiley and Sons

Friedreich's Ataxia: ^{1}H MRS in Pons and Spinal Cord

In addition to the cerebellum, a few of the abovementioned studies also reported results from ^{1}H MRS in the pons, a challenging region for MRS due to the small voxel size and the difficulty to obtain good B_0 shimming in this region. Mascalchi et al. reported lower tNAA/tCr ratio in the pons in patients versus controls [54]. Subsequent studies also reported lower tNAA, as well as higher mIns, although these changes did not reach statistical significance [55, 58].

In contrast, no MRS study has been reported in the spinal cord in FRDA. Spinal cord degeneration occurs early in FRDA, and spinal cord atrophy measured on anatomical MRI images was recently shown to correlate with FARS scores [60]. Recent preliminary results show large differences in tNAA (−34 %) and mIns (+30 %) in patients with FRDA relative to controls, consistent with neuronal damage and gliosis in the spinal cord [61].

Friedreich's Ataxia: ^{31}P MRS in Heart and Muscle

In FRDA, the mutation in the Frataxin gene leads to a reduction in levels of the mitochondrial protein Frataxin, leading to impaired mitochondrial respiration. As a result, FRDA is associated with impaired muscle metabolism and cardiomyopathy. A number of studies have investigated muscle and heart energy metabolism in patients with FRDA using ^{31}P MRS, which allows measurement of steady-state ATP, phosphocreatine (PCr), and inorganic phosphate (Pi) levels. Dynamic measurements of PCr and Pi concentrations in muscle during exercise can also be used to calculate kinetic parameters such as the maximum rate of ATP production Vmax (ATP).

In muscle, Vmax(ATP) was found to be lower in patients with FRDA than in control subjects [62]. There was also a strong correlation between Vmax and the number of GAA repeats in the small allele. Consistent with these findings, PCr recovery after ischemic exercise was dramatically delayed (time constant nearly tripled) in patients with FRDA compared to controls [63–65]. Time constants of recovery correlated with the number of repeats in the smaller allele, the age of the patients, and disease duration [63].

In the heart, cardiac PCr/ATP ratio was shown to be reduced by 40 % in patients with FRDA relative to controls [66, 67]. The decrease in PCr/ATP was accompanied by an increase in Pi/PCr [67]. Both ratios correlated with the degree of heart hypertrophy measured by septal wall thickness [67].

In light of those findings, ^{31}P MRS was used in a number of clinical trials in patients with FRDA. For example, treatment with Co-Q10 and vitamin E or E2 resulted in increased cardiac PCr/ATP ratio, and increased Vmax(ATP) in muscle in patients with FRDA relative to controls [68, 69]. Those improvements, however, were not associated with clear clinical benefits, although one study suggested improvement in ICARS scores in some patients [69]. Similarly, treatment with antioxidant idebenone [64] or with recombinant human erythropoietin (rhuEPO) had no effect on the rate of PCr recovery after exercise and did not improve clinical scores [65].

Although clinical trials have not identified effective treatments for FRDA so far, ^{31}P MRS may prove useful to assess heart and muscle energy metabolism in future clinical trials as new treatments become available.

Ataxia-Telangiectasia (A-T) and Ataxia-Telangiectasia-Like Disorder (ATLD)

One study reported decreased tNAA/tCho and increased tCho/tCr in the cerebellar white matter of patients with A-T compared to controls, with no change in tNAA/tCr [70]. This suggests elevated tCho, although the data are also consistent with a simultaneous decrease in tCr and tNAA. Proton MRS using water as an internal concentration reference would permit discrimination between those two possibilities.

In ATLD, tNAA/tCr was markedly reduced in two patients (−23 % and −40 %, respectively) in the cerebellum compared to known control values in the same region [71]. Other metabolite ratios were not reported.

Ataxia with Oculomotor Apraxia Type 2 (AOA2)

Only one MRS study has been reported in AOA2 [58]. Patients had lower tNAA and higher mIns concentrations in the vermis and pons relative to controls, as well as lower glutamate in the vermis, suggesting altered glutamatergic neurotransmission. tNAA levels in the vermis strongly correlated with FARS scores.

Autosomal Recessive Spastic Ataxia of Charlevoix-Saguenay (ARSACS)

Only one MRS study has been reported in ARSACS and showed decreased tNAA and increased mIns in the vermis and cerebellar hemispheres of patients versus controls [55].

Dominantly Inherited Ataxias

The majority of MRS studies in patients with dominantly inherited ataxias thus far showed a reduction in tNAA or tNAA/tCr at 1.5T, as is common in neurodegenerative diseases [17, 72]. Evidence that these changes are detectable very early in the disease course came from a report of lower tNAA/tCr and tCho/tCr in the pons of carriers of the SCA1 mutation, including two asymptomatic carriers [73]. While metabolite level/ratio alterations are detectable prior to symptoms, MRS-measured metabolites have also been found to reflect the symptomatic progression in ataxias. Namely, correlations between tNAA and clinical scores were shown in patients with cerebellar degeneration starting with the earliest studies [74]. Correlations were observed between pontine tNAA/tCr and clinical disability in a combined group of patients with OPCA, including SCA1 and SCA2 [54], between pontine tNAA and ataxia scores in SCA1 [26] and between tNAA, tCho, and disease duration in SCA2 [55]. Furthermore, a recent longitudinal pilot study at 3T indicated that the pontine tNAA/mIns ratio is more sensitive to disease progression than the SARA score [75].

Several investigations focused on neurochemical differences between different dominant ataxia subtypes. A pilot study reported lower cerebellar tNAA/tCr and tCho/tCr ratios in SCA2 than SCA6, suggesting a higher degree of neuronal dysfunction or loss in SCA2 [25]. This was consistent with a larger study that reported lower cerebellar tNAA/tCr and tCho/tCr ratios in SCA2 than both SCA3 and SCA6 [76]. A multi-modal MR study reported lower tNAA levels in the pons and cerebellar white matter of patients with SCA1 and SCA2 relative to controls, with no differences between the two disease entities [26]. Therefore distinguishing SCA2 from SCA3 and SCA6 seems feasible by the tNAA levels or ratios; however distinguishing SCA1 and SCA2 has been more challenging at 1.5T.

Studies at high field (3T, 4T) utilizing optimized MRS methodology reported more extensive changes in neurochemical profiles consisting of 10–15 metabolite concentrations [29, 38, 45]. These included altered levels of tNAA, glutamate, glutamine, mIns, and tCr, likely indicating neuronal dysfunction/loss, gliotic activity, alterations in glutamate–glutamine cycling, and deficits in energy metabolism in SCA1, SCA2, SCA3, SCA6, and SCA7 [29, 38, 45]. Interestingly, the same ranking was observed in the severity of neurochemical abnormalities and known levels of synaptic loss between SCA1, SCA2, and SCA6, with the most severe involvement in SCA2, followed by SCA1 and SCA6 (Fig. 9.3), raising the possibility that neurochemical alterations may reflect synaptic density [29]. The higher data quality and the ability to quantify extended neurochemical profiles also allowed the accurate classification of these dominant ataxias, even SCA1 and SCA2 [29], which has been challenging previously. In SCA1, SCA3, and SCA7, strong correlations were detected between select metabolites (tNAA, mIns, tCr, glutamate) and a validated ataxia score [38, 45]. These neurochemicals also allowed the separation of patients from controls without any overlap, suggesting that high-field MRS may be used for clinical decision making on an individual subject basis.

MRS was further used to noninvasively demonstrate the pathological involvement of regions other than the cerebellum and brainstem in dominant ataxias, e.g., the deep cerebral white matter in SCA3 [77] and cerebral cortical and white matter regions in SCA1 [78]. Finally, the cerebellar tNAA/tCr ratio was recently shown to accurately predict the age of disease onset in SCA2 and SCA3 [79].

Sporadic Ataxias

MRS studies in sporadic ataxias primarily tested the diagnostic utility of the technique, which may be very valuable in the absence of genetic testing. For example, distinguishing MSA-C from other forms of sporadic cerebellar degeneration is important as prognoses are quite different between these disease entities.

An early study reported reductions in cerebral and cerebellar tNAA/tCr in patients with MSA and sporadic CCA and suggested that the two groups could be distinguished based on putaminal tNAA/tCr, which was low only in MSA [53]. This study also showed significant correlations between the cerebellar tNAA/tCr and a functional ataxia scale and between frontal cortex tNAA/tCr and mini-mental state examination scores in both groups.

Others attempted to distinguish MSA-C from MSA-P. tNAA/tCr was lower than controls both in the pons and putamen in MSA-P, while only the pontine tNAA/tCr was altered in MSA-C [80]. In addition, tNAA/tCr alterations were present prior to ataxic symptoms and MRI abnormalities in this study.

Due to the similarities in clinical presentation and ages of onset, distinguishing MSA-C from the SCAs using neuroimaging is also expected to be valuable in the clinic. Among the most common SCAs, SCA2 has the most similar cerebellar metabolite ratios (tNAA/tCr and tCho/tCr) to those in MSA-C [76], and metabolite ratios have been largely insufficient to distinguish MSA-C from SCA2 [76, 81]. The similarity between neurochemical abnormalities in MSA-C and SCA2 was also demonstrated at 4T (Fig. 9.3); however the access to extended neurochemical profiles, as well as the better spectral quality, allowed the separation of these groups [29]. Specifically, utilizing selected neurochemical levels from the cerebellum and pons allowed 89 % accurate classification of a cohort consisting of patients with SCA1, SCA2, SCA6, MSA-C, and controls.

Future Perspectives

While the existing literature clearly has demonstrated the sensitivity of MRS to early and progressive neurochemical changes in ataxias, the majority of the work has been limited to metabolite ratios at 1.5T. With the wider availability of high-magnetic-field scanners (3T and above) and advanced MRS methods, studies with high MRS data quality are expected and needed in the ataxia field. The acquisition of neurochemical profiles is also expected to provide ataxia-subtype-specific metabolite profiles and insights into differences in pathology, as was demonstrated in pilot work [29]. Furthermore, future studies at ultra-high magnetic fields (e.g., 7T) are expected to bring higher quantification precision, especially for J-coupled metabolites such as glutamate and glutamine. In addition, increased sensitivity and the resulting spatial resolution at ultra-high fields are expected to enable future studies to better focus on smaller nuclei such as the dentate nucleus, a primary site of pathological involvement in FRDA, with less partial volume effects.

Longitudinal investigations, starting at the earliest disease stages, are critical in this endeavor to determine how early the neurochemical alterations can be detected, as well as the sensitivity of MRS-measured metabolites to disease progression. While this has already been demonstrated in animal model studies [44] and a number of cross-sectional clinical studies have shown correlations between MRS measures and clinical status [26, 45, 53, 54, 73, 74], it will be important to demonstrate the sensitivity of MRS to disease onset and progression in longitudinal studies in ataxias [75].

Finally, based on the promising data in mouse models showing the sensitivity of the technique to reversal of pathology [46, 47], there is great motivation to start using MRS as an outcome measure in preclinical and clinical trials. Standardization of advanced MRS methodology for multi-site investigations is critical in this respect, especially considering the rarity of degenerative ataxias and the difficulty of obtaining data from large patient cohorts in single-site investigations. Recent standardization efforts at high field have been very promising [37, 82] and robust multi-site data in ataxias are expected in the near future.

Acknowledgements The preparation of this chapter was in part supported by the National Institute of Neurological Disorders and Stroke (NINDS) grant R01 NS070815. The Center for MR Research is supported by the National Institute of Biomedical Imaging and Bioengineering (NIBIB) grant P41 EB015894 and the Institutional Center Cores for Advanced Neuroimaging award P30 NS076408.

References

1. Trouillas P, Takayanagi T, Hallett M, Currier RD, Subramony SH, Wessel K, Bryer A, Diener HC, Massaquoi S, Gomez CM, Coutinho P, Ben Hamida M, Campanella G, Filla A, Schut L, Timann D, Honnorat J, Nighoghossian N, Manyam B (1997) International cooperative ataxia rating scale for pharmacological assessment of the cerebellar syndrome. The ataxia neuropharmacology committee of the world federation of neurology. J Neurol Sci 145(2):205–211
2. Schmitz-Hübsch T, du Montcel ST, Baliko L, Berciano J, Boesch S, Depondt C, Giunti P, Globas C, Infante J, Kang JS, Kremer B, Mariotti C, Melegh B, Pandolfo M, Rakowicz M, Ribai P, Rola R, Schöls L, Szymanski S, van de Warrenburg BP, Dürr A, Klockgether T, Fancellu R (2006) Scale for the assessment and rating of ataxia: development of a new clinical scale. Neurology 66(11):1717–1720
3. Subramony SH, May W, Lynch D, Gomez C, Fischbeck K, Hallett M, Taylor P, Wilson R, Ashizawa T (2005) Measuring Friedreich ataxia: Interrater reliability of a neurologic rating scale. Neurology 64(7):1261–1262
4. Lynch DR, Farmer JM, Tsou A, Perlman S, Subramony SH, Gomez CM, Ashizawa T, Wilmot GR, Wilson RB, Balcer LJ (2006) Measuring Friedreich ataxia: complementary features of examination and performance measures. Neurology 66(11):1711–1716
5. Rüb U, Schöls L, Paulson H, Auburger G, Kermer P, Jen JC, Seidel K, Korf HW, Deller T (2013) Clinical features, neurogenetics and neuropathology of the polyglutamine spinocerebellar ataxias type 1, 2, 3, 6 and 7. Prog Neurobiol 104:38–66
6. Koeppen AH, Davis AN, Morral JA (2011) The cerebellar component of Friedreich's ataxia. Acta Neuropathol 122(3):323–330
7. Koeppen AH, Mazurkiewicz JE (2013) Friedreich ataxia: neuropathology revised. J Neuropathol Exp Neurol 72(2):78–90
8. Boder E (1985) Ataxia-telangiectasia: an overview. Kroc Found Ser 19:1–63
9. Criscuolo C, Chessa L, Di Giandomenico S, Mancini P, Sacca F, Grieco GS, Piane M, Barbieri F, De Michele G, Banfi S, Pierelli F, Rizzuto N, Santorelli FM, Gallosti L, Filla A, Casali C (2006) Ataxia with oculomotor apraxia type 2: a clinical, pathologic, and genetic study. Neurology 66(8):1207–1210

10. Bouchard J-P (1991) Recessive ataxia of Charlevoix-Saguenay. Handb Clin Neurol 16:451–459
11. Gouw LG, Digre KB, Harris CP, Haines JH, Ptacek LJ (1994) Autosomal dominant cerebellar ataxia with retinal degeneration: clinical, neuropathologic, and genetic analysis of a large kindred. Neurology 44(8):1441–1447
12. Robitaille Y, Schut L, Kish SJ (1995) Structural and immunocytochemical features of olivopontocerebellar atrophy caused by the spinocerebellar ataxia type 1 (SCA-1) mutation define a unique phenotype. Acta Neuropathol 90(6):572–581
13. Gilman S, Little R, Johanns J, Heumann M, Kluin KJ, Junck L, Koeppe RA, An H (2000) Evolution of sporadic olivopontocerebellar atrophy into multiple system atrophy. Neurology 55(4):527–532
14. Marie P, Foix C, Alajouanine T (1922) De l'atrophie cerebelleuse tardive a predominance corticale. Revue Neurol 38(849-885):1082–1111
15. Duarte JM, Lei H, Mlynárik V, Gruetter R (2012) The neurochemical profile quantified by in vivo ^{1}H NMR spectroscopy. Neuroimage 61(2):342–362. doi:10.1016/j.neuroimage.2011.12.038
16. Emir UE, Auerbach EJ, Moortele PF, Marjańska M, Ugurbil K, Terpstra M, Tkáč I, Öz G (2012) Regional neurochemical profiles in the human brain measured by ^{1}H MRS at 7 T using local B_1 shimming. NMR Biomed 25(1):152–160. doi:10.1002/nbm.1727
17. Öz G (2013) MR spectroscopy in health and disease. In: Manto M, Gruol DL, Schmahmann JD, Koibuchi N, Rossi F (eds) Handbook of the cerebellum and cerebellar disorders, vol 1. Springer, Dordrecht, pp 713–733
18. Zu T, Duvick LA, Kaytor MD, Berlinger MS, Zoghbi HY, Clark HB, Orr HT (2004) Recovery from polyglutamine-induced neurodegeneration in conditional SCA1 transgenic mice. J Neurosci 24(40):8853–8861
19. Xia H, Mao Q, Eliason SL, Harper SQ, Martins IH, Orr HT, Paulson HL, Yang L, Kotin RM, Davidson BL (2004) RNAi suppresses polyglutamine-induced neurodegeneration in a model of spinocerebellar ataxia. Nat Med 10(8):816–820
20. Klockgether T (2011) Update on degenerative ataxias. Curr Opin Neurol 24(4):339–345. doi:10.1097/WCO.0b013e32834875ba
21. Voncken M, Ioannou P, Delatycki MB (2004) Friedreich ataxia-update on pathogenesis and possible therapies. Neurogenetics 5(1):1–8
22. Zoghbi HY, Orr HT (2009) Pathogenic mechanisms of a polyglutamine-mediated neurodegenerative disease, spinocerebellar ataxia type 1. J Biol Chem 284(12):7425–7429
23. Mueller SG, Schuff N, Weiner MW (2006) Evaluation of treatment effects in Alzheimer's and other neurodegenerative diseases by MRI and MRS. NMR Biomed 19(6):655–668
24. Klockgether T, Skalej M, Wedekind D, Luft AR, Welte D, Schulz JB, Abele M, Burk K, Laccone F, Brice A, Dichgans J (1998) Autosomal dominant cerebellar ataxia type I. - MRI-based volumetry of posterior fossa structures and basal ganglia in spinocerebellar ataxia types 1, 2 and 3. Brain 121(Pt 9):1687–1693
25. Boesch SM, Schocke M, Burk K, Hollosi P, Fornai F, Aichner FT, Poewe W, Felber S (2001) Proton magnetic resonance spectroscopic imaging reveals differences in spinocerebellar ataxia types 2 and 6. J Magn Reson Imaging 13(4):553–559
26. Guerrini L, Lolli F, Ginestroni A, Belli G, Nave RD, Tessa C, Foresti S, Cosottini M, Piacentini S, Salvi F, Plasmati R, De Grandis D, Siciliano G, Filla A, Mascalchi M (2004) Brainstem neurodegeneration correlates with clinical dysfunction in SCA1 but not in SCA2. A quantitative volumetric, diffusion and proton spectroscopy MR study. Brain 127 (Pt 8):1785–1795
27. Prakash N, Hageman N, Hua X, Toga AW, Perlman SL, Salamon N (2009) Patterns of fractional anisotropy changes in white matter of cerebellar peduncles distinguish spinocerebellar ataxia-1 from multiple system atrophy and other ataxia syndromes. Neuroimage 47(Suppl 2):T72–T81

28. Schulz JB, Borkert J, Wolf S, Schmitz-Hübsch T, Rakowicz M, Mariotti C, Schöls L, Timmann D, van de Warrenburg B, Dürr A, Pandolfo M, Kang JS, Mandly AG, Nagele T, Grisoli M, Boguslawska R, Bauer P, Klockgether T, Hauser TK (2010) Visualization, quantification and correlation of brain atrophy with clinical symptoms in spinocerebellar ataxia types 1, 3 and 6. Neuroimage 49(1):158–168
29. Öz G, Iltis I, Hutter D, Thomas W, Bushara KO, Gomez CM (2011) Distinct neurochemical profiles of spinocerebellar ataxias 1, 2, 6, and cerebellar multiple system atrophy. Cerebellum 10(2):208–217. doi:10.1007/s12311-010-0213-6
30. Rudnicki DD, Margolis RL (2003) Repeat expansion and autosomal dominant neurodegenerative disorders: consensus and controversy. Expert Rev Mol Med 5(21):1–24
31. Schöls L, Bauer P, Schmidt T, Schulte T, Riess O (2004) Autosomal dominant cerebellar ataxias: clinical features, genetics, and pathogenesis. Lancet Neurol 3(5):291–304
32. Schmitz-Hübsch T, Coudert M, Bauer P, Giunti P, Globas C, Baliko L, Filla A, Mariotti C, Rakowicz M, Charles P, Ribai P, Szymanski S, Infante J, van de Warrenburg BP, Dürr A, Timmann D, Boesch S, Fancellu R, Rola R, Depondt C, Schöls L, Zdienicka E, Kang JS, Döhlinger S, Kremer B, Stephenson DA, Melegh B, Pandolfo M, di Donato S, du Montcel ST, Klockgether T (2008) Spinocerebellar ataxia types 1, 2, 3, and 6: disease severity and nonataxia symptoms. Neurology 71(13):982–989
33. Garwood M, DelaBarre L (2001) The return of the frequency sweep: designing adiabatic pulses for contemporary NMR. J Magn Reson 153(2):155–177
34. Öz G, Tkáč I (2011) Short-echo, single-shot, full-intensity proton magnetic resonance spectroscopy for neurochemical profiling at 4 T: Validation in the cerebellum and brainstem. Magn Reson Med 65(4):901–910. doi:10.1002/mrm.22708
35. Scheenen TW, Klomp DW, Wijnen JP, Heerschap A (2008) Short echo time ^{1}H-MRSI of the human brain at 3T with minimal chemical shift displacement errors using adiabatic refocusing pulses. Magn Reson Med 59(1):1–6
36. Mlynárik V, Gambarota G, Frenkel H, Gruetter R (2006) Localized short-echo-time proton MR spectroscopy with full signal-intensity acquisition. Magn Reson Med 56(5):965–970
37. Deelchand DK, Adanyeguh IM, Emir UE, Nguyen TM, Valabregue R, Henry PG, Mochel F, Öz G (2015) Two-site reproducibility of cerebellar and brainstem neurochemical profiles with short-echo, single voxel MRS at 3 T. Magn Reson Med 73(5):1718–1725. doi:10.1002/mrm.25295
38. Adanyeguh IM, Henry PG, Nguyen TM, Rinaldi D, Jauffret C, Valabregue R, Emir UE, Deelchand DK, Brice A, Eberly LE, Öz G, Durr A, Mochel F (2015) In vivo neurometabolic profiling in patients with spinocerebellar ataxia types 1, 2, 3, and 7. Mov Disord 30 (5):662–670. doi:10.1002/mds.26181
39. Guerrini L, Belli G, Mazzoni L, Foresti S, Ginestroni A, Della Nave R, Diciotti S, Mascalchi M (2009) Impact of cerebrospinal fluid contamination on brain metabolites evaluation with ^{1}H-MR spectroscopy: a single voxel study of the cerebellar vermis in patients with degenerative ataxias. J Magn Reson Imaging 30(1):11–17
40. Ernst T, Kreis R, Ross BD (1993) Absolute quantitation of water and metabolites in the human brain. I Compartments and water. J Magn Reson 102:1–8
41. Hetherington HP, Pan JW, Mason GF, Adams D, Vaughn MJ, Twieg DB, Pohost GM (1996) Quantitative ^{1}H spectroscopic imaging of human brain at 4.1 T using image segmentation. Magn Reson Med 36(1):21–29
42. Deelchand DK, Iltis I, Henry PG (2014) Improved quantification precision of human brain short echo-time ^{1}H magnetic resonance spectroscopy at high magnetic field: a simulation study. Magn Reson Med 72(1):20–25. doi:10.1002/mrm.24892
43. Griffin JL, Cemal CK, Pook MA (2004) Defining a metabolic phenotype in the brain of a transgenic mouse model of spinocerebellar ataxia 3. Physiol Genomics 16(3):334–340
44. Öz G, Nelson CD, Koski DM, Henry PG, Marjanska M, Deelchand DK, Shanley R, Eberly LE, Orr HT, Clark HB (2010) Noninvasive detection of presymptomatic and progressive neurodegeneration in a mouse model of spinocerebellar ataxia type 1. J Neurosci 30 (10):3831–3838

45. Öz G, Hutter D, Tkáč I, Clark HB, Gross MD, Jiang H, Eberly LE, Bushara KO, Gomez CM (2010) Neurochemical alterations in spinocerebellar ataxia type 1 and their correlations with clinical status. Mov Disord 25(9):1253–1261
46. Öz G, Vollmers ML, Nelson CD, Shanley R, Eberly LE, Orr HT, Clark HB (2011) In vivo monitoring of recovery from neurodegeneration in conditional transgenic SCA1 mice. Exp Neurol 232(2):290–298. doi:10.1016/j.expneurol.2011.09.021
47. Öz G, Kittelson E, Demirgöz D, Rainwater O, Eberly LE, Orr HT, Clark HB (2015) Assessing recovery from neurodegeneration in spinocerebellar ataxia 1: Comparison of in vivo magnetic resonance spectroscopy with motor testing, gene expression and histology. Neurobiol Dis 74:158–166. doi:10.1016/j.nbd.2014.11.011
48. Watase K, Weeber EJ, Xu B, Antalffy B, Yuva-Paylor L, Hashimoto K, Kano M, Atkinson R, Sun Y, Armstrong DL, Sweatt JD, Orr HT, Paylor R, Zoghbi HY (2002) A long CAG repeat in the mouse Sca1 locus replicates SCA1 features and reveals the impact of protein solubility on selective neurodegeneration. Neuron 34(6):905–919
49. Emir UE, Brent Clark H, Vollmers ML, Eberly LE, Öz G (2013) Non-invasive detection of neurochemical changes prior to overt pathology in a mouse model of spinocerebellar ataxia type 1. J Neurochem 127(5):660–668. doi:10.1111/jnc.12435
50. Armbrust KR, Wang X, Hathorn T, Cramer SW, Chen G, Zu T, Obu T, Zink AN, Öz G, Ebner TJ, Ranum LPW (2014) Mutant β-III spectrin causes mGluR1α mislocalization and functional deficits in a mouse model of spinocerebellar ataxia type 5. J Neurosci 34(30):9891–9904. doi:10.1523/JNEUROSCI.0876-14.2014
51. Harding AE (1983) Classification of the hereditary ataxias and paraplegias. Lancet 1 (8334):1151–1155. doi:S0140-6736(83)92879-9
52. Davie CA, Barker GJ, Webb S, Tofts PS, Thompson AJ, Harding AE, McDonald WI, Miller DH (1995) Persistent functional deficit in multiple sclerosis and autosomal dominant cerebellar ataxia is associated with axon loss. Brain 118(Pt 6):1583–1592
53. Terakawa H, Abe K, Watanabe Y, Nakamura M, Fujita N, Hirabuki N, Yanagihara T (1999) Proton magnetic resonance spectroscopy (^{1}H MRS) in patients with sporadic cerebellar degeneration. J Neuroimaging 9(2):72–77
54. Mascalchi M, Cosottini M, Lolli F, Salvi F, Tessa C, Macucci M, Tosetti M, Plasmati R, Ferlini A, Tassinari CA, Villari N (2002) Proton MR spectroscopy of the cerebellum and pons in patients with degenerative ataxia. Radiology 223(2):371–378
55. Viau M, Marchand L, Bard C, Boulanger Y (2005) ^{1}H magnetic resonance spectroscopy of autosomal ataxias. Brain Res 1049(2):191–202
56. Franca MC Jr, D'Abreu A, Yasuda CL, Bonadia LC, Santos da Silva M, Nucci A, Lopes-Cendes I, Cendes F (2009) A combined voxel-based morphometry and ^{1}H-MRS study in patients with Friedreich's ataxia. J Neurol 256(7):1114–1120
57. Hadjivassiliou M, Wallis LI, Hoggard N, Grunewald RA, Griffiths PD, Wilkinson ID (2012) MR spectroscopy and atrophy in Gluten, Friedreich's and SCA6 ataxias. Acta Neurol Scand 126(2):138–143. doi:10.1111/j.1600-0404.2011.01620.x
58. Iltis I, Hutter D, Bushara KO, Clark HB, Gross M, Eberly LE, Gomez CM, Öz G (2010) ^{1}H MR spectroscopy in Friedreich's ataxia and ataxia with oculomotor apraxia type 2. Brain Res 1358:200–210. doi:10.1016/j.brainres.2010.08.030
59. Mascalchi M (2013) The cerebellum looks normal in Friedreich ataxia. AJNR Am J Neuroradiol 34(2), E22. doi:10.3174/ajnr.A3480
60. Chevis CF, da Silva CB, D'Abreu A, Lopes-Cendes I, Cendes F, Bergo FP, Franca MC Jr (2013) Spinal cord atrophy correlates with disability in Friedreich's ataxia. Cerebellum 12 (1):43–47. doi:10.1007/s12311-012-0390-6
61. Henry PG, Deelchand DK, Iltis I, Hutter D, Bushara KO, Öz G, Lenglet C MRS and Diffusion MRI of the Spinal Cord in Friedreich's Ataxia. In: Proc Intl Soc Mag Reson Med, Milan, Italy, 2014. p 571
62. Lodi R, Cooper JM, Bradley JL, Manners D, Styles P, Taylor DJ, Schapira AH (1999) Deficit of in vivo mitochondrial ATP production in patients with Friedreich ataxia. Proc Natl Acad Sci U S A 96(20):11492–11495

63. Vorgerd M, Schols L, Hardt C, Ristow M, Epplen JT, Zange J (2000) Mitochondrial impairment of human muscle in Friedreich ataxia in vivo. Neuromuscul Disord 10(6):430–435
64. Schols L, Vorgerd M, Schillings M, Skipka G, Zange J (2001) Idebenone in patients with Friedreich ataxia. Neurosci Lett 306(3):169–172
65. Nachbauer W, Boesch S, Schneider R, Eigentler A, Wanschitz J, Poewe W, Schocke M (2013) Bioenergetics of the calf muscle in Friedreich ataxia patients measured by ^{31}P-MRS before and after treatment with recombinant human erythropoietin. PLoS One 8(7), e69229. doi:10.1371/journal.pone.0069229
66. Lodi R, Rajagopalan B, Blamire AM, Cooper JM, Davies CH, Bradley JL, Styles P, Schapira AH (2001) Cardiac energetics are abnormal in Friedreich ataxia patients in the absence of cardiac dysfunction and hypertrophy: an in vivo ^{31}P magnetic resonance spectroscopy study. Cardiovasc Res 52(1):111–119
67. Bunse M, Bit-Avragim N, Riefflin A, Perrot A, Schmidt O, Kreuz FR, Dietz R, Jung WI, Osterziel KJ (2003) Cardiac energetics correlates to myocardial hypertrophy in Friedreich's ataxia. Ann Neurol 53(1):121–123. doi:10.1002/ana.10419
68. Lodi R, Hart PE, Rajagopalan B, Taylor DJ, Crilley JG, Bradley JL, Blamire AM, Manners D, Styles P, Schapira AH, Cooper JM (2001) Antioxidant treatment improves in vivo cardiac and skeletal muscle bioenergetics in patients with Friedreich's ataxia. Ann Neurol 49(5):590–596
69. Hart PE, Lodi R, Rajagopalan B, Bradley JL, Crilley JG, Turner C, Blamire AM, Manners D, Styles P, Schapira AH, Cooper JM (2005) Antioxidant treatment of patients with Friedreich ataxia: four-year follow-up. Arch Neurol 62(4):621–626
70. Wallis LI, Griffiths PD, Ritchie SJ, Romanowski CA, Darwent G, Wilkinson ID (2007) Proton spectroscopy and imaging at 3T in ataxia-telangiectasia. AJNR Am J Neuroradiol 28(1):79–83
71. Palmeri S, Rufa A, Pucci B, Santarnecchi E, Malandrini A, Stromillo ML, Mandala M, Rosini F, De Stefano N, Federico A (2013) Clinical course of two Italian siblings with ataxia-telangiectasia-like disorder. Cerebellum 12(4):596–599. doi:10.1007/s12311-013-0460-4
72. Viau M, Boulanger Y (2004) Characterization of ataxias with magnetic resonance imaging and spectroscopy. Parkinsonism Relat Disord 10(6):335–351
73. Mascalchi M, Tosetti M, Plasmati R, Bianchi MC, Tessa C, Salvi F, Frontali M, Valzania F, Bartolozzi C, Tassinari CA (1998) Proton magnetic resonance spectroscopy in an Italian family with spinocerebellar ataxia type 1. Ann Neurol 43(2):244–252
74. Tedeschi G, Bertolino A, Massaquoi SG, Campbell G, Patronas NJ, Bonavita S, Barnett AS, Alger JR, Hallett M (1996) Proton magnetic resonance spectroscopic imaging in patients with cerebellar degeneration. Ann Neurol 39(1):71–78
75. Deelchand DK, Emir UE, Hutter D, Gomez CM, Eberly LE, Bushara KO, Öz G High field MRS is more sensitive to progression of neurodegeneration than clinical decline in spinocerebellar ataxia type 1 (SCA1). In: Proc Intl Soc Mag Reson Med, Milan, Italy, 2014. p 64
76. Lirng JF, Wang PS, Chen HC, Soong BW, Guo WY, Wu HM, Chang CY (2012) Differences between spinocerebellar ataxias and multiple system atrophy-cerebellar type on proton magnetic resonance spectroscopy. PLoS One 7(10), e47925. doi:10.1371/journal.pone.0047925
77. D'Abreu A, Franca M Jr, Appenzeller S, Lopes-Cendes I, Cendes F (2009) Axonal dysfunction in the deep white matter in Machado-Joseph disease. J Neuroimaging 19(1):9–12. doi:10.1111/j.1552-6569.2008.00260.x
78. Doss S, Brandt AU, Oberwahrenbrock T, Endres M, Paul F, Rinnenthal JL (2014) Metabolic evidence for cerebral neurodegeneration in spinocerebellar ataxia type 1. Cerebellum 13 (2):199–206. doi:10.1007/s12311-013-0527-2
79. Wang PS, Chen HC, Wu HM, Lirng JF, Wu YT, Soong BW (2012) Association between proton magnetic resonance spectroscopy measurements and CAG repeat number in patients with spinocerebellar ataxias 2, 3, or 6. PLoS One 7(10), e47479. doi:10.1371/journal.pone.0047479

80. Watanabe H, Fukatsu H, Katsuno M, Sugiura M, Hamada K, Okada Y, Hirayama M, Ishigaki T, Sobue G (2004) Multiple regional ^{1}H-MR spectroscopy in multiple system atrophy: NAA/Cr reduction in pontine base as a valuable diagnostic marker. J Neurol Neurosurg Psychiatry 75(1):103–109
81. Boesch SM, Wolf C, Seppi K, Felber S, Wenning GK, Schocke M (2007) Differentiation of SCA2 from MSA-C using proton magnetic resonance spectroscopic imaging. J Magn Reson Imaging 25(3):564–569
82. van de Bank BL, Emir UE, Boer VO, van Asten JJ, Maas MC, Wijnen JP, Kan HE, Öz G, Klomp DW, Scheenen TW (2015) Multi-center reproducibility of neurochemical profiles in the human brain at 7 T. NMR Biomed 28(3):306–316. doi:10.1002/nbm.3252

Chapter 10
Magnetic Resonance Spectroscopy in Prion Diseases

Damien Galanaud

Abstract Prion diseases are uncommon dementias linked to the accumulation in the brain of humans and other mammals of an "infectious" autologous protein. The rapid clinical deterioration of the patients makes them difficult to study with MRS. Their spectroscopic profile includes decreased *N*-acetylaspartate and increased myo-inositol, reflecting the neuronal loss and gliosis characteristics of these pathologies.

Keywords Prion-related protein • Creutzfeldt–Jakob disease • Fatal familial insomnia • Gerstmann–Straussler–Scheinker disease • Scrapie

Introduction

Prion diseases are an uncommon group of neurodegenerative disorders characterized by the presence of an "infectious" endogenous protein. In addition to humans, they affect a wide range of animals, such as cows, elks, mice, sheep, and felids [1, 2].

The pathogenicity of this protein is linked to its misfolding, resulting from an alteration in its sequence (in genetic prion diseases), interaction with an exogenous altered protein (in "mad cow" disease and growth hormone-related Creutzfeldt–Jakob disease), or unknown factors. This misfolding makes the prion-related protein (PrP) highly resistant to proteases, leading to its accumulation and to the development of secondary neurodegenerative processes. While the PrP is not observable with MRS, these secondary changes are easily detectable.

D. Galanaud (✉)
Department of Neuroradiology & Institut du Cerveau et de la Moëlle, Pitié Salpêtrière Hospital, Paris, France
e-mail: Damien.galanaud@psl.aphp.fr

G. Öz (ed.), *Magnetic Resonance Spectroscopy of Degenerative Brain Diseases*, Contemporary Clinical Neuroscience, DOI 10.1007/978-3-319-33555-1_10

Clinical Variants, Symptoms, and Radiological Signs

The classical clinical presentation of Creutzfeldt–Jakob disease, both sporadic (s-CJD) and genetic (g-CJD), is a dementia of rapid onset, with usually just a few weeks between the first symptoms and the fully developed clinical picture. Cerebellar syndrome and myoclonus are strongly suggestive of the disease. The median survival time is a few months. MRI shows in most of the cases areas of increased signal on FLAIR in the cortex and/or basal ganglia (putamen and pallidum). These signal changes are associated with a decrease in apparent diffusion coefficient (ADC) [3, 4]. They are rapidly evolving, with changes in both the pattern of signal and ADC changes, and accelerated brain atrophy [5].

Iatrogenic Creutzfeldt–Jakob disease (i-CJD) has a clinical pattern very similar to s-CJD and occurs 1.3 to 40+ years after contamination with prion-laden somatotropic hormone or dura matter grafts. Bulimia is sometimes also present. Evolution is somewhat slower, with disease duration of about 6 months. Its neuroimaging pattern is similar to s-CJD.

Variant Creutzfeldt–Jakob disease (v-CJD) usually affects younger patients, and usually starts with atypical depression, often associated with diffuse pain in the lower limbs, followed by dementia. Disease duration is between 1 and 2 years. Neuroimaging is a major diagnostic criterion, with a typical pattern of areas of high signal/low ADC in the pulvinar and dorso medial nuclei of the thalamus.

The two less common, genetic, prion diseases are fatal familial insomnia (FFI), a rapidly progressive pattern of sleep disturbance, psychiatric symptoms, ataxia and dementia, and Gerstmann–Straussler–Scheinker (GSS) disease, a slowly evolving ataxia later followed by dementia. Neuroimaging in these two diseases is usually normal or only shows nonspecific findings.

MRS Protocol for Prion Diseases

MRS in prion diseases is performed using single voxel acquisition. Since these patients usually suffer from advanced dementia when they are studied, MRSI acquisitions cannot reasonably be included, except in very special circumstances (the rare slowly evolving forms) [6]. Short echo time sequences should be utilized when studying prion diseases by MRS to reliably assess myo-inositol. The lentiform and caudate nuclei, the thalamus, the cerebellum, the bifrontal or bioccipital cortices are most commonly affected by prion diseases and therefore typically examined by MRS.

MRS should be ideally performed as part of a global neuroimaging protocol, including at least FLAIR, DWI, and 3D T1 acquisitions. This, however, is very often difficult due to the altered neurological status of the patients.

MRS Changes in Prion Diseases and Their Relationship to the Pathology

The abnormal PrP accumulates in the intra and extracellular spaces. This leads first to neuronal suffering, then to neuronal vacuolation of the axonal and dendritic processes (the "spongiform" changes), eventually leading to neuronal death [7]. This neuronal suffering and death leads to a decrease in *N*-acetylaspartate (NAA). Interestingly, despite the rapidity of the clinical deterioration and brain atrophy, free lipids are not present in significant amounts. Lactate is usually also not detectable but has been reported by some authors [8]. Astrogliosis may be present even in the earliest stages of the disease, leading to an increase in myo-inositol (Fig. 10.1). Experimental studies in the mouse [9] and hamster [10] models of prion diseases have shown similar results. No metabolite usually not detected in the brain has ever been reported. Since the evolution of the disease is rather fast, correlation with pathological studies in humans must be taken cautiously. However, the few reports that studied patients in MRI very shortly before their death showed a very good correlation between the pathological findings and MRS results [8, 11].

Temporal Evolution of MRS Changes

Due to the rapid evolution of most cases of prion diseases, repeated scanning of these patients is usually not practical, and reports of temporal changes of MRS in prion diseases are scant. Moreover, since they are almost always performed in patients with atypical, slowly evolving diseases they may not be representative of the "classical" evolution of this pathology. In our prospective study of 40 patients with prion disease, rescanning was not possible at 1 month except for 3 patients, and the resulting spectra were of very poor quality likely due to severe atrophy and motion artifacts (unpublished results). A report by Fujita et al., on two patients with s-CJD rescanned at 1 month showed a trend for a decrease in NAA in the caudate nucleus [8]. Interestingly, an increase in myo-inositol may be present even before the onset of symptoms in presymptomatic carriers of mutations in the PrP gene [12], while hamsters infected with PrP did not exhibit changes in NAA and myo-inositol until the clinical phase of the disease [8]. In a study of the mouse model of scrapie, a decrease in NAA appeared near the symptomatic phase of the disease [9]. This decrease was stable over the course of the disease, with the exception of a trend toward additional reduction in NAA at end-stage disease.

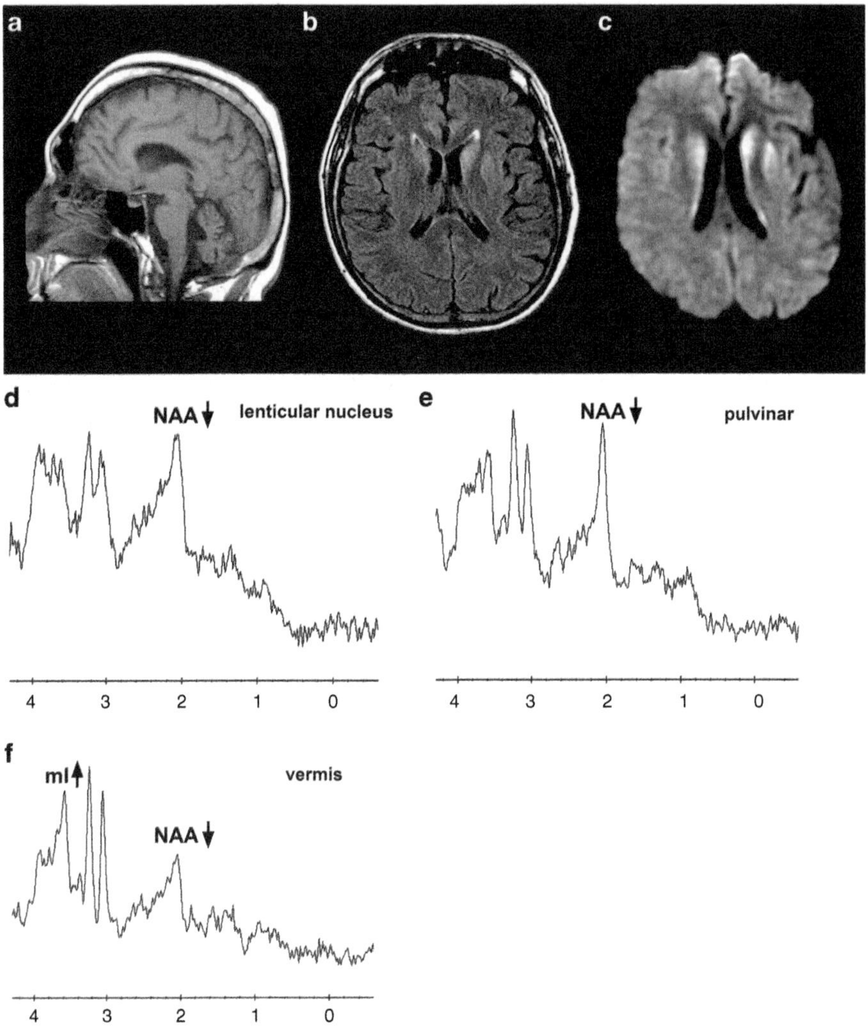

Fig. 10.1 Imaging and spectroscopy (STEAM acquisition, TR = 1500 ms, TE = 20 ms) in a case of sporadic Creutzfeldt–Jakob disease. Sagittal T1 acquisition (**a**) showing marked cerebellar atrophy. Marked hyperintensity is detected in the lentiform and caudate nuclei on FLAIR (**b**) and diffusion (**c**) images. MRS on the lentiform nucleus (**d**) and pulvinar (**e**) shows isolated decrease in NAA, associated with an increase in myo-inositol in the vermis (**f**)

Potential Diagnostic Value of MRS in Prion Diseases

Multivariate analysis of myo-inositol/NAA ratios measured in the putamen, cerebellum, and frontal cortex clearly separate patients with various prion diseases from controls (Fig. 10.2, Galanaud et al., unpublished). In addition, MRS changes can be detected in patients without any signal changes obvious on FLAIR and DWI

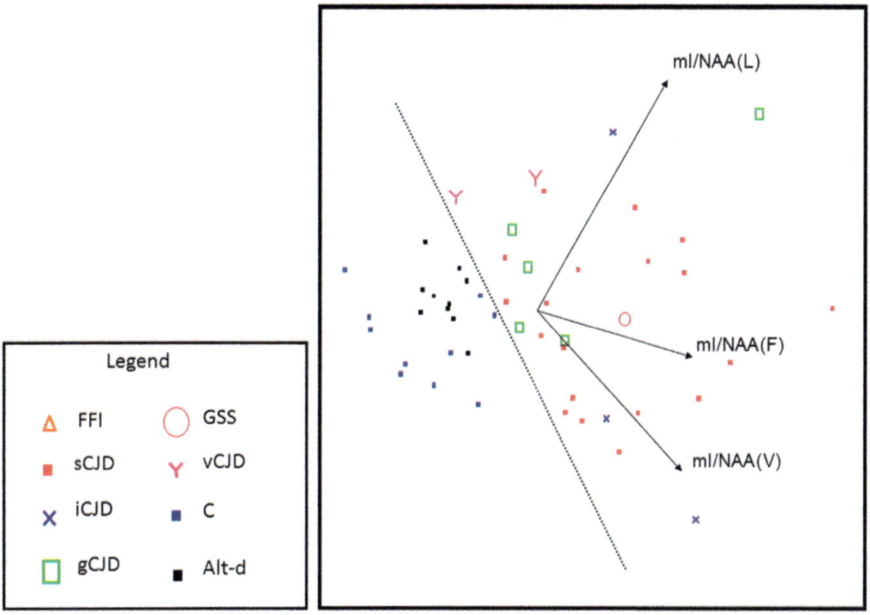

Fig. 10.2 Principal component analysis of the mI/NAA ratio in 3 STEAM single voxel MRS acquisitions (TR = 1500 ms, TE = 20 ms) performed on the vermis (V), frontal cortex (F), and lentiform nucleus (L) in 30 patients with various prion diseases, 10 controls (*blue squares*) and 9 other suspected cases of prion diseases, for which this diagnosis was eventually ruled out (*black boxes*). The hyperplane (*dotted line*) clearly separates the populations, showing the potential interest of MRS for the positive diagnosis of prion diseases

[11, 12]. However, as these authors stressed, these findings are nonspecific, and reduced NAA and increased myo-inositol is a frequent finding in most neurodegenerative diseases. The presymptomatic decline in NAA and increase in myo-inositol has also been reported in Alzheimer's disease [13] and is also present in frontotemporal and vascular dementias [14]. Assessing the precise diagnostic value of MRS in prion diseases would thus require multivoxel MR spectroscopy protocols, including a large number of patients with prion diseases and the most common neurodegenerative diseases. However, the implementation of such a study is very challenging.

In a series of 14 patients with s-CJD, Kim et al. showed that the value of NAA/Cr at diagnosis, measured in the parietooccipital cortices was correlated with disease duration, giving it a somewhat prognostic value [15].

While without clear diagnostic value for individual patients, some form-specific patterns can be described in prion diseases. For example, v-CJD is dominated by an increase in myo-inositol and a decrease in NAA in the pulvinar [16, 17], and i-CJD is characterized by a marked decrease in NAA in the cerebellum [18]. The use of MRS methods that provide quantification of a larger number of neurochemicals is expected to be of value for specificity.

Association of MRS with Findings on Other MR Modalities

Understanding the MRS findings in prion diseases is relatively straightforward, since they correspond well to the histological findings in these pathologies. Spectroscopy has also been suggested as an interesting tool to understand the signal changes on structural MRI and diffusion-weighted imaging observed in these pathologies, which still remain subject to intense debate. In a case of FFI, we have shown that an isolated increase in myo-inositol was associated with an increase in ADC of the corresponding region (the thalamus) [11]. This was attributed to isolated gliosis, and this finding was confirmed by postmortem analysis. It should also be noted that the cerebellum, frequently involved in CJD, does not exhibit any signal changes on both FLAIR and DWI. However, metabolic alterations can often be detected with MRS on this structure [19].

Fulbright et al. reported both decreased ADC and NAA/Cr ratios on the putamen on an analysis of seven patients with the E200K but did not draw any correlation between these findings [6]. In a series of 31 patients with various prion diseases, we did not find any correlation between ADC values and any metabolite ratio in the putamen, cerebellum, frontal cortex, and pulvinar [18]. This suggests that a complex mechanism, combining neuronal suffering/loss, PrP deposits, and astrogliosis is responsible for the ADC changes observed in these pathologies [18].

Conclusion

MRS can help better understand the physiological processes underlying prion diseases. However, its use is hampered by the scarcity of these pathologies and the altered clinical status of the patients.

References

1. Imran M, Mahmood S (2011) An overview of animal prion diseases. Virol J 8:493
2. Imran M, Mahmood S (2011) An overview of human prion diseases. Virol J 8:559
3. Meissner B, Körtner K, Bartl M, Jastrow U, Mollenhauer B, Schröter A, Finkenstaedt M et al (2004) Sporadic Creutzfeldt-Jakob disease: magnetic resonance imaging and clinical findings. Neurology 63:450–456
4. Zerr I, Kallenberg K, Summers DM, Romero C, Taratuto A, Heinemann U, Breithaupt M et al (2009) Updated clinical diagnostic criteria for sporadic Creutzfeldt-Jakob disease. Brain 132:2659–2668
5. Letourneau-Guillon L, Wada R, Kucharczyk W (2012) Imaging of prion diseases. J Magn Reson Imaging 35:998–1012
6. Fulbright RK, Kingsley PB, Guo X, Hoffmann C, Kahana E, Chapman JC, Prohovnik I (2006) The imaging appearance of Creutzfeldt-Jakob disease caused by the E200K mutation. Magn Reson Imaging 24:1121–1129

7. Hauw JJ, Sazdovitch V, Laplanche JL, Peoc'h K, Kopp N, Kemeny J, Privat N et al (2000) Neuropathologic variants of sporadic Creutzfeldt-Jakob disease and codon 129 of PrP gene. Neurology 54:1641–1646
8. Fujita K, Harada M, Yuasa T, Sasaki M, Izumi Y, Kaji R (2011) Temporal evolution of sporadic Creutzfeldt–Jakob disease monitored by 3-Tesla MR spectroscopy. J Neurol 258:1368–1370
9. Bell JD, Cox IJ, Williams SC, Belton PS, McConnell I, Hope J (1991) In vivo detection of metabolic changes in a mouse model of scrapie using nuclear magnetic resonance spectroscopy. J Gen Virol 72(Pt 10):2419–2423
10. Behar KL, Boucher R, Fritch W, Manuelidis L (1998) Changes in N-acetylaspartate and myo-inositol detected in the cerebral cortex of hamsters with Creutzfeldt-Jakob disease. Magn Reson Imaging 16:963–968
11. Haïk S, Galanaud D, Linguraru MG, Peoc'h K, Privat N, Faucheux BA, Ayache N et al (2008) In vivo detection of thalamic gliosis: a pathoradiologic demonstration in familial fatal insomnia. Arch Neurol 65:545–549
12. Waldman AD, Cordery RJ, MacManus DG, Godbolt A, Collinge J, Rossor MN (2006) Regional brain metabolite abnormalities in inherited prion disease and asymptomatic gene carriers demonstrated in vivo by quantitative proton magnetic resonance spectroscopy. Neuroradiology 48:428–433
13. Miller BL, Moats RA, Shonk T, Ernst T, Woolley S, Ross BD (1993) Alzheimer disease: depiction of increased cerebral myo-inositol with proton MR spectroscopy. Radiology 187:433–437
14. Akhvlediani T, Henning A, Sándor PS, Boesiger P, Jung HH (2010) Adaptive metabolic changes in CADASIL white matter. J Neurol 257:171–177
15. Kim JH, Choi BS, Jung C, Chang Y, Kim S (2011) Diffusion-weighted imaging and magnetic resonance spectroscopy of sporadic Creutzfeldt-Jakob disease: correlation with clinical course. Neuroradiology 53:939–945
16. Cordery RJ, MacManus D, Godbolt A, Rossor MN, Waldman AD (2006) Short TE quantitative proton magnetic resonance spectroscopy in variant Creutzfeldt-Jakob disease. Eur Radiol 16:1692–1698
17. Galanaud D, Dormont D, Grabli D, Charles P, Hauw JJ, Lubetzki C, Brandel JP et al (2002) MR spectroscopic pulvinar sign in a case of variant Creutzfeldt-Jakob disease. J Neuroradiol 29:285–287
18. Galanaud D, Haik S, Linguraru MG, Ranjeva JP, Faucheux B, Kaphan E, Ayache N et al (2010) Combined diffusion imaging and MR spectroscopy in the diagnosis of human prion diseases. Am J Neuroradiol 31:1311–1318
19. Oppenheim C, Zuber M, Galanaud D, Detilleux M, Bolgert F, Mas JL, Chiras J et al (2004) Spectroscopy and serial diffusion MR findings in hGH-Creutzfeldt-Jakob disease. J Neurol Neurosurg Psychiatry 75:1066–1069

Chapter 11
Magnetic Resonance Spectroscopy in HIV-Associated Neurocognitive Disorders: HAND

Eva-Maria Ratai

Abstract A significant number of HIV-infected patients develop neurological symptoms ranging from minor cognitive impairment to severe dementia (known HIV-associated dementia, HAD). Without combination antiretroviral therapy (cART) HAD occurs in 20–40 % of HIV positive subjects, with the advent of cART the incident has decreased to 5–10 %, although milder forms of cognitive deficits may occur in 30–50 % of those infected with HIV. It is believed that HIV enters the CNS during the early stages of infection by infected immune cells which initiate an inflammatory cascade which results in neuronal injury and loss. Imaging has been widely used to evaluate the effect of HIV to the brain. Specifically, proton magnetic resonance spectroscopy (^{1}H MRS) is one of the most informative methods employed in patients suffering from HIV-associated neurocognitive disorders (HAND). MRS is able to noninvasively measure metabolic changes pertaining to neuronal injury and inflammation, thus, it can assist in the diagnosis of the disease and measure the severity of injury. In animal models of neuroAIDS, MRS has been proven extremely powerful to assess disease progression and response to treatment. Here we review the literature of preclinical models as well as MRS studies of HIV+ adults and children before and after the advent of cART regiments. In addition, we discuss technical considerations related to the disease and finally talk about future direction in HAND using MRS.

Keywords AIDS (acquired immune deficiency syndrome) • HIV (Human immunodeficiency Virus) • NeuroAIDS • HAND (HIV-associated neurocognitive disorders) • SIV (Simian immunodeficiency virus) • Rhesus Macaque • Brain • MRS (magnetic resonance spectroscopy) • MRSI (magnetic resonance spectroscopic imaging) • cART (combination antiretroviral therapy) • Minocycline • Neuroinflammation

---

E.-M. Ratai, Ph.D. (✉)
Department of Radiology, Neuroradiology Division, Massachusetts General Hospital, Harvard Medical School and Athinuola A. Martinos Center for Biomedical Imaging, 149 13th Street, Room 2301, Boston, MA 02129, USA
e-mail: ratai@nmr.mgh.harvard.edu

G. Öz (ed.), *Magnetic Resonance Spectroscopy of Degenerative Brain Diseases*, Contemporary Clinical Neuroscience, DOI 10.1007/978-3-319-33555-1_11

Background on Disease Pathology

Human immunodeficiency virus (HIV) affects more than 1 million people in the United States of America and ~34 million worldwide. Approximately 1.7 million people have died from AIDS and 2.5 million people were newly infected with the virus in 2011 [1]. Figure 11.1 showed the natural disease progression of HIV/AIDS: During the early phase of infection, the virus replicates extensively, reaching levels of 10^6–10^8 copies of HIV RNA/mL in blood plasma due to the absence of discernible immune responses [2]. Once seroconversion occurs (as determined by ELISA/Western Blot), it is thought that the host's immune response begins to control viral replication, causing the high levels of virus to be quickly reduced within weeks [2]. During acute HIV-1 infection, there is frequently a marked decrease in the $CD4^+$ T cell count [3, 4], followed by an increase toward the end of the acute stage, which typically does not return to preinfection levels in the absence of antiretroviral therapy (ART). Throughout the latent period, there is usually a steady decline in the numbers of $CD4^+$ lymphocytes. A diagnosis of AIDS is made if the $CD4^+$ T-cell count falls below 200 per μL. At these later stages of the disease, viral loads are starting to increase once more, leading to symptoms, opportunistic infections, and death.

Early in the AIDS epidemic, HIV-associated neurocognitive disorders (HAND) were recognized as an important clinical manifestation of the disease [5–9]. Prior to

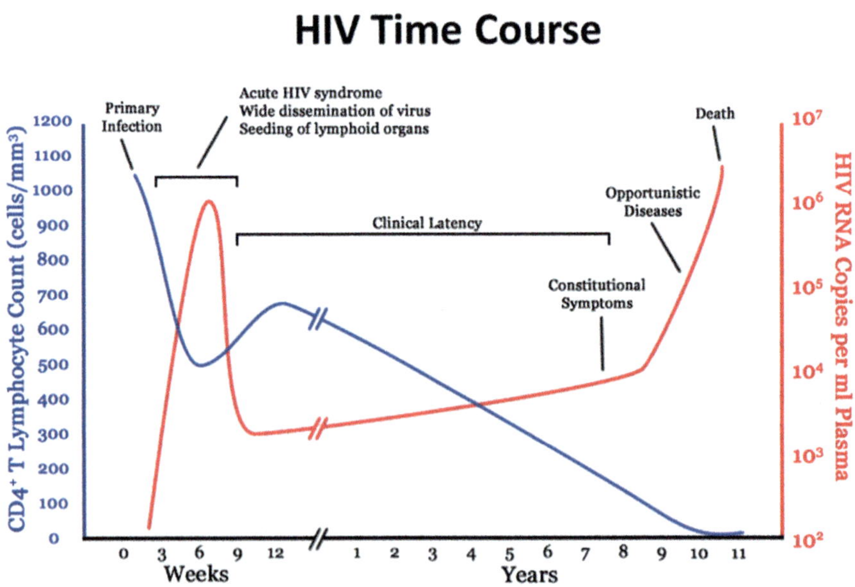

Fig. 11.1 A generalized graph of the relationship between HIV copies (viral load) and $CD4^+$ T cell counts over the average course of untreated HIV infection. This is a file from the Wikimedia Commons. Commons is a freely licensed media file repository. Adapted from Fig. 1 in Pantaleo, G et al. (February 1993). "New concepts in the immunopathogenesis of human immunodeficiency virus infection". *New England Journal of Medicine* 328 (5): 327–335. PMID 8093551

the advent of combination antiretroviral therapy (cART), 20–40 % of HIV-infected patients developed neurological symptoms ranging from minor cognitive impairment to severe dementia (known as HIV-associated dementia, HAD or AIDS dementia complex, ADC). ADC is characterized clinically by severe cognitive, motor, and behavioral abnormalities in the absence of opportunistic infections. Physical symptoms include headaches, generalized seizures, and ataxia. Cognitive symptoms comprise forgetfulness, slowing of thought processes, global dementia, confusion, and disorientation. Abnormalities in motor functions are characterized by unsteady gait, clumsiness, tremor, limb weakness, loss of coordination, and fine motor control. Finally, behavioral changes such as social withdrawal, lethargy, personality change, and hallucinations have been perceived in patients with ADC [7, 10].

Although cART has reduced AIDS-related mortality, HAND continues to be a major problem in patients with HIV. Antiretroviral drugs suppress virus burden in the plasma, CSF, and lymphoid tissue of HIV+ individuals; however, due to their restricted CNS penetration, these drugs exhibit only limited efficacy in the treatment of HAND. While the incidence of more severe neurological symptoms of ADC has been seen to decrease with cART, less severe versions of the disease persist among the infected population [11, 12]. The overall prevalence of HAND and associated morbidity remains high at approximately ~50 % [13–15]. It is projected that the prevalence of HAND is again increasing as life expectancy among patients increases. As the virus gains resistance to ART, the incidence of severe neurodegeneration is predicted to increase [13, 16, 17] and remains a significant independent risk factor for death [18, 19].

There is a consensus that HIV enters the central nervous system (CNS) during the early stages of infection primarily through virally infected/activated monocytes from the blood across the blood–brain barrier (BBB) by a so-called Trojan horse mechanism [20, 21]. Once in the brain, infected monocytes/macrophages and microglia release viral envelope proteins and cytokines, which in turn activate uninfected macrophages and microglia [22–29]. These activated macrophages and microglia release several neurotoxic substances, such as free radicals and glutamate, which lead to neuronal injury and apoptosis [30–34]. In summary, the virus does not directly infect neurons, yet neurons suffer injury due to indirect mechanisms mediated by host proinflammatory and viral proteins [26, 31, 35–37].

Post mortem neuropathological studies have demonstrated that the virus variably affects different regions of the brain. Pre-ART, abnormalities were predominantly found in the white matter and in subcortical structures, with relative sparing of the cortex [5]. Furthermore, higher levels of virus were found in the basal ganglia and hippocampus compared to cerebellar cortex and midfrontal cortical gray matter [38]. However, it appears that the anatomic distribution and temporal progression of neuropathologic changes differ across individuals, thus, it is important to consider both cortical and subcortical brain regions in studies of neuropathogenesis and treatment of HIV-related brain disease [39].

Potential Clinical Utility of MRS in HAND

Imaging has been used widely to evaluate the effect of HIV infection on the brain. Structural neuroimaging methods such as computerized tomography (CT) and magnetic resonance imaging (MRI) are important for the diagnosis of cerebral opportunistic infections such as Progressive Multifocal Leukoencephalopathy (PML), Toxoplasmosis, Neurosyphilis, and Cytomegalovirus (CMV) Encephalitis. Morphological alterations detected by MRI include cortical atrophy at later stages of HIV-associated dementia (HAD). Furthermore, hyperintensities in the white matter (WM) and basal ganglia (BG) were observed on T2-weighted images of HIV+ patients. Diffusion tensor imaging (DTI) has been used to study the white matter structural integrity as it is typically more sensitive compared to T2-weighted imaging [40]. More recently, also quantitative morphological MRI has been used for assessing the severity of HAND [41–43]. However, structural imaging techniques were found to be limited in the evaluation of early cognitive changes [44–47]. Thus, neuroimaging techniques that provide functional and biochemical information have proven to be more useful in the evaluation of HAND [48].

SPECT and PET are sensitive to changes related to cerebral perfusion and metabolism. Specifically, PET has been used in patients with HAND and in animal models of neuroAIDS to study glucose metabolism, neuroinflammation, microglial activation, fibrillar amyloid, and dopamine transporters [49–53]. However, nuclear medicine techniques expose the subject to radiation and have therefore limited use for monitoring diseases progression or treatment effects when repeat measurements are needed [54].

Magnetic resonance spectroscopy (MRS) offers the unique ability to measure a wide range of cerebral metabolite levels in a noninvasive manner. Specifically, proton MRS (^{1}H MRS) is one of the most informative methods employed in neuroAIDS research. MRS has been used in HIV-infected patients to understand the cognitive deficits and to establish biomarkers in HAND. To date, more than 80 original studies have been employed to evaluate the effects of HIV infection using MRS. The resonances seen in the brain by ^{1}H MRS are typically low weight molecules related to neuronal injury/loss and inflammation. Specifically, *N*-Acetylaspartate (NAA), an established marker for neuronal density and viability is decreased in patients with advanced neurocognitive symptoms and is of particular value for the in vivo assessment of neuronal integrity [55–58]. The earliest MRS studies of neuroAIDS, published in the early 1990s, reported decreases in NAA/Creatine (NAA/Cr) in patients with advanced neurocognitive symptoms and the decrease in NAA/Cr was found to be associated with disease severity [59–61].

Choline-containing compounds are related to cell/lipid membrane metabolism, thus an increase in Cho or Cho/Cr in HIV-infected patients is possibly associated with an immune response that includes cerebral inflammation or gliosis [61–63]. *Myo*-inositol (mI) is an organic osmolyte which is primarily located in glial cells, thus an increase in mI is associated with gliosis or inflammation [64]. Since HAND is considered a neuroinflammatory disease mI is especially important in the evaluation of glial response. In fact, MRS studies have found an elevation in mI/Cr as a result of HIV-associated brain injury [65–67].

The total creatine (tCr) resonance consists of the sum of creatine (Cr) and phosphocreatine (PCr), a high energy reservoir for the generation of ATP. Since Cr and PCr are in equilibrium the tCr peak is commonly assumed to remain stable in size despite bioenergetic abnormalities that occur with many pathologies or with age [68]. Consequently, the tCr resonance is often used as an internal standard and is commonly referred to as simply creatine (Cr). However, elevated Cr levels have been reported in the white matter in ART naïve HIV+ ADC stage 3 patients [69] but also in the WM in simian immunodeficiency virus (SIV) infected rhesus macaques during acute infection [70] and in an accelerated macaque model of neuroAIDS a few weeks after infection [71]. In the context of HAND, an elevation in Cr may reflect enhanced high-energy phosphate turnover in activated astrocytes and microglia [71].

Other metabolites of interest include glutamate (Glu), glutamine (Gln), and Glx, which is the sum of Gln and Glu. Glu is an amino acid acting as excitatory neurotransmitter and Gln is its precursor. Even though activated microglia release neurotoxic substances, such as free radicals and glutamate, which result in excitotoxicity and finally to apoptosis [30–34], decreases in Glx levels have been reported during early and chronic HIV infection [72–76]. The observed decrease in Glx may indicate neuronal dysfunction. Furthermore, levels of lactate (Lac), a marker of anaerobic glycolysis and inflammation and lipids, indicators of cell membrane turnover, have been assessed in HIV+ patients. The Lac/Cr ratio in the lenticular nuclei was significantly greater in HIV+ patients with moderate to severe impairment compared to seronegative controls. Moreover, the (Lipids + Lac)/Cr ratio was significantly elevated in both mild and moderate to severe HIV+ patients compared to seronegative controls [77].

Consequently, quantification of these abovementioned metabolite levels can assist in the diagnosis of the disease and measuring the severity of injury. Especially, in animal models where longitudinal scans are common, MRS has been proven extremely powerful to assess disease progression and response to treatment.

Technical Considerations for HAND

Echo Time

As with MRI, the choice of echo time (TE) can have an enormous effect on the appearance of the information obtained in a ^{1}H MRS study. MR spectra obtained with shorter echo times (~30 ms) allow the detection of more metabolites including glutamate, glutamine, and *myo*-inositol. However, the baseline is typically more distorted due to increased lipids and macromolecular background signals. Spectra obtained with longer echo times (135–144 ms) show reduced number of

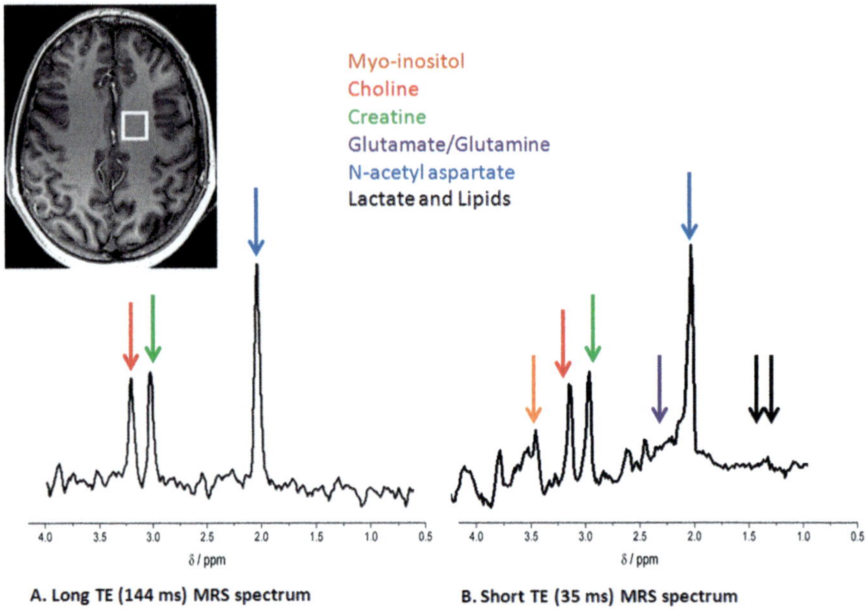

Fig. 11.2 MRS of white matter in a normal brain. (**a**) Long TE spectra have less baseline distortion and are easy to process and analyze but show fewer metabolites than short TE spectra. (**b**) Short TE demonstrates peaks attributable to more metabolites, including lipids and macromolecules, glutamine and glutamate, and *myo*-inositol

metabolites. However, spectra are easier to process and analyze due to the relatively flat baseline. In addition, the lactate (1.3 ppm) doublet is inverted, therefore allowing better differentiation between Lac and lipids. On the other hand, longer TE attenuates the signal-to-noise ratio (SNR) and some metabolites, e.g., mI, Glu, and Gln, which can only be detected at short echo times, would be missed, thus short TE is typically preferred (Fig. 11.2).

Single Voxel vs. MRSI

To measure MR spectra in vivo one has to be able to define the spatial origin of the detected signal. Basically, two methods exist to obtain the spatially localized metabolic information in vivo:

1. Single voxel spectroscopy (SVS) uses selective excitation pulses to localize a voxel of typically 3–8 cm^3. SVS has the advantage of higher SNR and typically shorter acquisition times. However, the accumulative scan time permits acquisition of only a few locations (Fig. 11.3).
2. Magnetic resonance spectroscopic imaging (MRSI) can be obtained in two or three dimensions. MRSI allows one to collect the spectral information from a

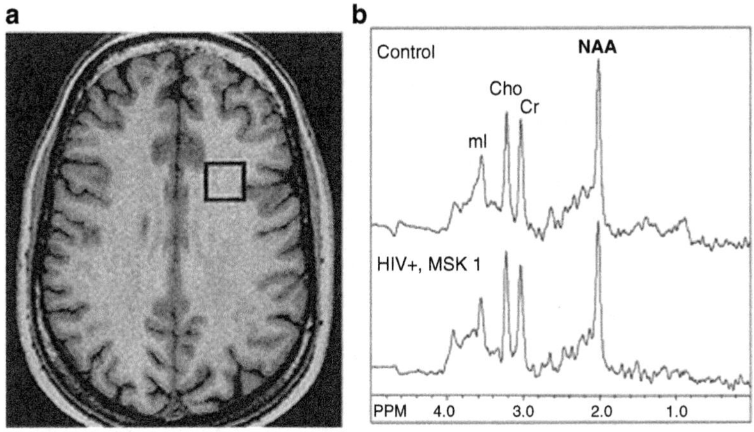

Fig. 11.3 Single-voxel proton magnetic resonance spectroscopy (SV-MRS). (**a**) Voxel locations the left frontal white matter region selected for quantitative spectroscopic analysis using SV MRS. (**b**) Sample spectra obtained in the frontal white matter in an HIV seronegative control and an HIV seropositive individual with mild dementia using SV-MRS. The NAA peak is decreased in the HIV dementia patient. *Reprinted with permission from* [80]

volume consisting of many voxels with individual voxel sizes of typically 0.5–3 cm^3 and most importantly makes it possible to cover large brain areas although spectra are typically characterized by lower SNR compared to SVS (Fig. 11.4).

MRSI is preferred for clinical studies where it is indicated to obtain metabolic information of a large and heterogeneous lesions (e.g., in tumors), and when additional simultaneous spectral information from control regions needs to be obtained. However, HIV is considered a predominantly "global disease" and therefore, most studies have acquired SVS from 1 to 3 brain regions covering frontal white matter (FWM), basal ganglia (BG), and parietal cortex (PC) consistent with recommendations by the HIV MRS Consortium [78, 79]. In a study using both techniques, Sacktor et al. demonstrated that short TE SVS and long TE MRSI offer complementary roles in evaluating individuals with HIV dementia [80] (Fig. 11.4).

Field Strengths

MR spectroscopy inherently suffers from low SNR resulting in voxels with low spatial resolution or long acquisition times. The SNR can be improved by using higher magnetic field strengths. Physics predicts a linear increase in signal with field strength if T1 and T2 relaxation times, coil and system losses, and radiofrequency (RF) penetration effects do not change significantly. In addition,

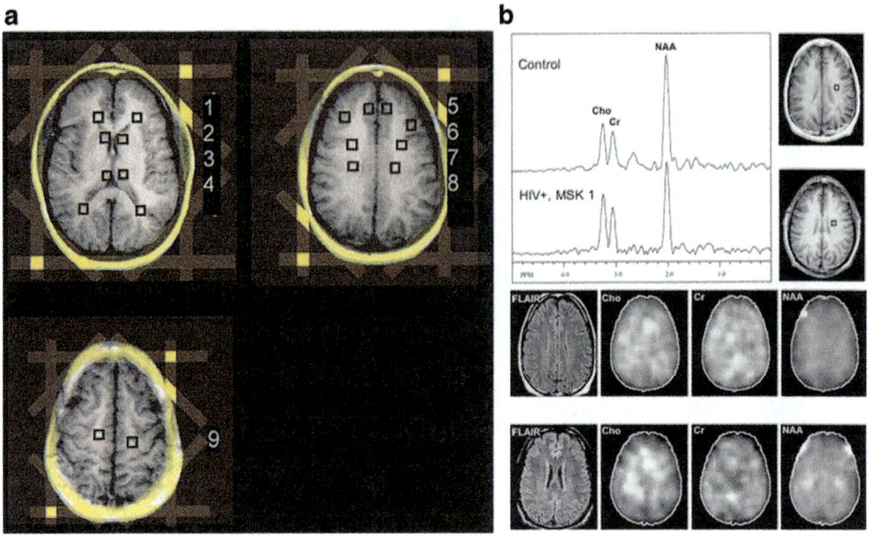

Fig. 11.4 Magnetic resonance spectroscopic imaging (MRSI). (**a**) The MRSI locations include bilateral regions examining the (*1*) forceps minor, (*2*) caudate nucleus, (*3*) thalamus, (*4*) forceps major, (*5*) mesial frontal gray matter, (*6*) dorso-lateral gray matter, (*7*) frontal white matter, (*8*) centrum semiovale, and (*9*) subcortical white matter. The location of octagonal outer-volume saturation bands for lipid suppression is indicated on localizer images in *yellow*. (**b**) Sample spectra, T1-weighted and FLAIR MR images, and spectroscopic images for Cho, Cr, and NAA obtained in the frontal white matter in an HIV seronegative control and an HIV seropositive individual with mild dementia using MRSI. The NAA peak is decreased in the HIV dementia patient. FLAIR MRI scans show no abnormalities; no obvious spatial variations in the metabolic images are apparent between the HIV seronegative control and an HIV seropositive individual. Focal hyperintensity at the edges of the brain in the NAA images is due to lipid contamination from the peri-cranial tissues. *Reprinted with permission from* [80]

the increased chemical shift range at 3 and 7 T relative to 1.5 T results in greater separation of the resonance peaks, and consequently, allows for better quantification of those metabolites that generally overlap with others such as Glu, Gln, and mI. Thus, MRS is preferred at 3 T over 1.5 T. Figure 11.5 shows the spectra from the parietal cortex of a healthy rhesus macaque obtained at three different field strengths, 1.5, 3, and 7 T. The SNR increases at higher fields. In addition, improvements in spectral resolution are evident.

Higher field strength may also result in better reproducibility. For example, in a study at Massachusetts General Hospital, four rhesus macaques underwent repeated (>4 times) SVS scans before infection to access reproducibility of the technique. Data at 1.5 T revealed coefficients of variance (CVs) within the same animals of 7–11 % for NAA/Cr, 13–16 % for Cho/Cr, and 8–12 % for mI/Cr in various brain regions including the FC, WM, and BG [81]. Four additional macaques were scanned a few years later at 3 T and revealed CVs of only 4 % for NAA/Cr, 2–8 % for Cho/Cr, and 3–7 % for mI/Cr in the FC, WM, and BG, respectively.

Fig. 11.5 Comparison between MR spectra of rhesus macaques obtained at three different field strengths. (**a**) 1.5 T, (**b**) 3 T, (**c**) 7 T

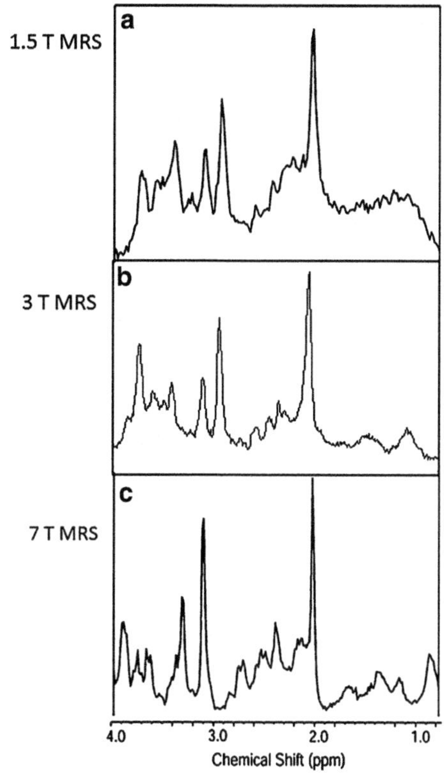

Ratios vs. Absolute Concentration

There are long-standing discussions on the best way to quantify MRS signals, specifically relative versus absolute quantification. The easiest method is to report metabolite ratios such as NAA/Cr. However, using Cr as an internal standard is often based on the assumption that the Cr concentration does not change during the disease process, which is sometimes, but not always true. Ratios are relatively easy to determine and are more reproducible than absolute concentrations. However, when a change is detected, it may not be possible to determine whether it is due to a change in the numerator, denominator, or both.

Changes in creatine have been reported in patients and an animal model of HAND [69–71]; increasing creatine levels may mask increases in Cho and mI. Absolute quantification of brain metabolites by MRS is more difficult to obtain. They are generally expressed in units of mmol/kg. Methods used for absolute quantification include: (a) phantom replacement techniques [82–84], (b) using the unsuppressed water signal as a reference [85–87], or (c) the use of an external reference [88, 89]. Most studies on HIV-infected patients that reported metabolite concentrations have used the water signal as reference [85].

Overview of MRS in Animal Models of HAND

In the previous chapter, we have established that in vivo MRS is a powerful tool in studying the effect of HIV infection on the brain. Due to its noninvasive nature, in vivo MRS is suitable for both cross-sectional and longitudinal studies. However, studies of HIV-infected individuals present difficulties in interpretation: the time of infection is often unknown; a number of individuals are infected with more than one strain of HIV; alcohol and substance abuse result in comorbidities. Animal models allow for greater flexibility to explore and address the questions concerning HIV neuropathogenesis in a controlled manner. Furthermore, animal models allow access to subsequent *post mortem* neuropathology studies helping define relationships between in vivo biomarkers of neuronal function and inflammation by MRS and *postmortem* histopathology.

Crucial insights into neuroAIDS have emerged from several excellent animal models, including a transgenic HIV mouse model that expresses viral surface protein gp120 in the brain [90]; injection of HIV-1-infected macrophages into the brains of severe combined immunodeficient (SCID) mice [91], which results in the development of pathologic hallmarks for HIV encephalitis (HIVE) and neuroinflammation [92–94]; a feline immunodeficiency virus (FIV) infection of cats [95–102]; and the SIV macaque models [103–107].

MRS Studies of Rodent Models

Rodent models have been applied in elucidating neuropathogenic pathways associated with HIV infection, such as the migration and distribution of monocytes in brain tissue and the neurotoxic effects of specific viral proteins [108–110]. In a SCID mouse model, HIVE was induced focally into the striatum by the unilateral injection of HIV-1ADA-infected human monocyte-derived macrophages. Histology revealed focal giant cell encephalitis, with reactive astrocytes, microgliosis, and neuronal dropout. Metabolite concentrations using in vivo MRSI at 7 T were measured 7 days after injection in infected mice compared to sham-operated and unmanipulated mice. These studies showed significant decreases in NAA in both the ipsilateral and contralateral site of injection in HIVE mice compared to sham-operated mice indicating that a highly focal encephalitis can produce global deficits for neuronal function and metabolism [111].

MRS Studies of FIV Models

Like HIV, FIV is known to infect $CD4^+$ T lymphocytes causing immunosuppression, encephalopathy, and neurobehavioral deficits [112, 113]. The neurological abnormalities are typically more profound when FIV is induced during the developmental period of brain maturation, making this an ideal model for understanding neurologic complications observed in children infected with HIV [102, 114]. Cats infected with Maryland isolate (FIV-MD) have demonstrated reductions in NAA and NAA/Cho between 8 and 14 months of infection. FIV-infected cats also exhibited a higher proportion of quantitative electroencephalographic relative slow wave activity that correlated to lower NAA in the frontal cortex [102]. Furthermore, cats infected with high viral titers of V1CSF (a neurovirulent FIV strain which was derived from the CSF of a cat with encephalopathy) had increased Cho/Cr and reduced NAA/Cr in the frontal cortex compared to those that received low titers indicating that infectious titers in the brain during the early stages of infection determine the severity of neurovirulence [115]. Like many HIV studies, FIV infection also resulted in reduction in Glu and Glu/Cr in the cortex and white matter [115].

MRS Studies of SIV Macaque Models

The SIV macaque models are arguably the most comparable and informative of CNS disease progression [107]. The SIV-infected rhesus macaque shares very similar pathology with HIV-infected human patients, including the development of AIDS, disease of the CNS, and cognitive or behavioral deficits [103–107]. SIV is the closest known phylogenetic relative of HIV and also infects $CD4^+$ macrophages, lymphocytes, and microglia. Its acute infection parallels that of HIV, with a very rapid change in the blood viral load observed during the first month of infection, usually around 11–12 days after infection [116–118]. Neuroinvasion occurs early in infection for both HIV and SIV [119], and their neuropathology includes gliosis, perivascular cuffing, and neuronal injury [104, 106, 120, 121]. Furthermore, antiretroviral therapy has been successfully applied to SIV-infected macaques [122, 123].

Traditional SIV Model

Acute/Early SIV Infection

Acute HIV infection is the time period from viral infection through seroconversion. It is characterized by high peak in viremia and impairment/destruction of $CD4^+$ T lymphocytes [2, 124, 125] similar to events that are manifest in chronic infection when HIV-related dementia is more likely to occur [126]. In the SIV macaque model peak viremia is typically observed within 2 weeks after inoculation with SIVmac251 (Fig. 11.6a). We found that plasma SIV RNA peaked at 11–12 days postinfection (dpi) at $\sim 10^8$ copies/mL. Within days after the peak, the viral load reverted to lower levels $\sim 10^6$ copies/mL due to the immunologic response by the macaques. In these studies nearly all macaques had transient reductions in NAA/Cr

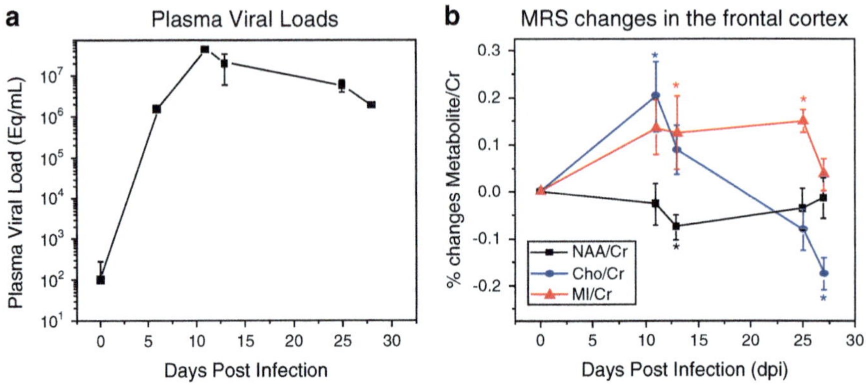

Fig. 11.6 Acute changes in the SIV macaque model of NeuroAIDS. Based on [118] and [70]. (**a**) Plasma viral loads: peak viremia can be detected at 11 days postinfection (dpi). (**b**) Metabolic changes in the frontal cortex: acutely SIV-infected macaques showed a transient decrease in NAA/Cr at 13 dpi; elevated mI/Cr at 11, 13, and 25 dpi; and a dynamic response in Cho/Cr with increases at 11 dpi and decreases below baseline at 27 dpi

by single voxel MRS within the first 2 weeks of infection in the frontal cortex (Fig. 11.6b) but not in the basal ganglia or white matter. Pathology confirmed neuronal dysfunction detected by significant decreases in synaptophysin, a marker for synaptic integrity, while the numbers of neurons were not affected by the virus suggesting that during early infection reversible neuronal injury occurs, but not neuronal loss [127].

Myo-inositol levels were significantly elevated in both the frontal cortex and white matter regions, suggesting stimulation of glial metabolism within the first month of infection [70]. In addition, dynamic changes in choline were observed in the same regions. At the time of peak viremia, both choline and Cho/Cr were significantly elevated; however, between 2 and 4 weeks postinfection (wpi), Cho/Cr ratios underwent a large reduction, resulting in levels below preinfection [118] (Fig. 11.6b). An increase in Cho reflects altered membrane metabolism [128] and in the context of neuroAIDS may be reflective of cerebral inflammation, cell proliferation, recruitment, microgliosis, and/or astrocytosis [61, 63]. Evidence of the association with astrocytosis is given by our observation that during the acute phase changes in Cho/Cr mirror the changes in the astrocyte response by glial fibrillary acidic protein (GFAP) [129]. Decreases in Cho below baseline have rarely been reported in diseases not involving necrosis and still warrant further investigations.

Furthermore, the SIV Model is characterized by elevations in creatine concentrations in the white matter during the acute phase of infection. During that stage, the virus enters the brain through infected monocytes [31], resulting in glial-cell activation and proliferation. These processes may manifest themselves in high tissue metabolism, which would explain an increase in Cr. Energy change as a function of enhanced glial activation is supported by the finding that the greatest increases in mI and Cho are in the same regions as the highest increase in Cr [70].

Chronic SIV Infection

Eight rhesus macaques were infected with SIVmac251 and serially imaged with MRI and ^{1}H MRS to terminal AIDS or the endpoint of the study at 2 years. Due to the low incidence of SIV encephalitis (SIVE), possible diverse viral mutation and the differences in host factors the metabolic response over the time course of infection was variable. However, comparing the same animals we found a positive correlation between frontal cortex Cho/Cr and plasma viral load and a negative correlation between basal ganglia NAA/Cr and plasma viral load [130].

Rapid SIV Model

Untreated SIV^{+}/CD8^{-} Animals

Because of its parallels with HIV pathogenesis, the traditional SIV macaque model is hindered by the low rate of development of SIVE and the long time period for its evolution. Only approximately 25 % of infected macaques develop encephalitis and progression to terminal AIDS may take several years [107, 131]. These factors make it difficult for use in testing potential drug therapies. Therefore, attention has focused on two rapidly progressing SIV macaque models. One model employs pig-tailed macaques that are coinoculated with two SIV strains (SIV/17E-Fr and SIV/DeltaB670), which accelerates SIV CNS disease, producing SIVE in over 90 % of these animals within 3 months [132–135]. The second model retains the use of the SIV-infected rhesus macaques, but uses a monoclonal antibody to deplete the animal of CD8^{+} lymphocytes [136, 137]. In this model, 80 % of persistently CD8-depleted animals develop SIVE, with a course of progression to terminal AIDS within 12 weeks [33, 123, 138]. Modifications in applying the antibody have resulted in >90 % of macaques becoming persistently CD8 depleted with nearly all of them developing SIVE. This model also produces profound neuronal injury detectable within weeks by in vivo ^{1}H MRS [123]. Thus, this accelerated model of neuroAIDS in combination with MRS provides an exceptional opportunity to efficiently explore potentially useful drug therapies that can control or reverse neuronal injury and damage.

In the SIV^{+}/CD8^{-} model plasma viral RNA was detectable 6 days after infection with ~10^{7} copies Eq./mL and approached a plateau by 4 weeks after infection at ~5×10^{8} copies Eq./mL (Fig. 11.7a). CSF viral loads were found to be approximately 3 orders of magnitude lower than plasma. SIV^{+}/CD8^{-} animals were euthanized at 4, 6, and 8 weeks after infection and the amount of viral RNA in the frontal cortex was found to be ~10^{6} copies Eq./g at all three time points (Fig. 11.7b).

The neuronal marker NAA/Cr steadily declined following SIV infection and CD8 depletion, reaching levels as low as 20 % below baseline by 8 wpi in untreated animals (Fig. 11.7c). The decrease in NAA/Cr following infection is due to both decreases in neuronal NAA and increases in creatine, which most likely reflects the

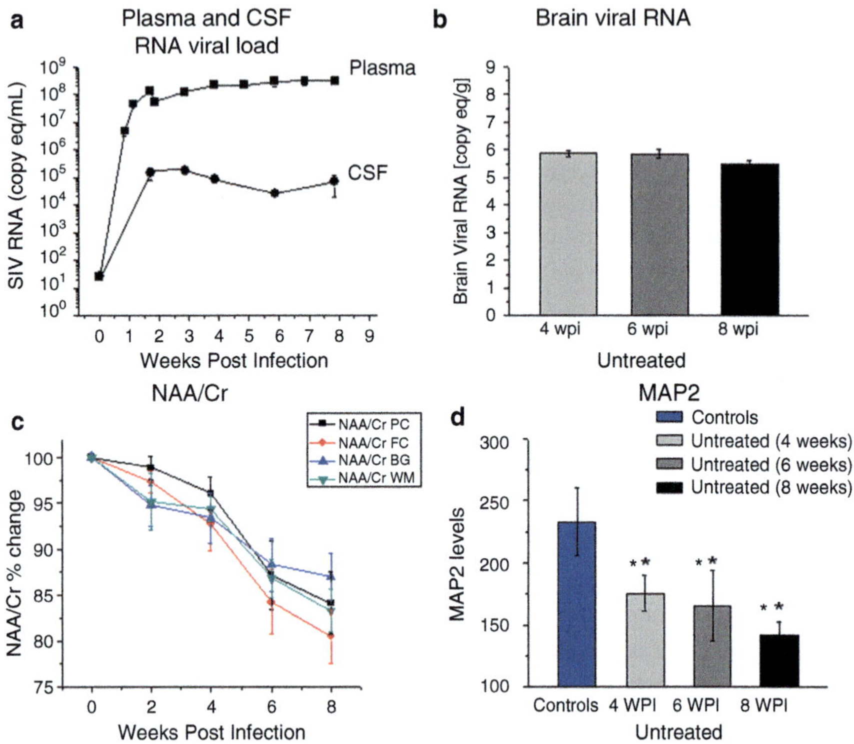

Fig. 11.7 (**a** and **b**) Viral RNA and neuronal injury following SIV infection and CD8 depletion. Based on [148] and [71]. (**a**) Viral RNA was detectable in plasma 6 days after infection (~10^7 copies Eq./mL) and the viral load approaches a plateau by 4 weeks after infection. The mean plasma viral load in infected animals was 4.9×10^8 copies Eq./mL 8 weeks after infection. CSF viral loads were found to be approximately 3 orders of magnitude lower than plasma in all animals. (**b**) Animals were euthanized at 4, 6, and 8 weeks after infection. The amount of viral RNA in the frontal cortex of untreated animals was 7.5×10^5, 9.2×10^5, and 3.0×10^5 copies Eq./g at 4, 6, and 8 weeks after infection, respectively. (**c**) The neuronal marker NAA/Cr steadily declined following SIV infection, reaching decreases as low as 20 % below baseline by 8 weeks postinfection (wpi) in the parietal cortex (PC). (**d**) Neuronal injury is confirmed by decreases in Microtubule-associated protein 2 (MAP2), a marker of dendritic integrity

cumulative effects of altered metabolic states of neurons and glial cells, respectively (for further discussion, see Ratai et al. [71]). Neuronal injury was confirmed by decreases in microtubule-associated protein 2 (MAP2), a marker of dendritic integrity (Fig. 11.7d).

CART and Minocycline-Treated SIV$^+$/CD8$^-$ Animals

To gain further insights into the neuropathogenesis of AIDS we modulated disease progression with treatments. In total, 11 SIV$^+$/CD8$^-$ rhesus macaques were treated. Seven received minocycline (MN), an antibiotic, which easily crosses the BBB, and

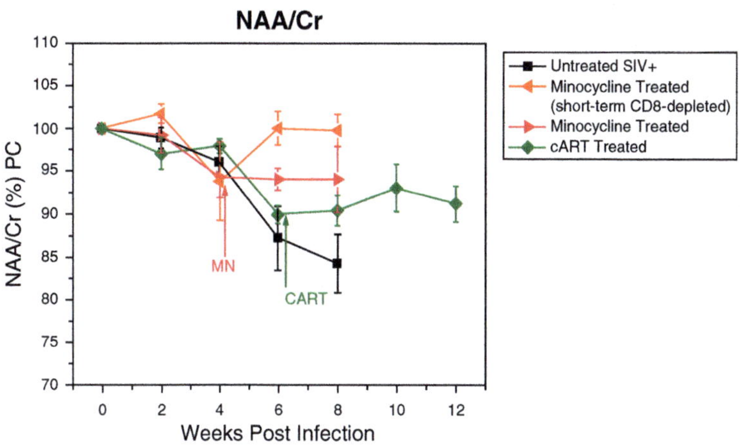

Fig. 11.8 Amelioration of neuronal injury with combination antiretroviral therapy (cART) and minocycline. Based on [147, 148]. NAA/Cr steadily declined following SIV infection in all animals, reaching levels as low as 20 % below baseline by 8 wpi in untreated animals (*black squares*). The decline in NAA/Cr was arrested with the cART (*green diamonds*) and minocycline treatment (*red triangles*) resulting in higher NAA/Cr levels when compared to untreated animals sacrificed at 8 wpi. Animals that were MN treated and had partial immune reconstitution of the $CD8^+$ T cell population had the most complete recovery (*orange triangles*)

has been found to have anti-inflammatory, neuroprotective, possibly antiviral effects [139–146]. Four of the seven MN-treated animals had persistent depletion of $CD8^+$ lymphocytes. However, three of the MN-treated animals had partial recovery of $CD8^+$ lymphocytes. This latter cohort was distinct from the long-term depleted cohorts in respect to its 1 order of magnitude lower viral loads and was thus grouped into a separate cohort of MN-treated short-term depleted animals. All seven animals were treated at 4 wpi for 4 weeks.

Four additional animals were treated with combination antiretroviral therapy (cART) consisting of 9-*R*-2-Phosphonomethoxypropyl adenine (PMPA), 5-Fluoro-1-[(2*R*,5*S*)-2-(hydroxymethyl)-[1,3]oxathiolan-5-yl]cytosine (FTC), and 2′-3-′-didehydro-2′-3′-dideoxythymidine (Stavudine, Zerit®) starting at 6 wpi for 6 weeks. Both treatments, MN and cART, resulted in a moderate decline of plasma, CSF, and brain viral levels [147, 148]. During minocycline as well as cART treatment, a significant reduction of circulating activated $CD14^+CD16^+$ monocytes was observed. Without treatment, SIV progression is characterized by the expansion of these monocytes in the periphery, which play a major role in trafficking virus across the BBB into the brain [149]. More importantly, the decline in NAA/Cr was arrested with MN and cART (Fig. 11.8, here shown for the parietal cortex, NAA/Cr) resulting in higher NAA/Cr levels when compared to untreated animals sacrificed at 8 wpi. Animals that were MN treated and had partial immune reconstitution of the $CD8^+$ T cell population had the most complete recovery. Thus, we hypothesize that possibly the best strategy to treat neuroAIDS is by the use of combination therapies targeting pathogenic factors in both the periphery and CNS [148].

Overview of Clinical MRS in HAND

Early HIV Infection

Until recently the effect of HIV to the brain during the early stages of HIV infection (less than 1 year of seroconversion) remained relatively unknown. However, a few studies show that soon after infection brain metabolism is affected by the virus as detected by in vivo MRS. In a study by Lentz et al., eight subjects were examined by SVS MRS within 3 months of their evolving Western blot and subsequently at

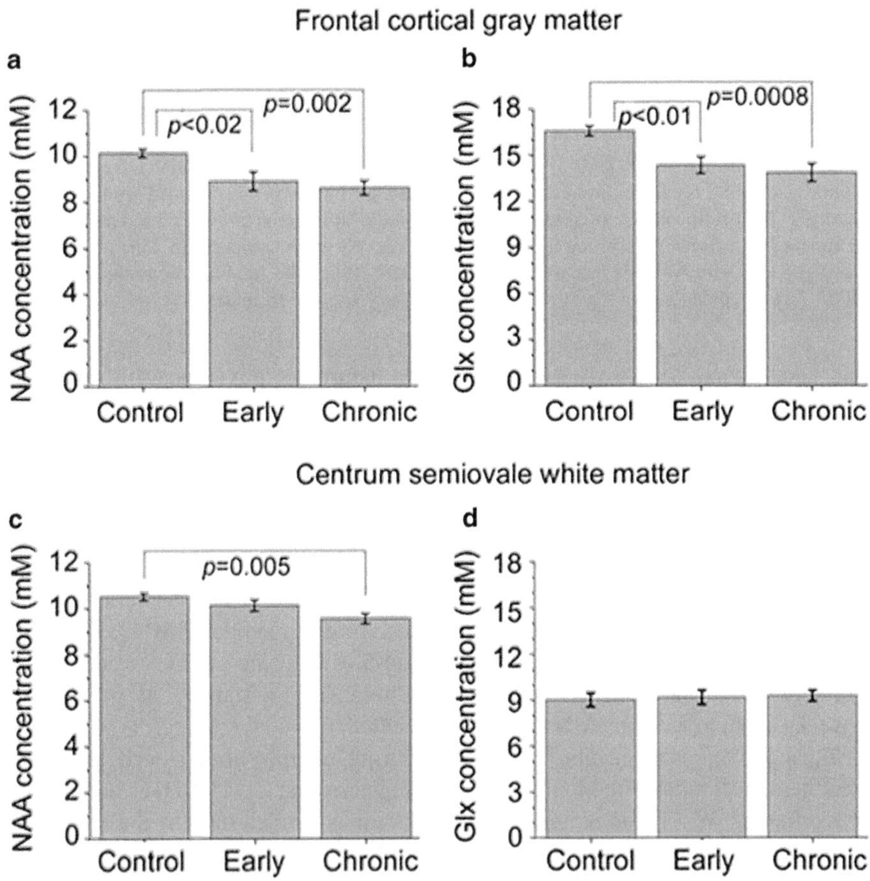

Fig. 11.9 Metabolic changes in NAA and Glx in the frontal cortex and white matter centrum semiovale during early and chronic infection. In the frontal cortical gray matter *N*-acetylaspartate (NAA) (**a**) and glutamate + glutamine (Glx) (**b**) levels are reduced in subjects during the first few months of infection, similar to subjects with chronic infection who are neurologically asymptomatic. Within the white matter of the centrum semiovale, NAA (**c**) was found to be reduced in chronically infected patients with HIV, but not in those during early infection. No changes from control levels for Glx in the white matter were observed (**d**). *Reprinted with permission from* [76]

2 and 6 months after their initial scan. Initially, a decrease in NAA and Glx were observed in the frontal cortex but not in the white matter semiovale compared to controls suggesting that frontal cortical gray matter to be more susceptible to early neuronal injury [76] (Fig. 11.9). Further longitudinal evaluation of this cohort showed increases in choline metabolism in the FC and WM indicative of lipid membrane turnover and glial metabolism [150]. These findings were confirmed by Sailasuta and Valcour who found increases in Cho/Cr in the WM and BG, respectively, in HIV+ patients being scanned during the first year of infection (median 2 weeks after infection) compared with HIV− controls [151, 152]. Recently, MRS markers have been correlated with markers of inflammation and neuronal injury suggesting early neuronal injury in a subset of participants with primary HIV infection through mechanisms involving central nervous system inflammation [153, 154].

Chronic HIV Infection

Observed MRS changes in chronically HIV-infected patients typically show reduction in NAA or NAA/Cr and elevations in Cho and mI or their respective ratios over Cr. The first case report of MRS studies in HIV-infected patients appeared in 1990 and discussed the strength of MRS in providing a marker of neuronal loss in patients with normal MRIs [155]. The first larger study of MRS in more than 100 HIV+ patients showed decreased NAA/Cr and elevated Cho/Cr [61]. Subsequently, it was found that Cho/Cr was elevated even in asymptomatic HIV+ patients and in patients with AIDS dementia complex while NAA/Cr was only decreased in symptomatic ADC patients [156] confirming that brain choline-containing compounds are elevated in HIV+ patients before the onset of ADC [157]. The decrease in NAA seems to correlate with the degree of neuronal dysfunction. Decreases in NAA/Cr of ~10 % have been observed in HIV+ patients with minor cognitive or motor dysfunction [67], while reductions of 15–28 % been reported in HIV+ patients with severe cognitive impairment or AIDS dementia complex compared to HIV− controls [63, 156, 158, 159].

Elevated mI or mI/Cr has been reported at various stages of HIV dementia and has been shown to increase with dementia severity [54, 65–67, 160]. Elevated Cr levels have been reported in the frontal white matter of chronic HIV patients [69]. Further changes in chronically infected patients include decreases in glutamate or Glx in the frontal white matter [72, 73]. Lower brain glutamate levels are associated with poorer cognitive deficits in HIV patients [74, 75].

Post Post-cART

As previously stated with the advent of ART, the incidence of severe neurological symptoms of HAD has been seen to decrease; however, less severe versions of the disease persist among the infected population [11, 12]. The characteristic involvement of the basal ganglia in pre-ART specimens is less commonly seen in post-ART specimens, which display neuroinflammation in the hippocampus and in adjacent parts of the entorhinal and temporal cortices [15, 161, 162].

Longitudinal studies in HIV+ patients with mild dementia showed improvement of CD4 counts and HIV Dementia Scale score when treated with cART. The initially increased Cho/Cr levels reversed in the midfrontal cortex and in the basal ganglia and the initially elevated mI/Cr and mI levels in the basal ganglia also decreased [163]. Another study that followed HIV+ patients before and after cART showed no significant improvement in Cho or mI after 3 months of treatment, but only after 6–9 months [54].

Clinical Trials Using MRS

A few clinical trials have included MRS as one of the outcome measures for neuroprotective drugs in HIV+ patients. In a Phase II randomized, double-blind, placebo-controlled, multicenter trial within the Adult AIDS Clinical Trials Group, 140 HIV-infected patients with mild to severe ADC receiving stable antiretroviral therapy were enrolled to assess the safety and efficacy of memantine, an *N*-methyl-D-aspartate receptor antagonist as treatment of HIV-associated cognitive impairment. While no significant differences in cognitive performance (neuropsychological z score (NPZ)-8) were shown 16 weeks later, the MRS data suggested that memantine may ameliorate neuronal metabolism, an important step to stabilizing or preventing neuronal injury. These results underscore the need for longer studies to assess the full potential of neuroprotective agents [164]. On the other hand, a study that assessed the effectiveness of selegiline transdermal system (STS) in reversing HIV-induced neurologic injury revealed no effect on either MRS metabolites, oxidative stress, or NPZ-8 scores [165].

Comorbidities

HAND is exacerbated by several comorbidities. Numerous studies have shown that neuronal injury is enhanced by both licit (alcohol) and illicit drugs (methamphetamine) [43, 166, 167]. As the HIV-infected population ages, other age-related neurological disorders will likely affect HIV+ patients. Several studies started to investigate the relationship between HIV and age [79].

Of note, increasing evidence suggests that certain antiretrovirals may be neurotoxic [168]. Nucleoside reverse transcriptase inhibitors (NRTIs) suppress HIV replication but are often associated with mitochondrial toxicity. Chronically infected HIV+ patients on long-term cART regimens that included NRTIs such as didanosine and/or stavudine revealed significant reductions in NAA in the FWM compared with HIV− controls while HIV+ individuals receiving alternative cART regimens that did not include NRTIs exhibited only intermediate decreases in NAA [169].

HIV-Infected Children

Overall, fewer MRS studies have been performed on the brains of HIV-infected children. Most HIV+ children under the age of 13 are infected during pregnancy, childbirth, or breastfeeding. A study by Keller et al. showed that while healthy children showed an increase in NAA with age in the frontal WM and hippocampus, HIV+ children did not show the age-related increase. Furthermore, HIV+ children with high plasma viral loads had lower Cho, mI, and Cr in the basal ganglia but higher Cho in the midfrontal gray matter compared to children with lower viral loads [170].

A study by Mbugua et al. imaged a cohort of 5-year-old HIV+ children in South Africa and found lower CD4/CD8 ratio in early infancy (mean age $= 8 \pm 2$ weeks old) to be significantly associated with lower NAA and Cho levels in the basal ganglia [171]. Furthermore, a longitudinal study by Holmes et al. compared HIV uninfected, exposed (HEU) and HIV unexposed, uninfected (HUU) young children; they found a significant increase in NAA levels with age in HUU only as well as significantly higher mean Cho levels in HEU children at age 7 years [172].

Summary and Future Perspectives

In summary, MRS offers a valuable imaging technique to assess disease status and response to therapeutics in patients with HAND. Since MRS changes are sensitive to early subtle neuropathogenic changes MRS studies certainly add information to conventional MRI exams. In animal models MRS has successfully been utilized to understand the neuropathogenesis of the disease. Specifically, the combination of MRS with immunology, virology, and pathology in animal models has led to a greater understanding of the inflammatory and neuronal events caused by the virus.

To date most human studies have been cross sectional; however, MRS is especially powerful when applied longitudinally to monitor individual patients. Thus, the next step in neuroimaging studies should include standardized MRS protocols in multicenter clinical trials to evaluate potential treatments for patients suffering from HAND. Significant effort has been made by the AIDS Clinical Trial Group (ACTG) using the same imaging paradigm at multiple sites [154].

Furthermore, most studies performed single voxel spectroscopy on either 1.5 or 3 T. However, more novel MRS techniques have been developed to more accurately measure metabolites that show overlapping peaks in the regular one-dimensional (1D) MRS spectra. For instance, using 2D localized chemical shift correlated spectroscopy (L-COSY) enables converting a crowded, overlapping 1D MR spectrum to a better resolved 2D spectrum through the addition of a spectral dimension [173] and allows for better detection of metabolites at lower concentrations and delineation of the components of otherwise overlapping peaks such as glutamate/glutamine, *myo*-inositol, aspartate (Asp), gamma-aminobutyrate (GABA), taurine, glutathione (GSH), threonine, and macromolecules [174].

In addition, GABA and GSH can specifically be detected with so-called editing techniques either using double-quantum coherence (DQC) [175, 176] or difference editing techniques, e.g., MEGA PRESS [177, 178]. Specifically, the quantification of GSH levels may be of pertinence in HIV-infected patients. HIV is associated with substantial oxidative stress [179]. Glutathione is thought to be an extremely important antioxidant for HIV-infected patients, because it appears to interfere with HIV's entry into its target cells [180].

Greater signal-to-noise ratio, spatial and spectral resolution can also be gained from higher field strengths. 7 T MRS in combination with LCModel [181] allows for the quantification of 17 metabolites [182]. 7 T MRS has successfully been employed in the CD8-depleted macaque model of neuroAIDS [183]. In addition, Gonen et al., have successfully implemented a 3D MRSI sequence with 0.05 cm^3 isotropic spatial resolution at 7 T [184].

Other technical advances include motion- corrected MR spectroscopy exams. For children between 2 months and 7 years of age, motion artifact and patient cooperation often limit the feasibility of MRI and MRS [185]. Using image-based navigators, it is possible to correct motion in structural imaging, single-voxel and multivoxel spectroscopy prospectively [186–192].

Acknowledgements This work was partly funded by NIH grants R21NS059331, R01NS050041, R01NS040237, R01NS37654, R01MH62962, MH59754, MH62512, RR00168, R24 RR016001, N01 AI040101, and P41RR14075. The author wishes to thank Dr. R Gilberto Gonzalez for valuable discussions.

References

1. Worldwide HIV AIDS Statistics (2014) http://www.avert.org/worldwide-hiv-aids-statistics.htm
2. Daar ES, Moudgil T, Meyer RD, Ho DD (1991) Transient high levels of viremia in patients with primary human immunodeficiency virus type 1 infection. N Engl J Med 324 (14):961–964
3. Gupta KK (1993) Acute immunosuppression with HIV seroconversion. N Engl J Med 328 (4):288–289
4. Vento S, Di Perri G, Cruciani M, Garofano T, Concia E, Bassetti D (1993) Rapid decline of CD4+ cells after IFN alpha treatment in HIV-1 infection. Lancet 341(8850):958–959
5. Navia BA, Cho ES, Petito CK, Price RW (1986) The AIDS dementia complex: II. Neuropathology. Ann Neurol 19(6):525–535
6. Navia BA, Jordan BD, Price RW (1986) The AIDS dementia complex: I. Clinical features. Ann Neurol 19(6):517–524

7. Price RW, Brew B, Sidtis J, Rosenblum M, Scheck AC, Cleary P (1988) The brain in AIDS: central nervous system HIV-1 infection and AIDS dementia complex. Science 239 (4840):586–592
8. Gray F, Hurtrel M, Hurtrel B (1993) Early central nervous system changes in human immunodeficiency virus (HIV)-infection. Neuropathol Appl Neurobiol 19(1):3–9
9. Vaughan JT, Hetherington HP, Otu JO, Pan JW, Pohost GM (1994) High frequency volume coils for clinical NMR imaging and spectroscopy. Magn Reson Med 32(2):206–218
10. Price RW, Brew BJ (1988) The AIDS dementia complex. J Infect Dis 158(5):1079–1083
11. Brodt HR, Kamps BS, Gute P, Knupp B, Staszewski S, Helm EB (1997) Changing incidence of AIDS-defining illnesses in the era of antiretroviral combination therapy. AIDS 11 (14):1731–1738
12. McArthur JC (2004) HIV dementia: an evolving disease. J Neuroimmunol 157(1–2):3–10
13. Nath A, Schiess N, Venkatesan A, Rumbaugh J, Sacktor N, McArthur J (2008) Evolution of HIV dementia with HIV infection. Int Rev Psychiatry 20(1):25–31
14. Heaton RK, Clifford DB, Franklin DR Jr, Woods SP, Ake C, Vaida F, Ellis RJ, Letendre SL, Marcotte TD, Atkinson JH, Rivera-Mindt M, Vigil OR, Taylor MJ, Collier AC, Marra CM, Gelman BB, McArthur JC, Morgello S, Simpson DM, McCutchan JA, Abramson I, Gamst A, Fennema-Notestine C, Jernigan TL, Wong J, Grant I (2010) HIV-associated neurocognitive disorders persist in the era of potent antiretroviral therapy: CHARTER Study. Neurology 75 (23):2087–2096
15. Gannon P, Khan MZ, Kolson DL (2013) Current understanding of HIV-associated neurocognitive disorders pathogenesis. Curr Opin Neurol 24(3):275–283
16. Sacktor N (2002) The epidemiology of human immunodeficiency virus-associated neurological disease in the era of highly active antiretroviral therapy. J Neurovirol 8(Suppl 2):115–121
17. Sacktor N, McDermott MP, Marder K, Schifitto G, Selnes OA, McArthur JC, Stern Y, Albert S, Palumbo D, Kieburtz K, De Marcaida JA, Cohen B, Epstein L (2002) HIV-associated cognitive impairment before and after the advent of combination therapy. J Neurovirol 8(2):136–142
18. Ellis RJ, Deutsch R, Heaton RK, Marcotte TD, McCutchan JA, Nelson JA, Abramson I, Thal LJ, Atkinson JH, Wallace MR, Grant I (1997) Neurocognitive impairment is an independent risk factor for death in HIV infection. San Diego HIV Neurobehavioral Research Center Group. Arch Neurol 54(4):416–424
19. Liner KJ II, Hall CD, Robertson KR (2008) Effects of antiretroviral therapy on cognitive impairment. Curr HIV/AIDS Rep 5(2):64–71
20. Peluso R, Haase A, Stowring L, Edwards M, Ventura P (1985) A Trojan Horse mechanism for the spread of visna virus in monocytes. Virology 147(1):231–236
21. Hickey WF, Hsu BL, Kimura H (1991) T-lymphocyte entry into the central nervous system. J Neurosci Res 28(2):254–260
22. Bonwetsch R, Croul S, Richardson MW, Lorenzana C, Valle LD, Sverstiuk AE, Amini S, Morgello S, Khalili K, Rappaport J (1999) Role of HIV-1 Tat and CC chemokine MIP-1alpha in the pathogenesis of HIV associated central nervous system disorders. J Neurovirol 5 (6):685–694
23. Catani MV, Corasaniti MT, Navarra M, Nistico G, Finazzi-Agro A, Melino G (2000) gp120 induces cell death in human neuroblastoma cells through the CXCR4 and CCR5 chemokine receptors. J Neurochem 74(6):2373–2379
24. Cheng J, Nath A, Knudsen B, Hochman S, Geiger JD, Ma M, Magnuson DS (1998) Neuronal excitatory properties of human immunodeficiency virus type 1 Tat protein. Neuroscience 82 (1):97–106
25. Meucci O, Fatatis A, Simen AA, Bushell TJ, Gray PW, Miller RJ (1998) Chemokines regulate hippocampal neuronal signaling and gp120 neurotoxicity. Proc Natl Acad Sci U S A 95(24):14500–14505
26. Nath A (2002) Human immunodeficiency virus (HIV) proteins in neuropathogenesis of HIV dementia. J Infect Dis 186(Suppl 2):S193–S198

27. Nath A, Psooy K, Martin C, Knudsen B, Magnuson DS, Haughey N, Geiger JD (1996) Identification of a human immunodeficiency virus type 1 Tat epitope that is neuroexcitatory and neurotoxic. J Virol 70(3):1475–1480
28. Ohagen A, Ghosh S, He J, Huang K, Chen Y, Yuan M, Osathanondh R, Gartner S, Shi B, Shaw G, Gabuzda D (1999) Apoptosis induced by infection of primary brain cultures with diverse human immunodeficiency virus type 1 isolates: evidence for a role of the envelope. J Virol 73(2):897–906
29. D'Aversa TG, Eugenin EA, Berman JW (2005) NeuroAIDS: contributions of the human immunodeficiency virus-1 proteins Tat and gp120 as well as CD40 to microglial activation. J Neurosci Res 81(3):436–446
30. Crowe S, Zhu T, Muller WA (2003) The contribution of monocyte infection and trafficking to viral persistence, and maintenance of the viral reservoir in HIV infection. J Leukoc Biol 74(5):635–641
31. Gartner S (2000) HIV infection and dementia. Science 287(5453):602–604
32. Persidsky Y, Gendelman HE (2003) Mononuclear phagocyte immunity and the neuropathogenesis of HIV-1 infection. J Leukoc Biol 74(5):691–701
33. Williams KC, Hickey WF (2002) Central nervous system damage, monocytes and macrophages, and neurological disorders in AIDS. Annu Rev Neurosci 25:537–562
34. Roy S, Wainberg MA (1988) Role of the mononuclear phagocyte system in the development of acquired immunodeficiency syndrome (AIDS). J Leukoc Biol 43(1):91–97
35. Kaul M, Garden GA, Lipton SA (2001) Pathways to neuronal injury and apoptosis in HIV-associated dementia. Nature 410(6831):988–994
36. Kaul M, Zheng J, Okamoto S, Gendelman HE, Lipton SA (2005) HIV-1 infection and AIDS: consequences for the central nervous system. Cell Death Differ 12(Suppl 1):878–892
37. Kraft-Terry SD, Buch SJ, Fox HS, Gendelman HE (2009) A coat of many colors: neuroimmune crosstalk in human immunodeficiency virus infection. Neuron 64(1):133–145
38. Wiley CA, Soontornniyomkij V, Radhakrishnan L, Masliah E, Mellors J, Hermann SA, Dailey P, Achim CL (1998) Distribution of brain HIV load in AIDS. Brain Pathol 8 (2):277–284
39. Moore DJ, Masliah E, Rippeth JD, Gonzalez R, Carey CL, Cherner M, Ellis RJ, Achim CL, Marcotte TD, Heaton RK, Grant I (2006) Cortical and subcortical neurodegeneration is associated with HIV neurocognitive impairment. AIDS 20(6):879–887
40. Filippi CG, Ulug AM, Ryan E, Ferrando SJ, van Gorp W (2001) Diffusion tensor imaging of patients with HIV and normal-appearing white matter on MR images of the brain. AJNR Am J Neuroradiol 22(2):277–283
41. Bonnet F, Amieva H, Marquant F, Bernard C, Bruyand M, Dauchy FA, Mercie P, Greib C, Richert L, Neau D, Catheline G, Dehail P, Dabis F, Morlat P, Dartigues JF, Chene G (2013) Cognitive disorders in HIV-infected patients: are they HIV-related? AIDS 27(3):391–400
42. Ances BM, Ortega M, Vaida F, Heaps J, Paul R (2012) Independent effects of HIV, aging, and HAART on brain volumetric measures. J Acquir Immune Defic Syndr 59(5):469–477
43. Pfefferbaum A, Rosenbloom MJ, Sassoon SA, Kemper CA, Deresinski S, Rohlfing T, Sullivan EV (2012) Regional brain structural dysmorphology in human immunodeficiency virus infection: effects of acquired immune deficiency syndrome, alcoholism, and age. Biol Psychiatry 72(5):361–370
44. Aylward EH, Henderer JD, McArthur JC, Brettschneider PD, Harris GJ, Barta PE, Pearlson GD (1993) Reduced basal ganglia volume in HIV-1-associated dementia: results from quantitative neuroimaging. Neurology 43(10):2099–2104
45. Broderick DF, Wippold FJ II, Clifford DB, Kido D, Wilson BS (1993) White matter lesions and cerebral atrophy on MR images in patients with and without AIDS dementia complex. AJR Am J Roentgenol 161(1):177–181
46. Dooneief G, Bello J, Todak G, Mun IK, Marder K, Malouf R, Gorman J, Hilal S, Stern Y, Mayeux R (1992) A prospective controlled study of magnetic resonance imaging of the brain in gay men and parenteral drug users with human immunodeficiency virus infection. Arch Neurol 49(1):38–43

47. McArthur JC, Kumar AJ, Johnson DW, Selnes OA, Becker JT, Herman C, Cohen BA, Saah A (1990) Incidental white matter hyperintensities on magnetic resonance imaging in HIV-1 infection. Multicenter AIDS Cohort Study. J Acquir Immune Defic Syndr 3(3):252–259
48. Boska MD, Mosley RL, Nawab M, Nelson JA, Zelivyanskaya M, Poluektova L, Uberti M, Dou H, Lewis TB, Gendelman HE (2004) Advances in neuroimaging for HIV-1 associated neurological dysfunction: clues to the diagnosis, pathogenesis and therapeutic monitoring. Curr HIV Res 2(1):61–78
49. Hammoud DA, Endres CJ, Chander AR, Guilarte TR, Wong DF, Sacktor NC, McArthur JC, Pomper MG (2005) Imaging glial cell activation with [11C]-R-PK11195 in patients with AIDS. J Neurovirol 11(4):346–355
50. Wiley CA, Lopresti BJ, Becker JT, Boada F, Lopez OL, Mellors J, Meltzer CC, Wisniewski SR, Mathis CA (2006) Positron emission tomography imaging of peripheral benzodiazepine receptor binding in human immunodeficiency virus-infected subjects with and without cognitive impairment. J Neurovirol 12(4):262–271
51. Andersen AB, Law I, Krabbe KS, Bruunsgaard H, Ostrowski SR, Ullum H, Hojgaard L, Lebech A, Gerstoft J, Kjaer A (2010) Cerebral FDG-PET scanning abnormalities in optimally treated HIV patients. J Neuroinflammation 7:13
52. Chang L, Wang GJ, Volkow ND, Ernst T, Telang F, Logan J, Fowler JS (2008) Decreased brain dopamine transporters are related to cognitive deficits in HIV patients with or without cocaine abuse. Neuroimage 42(2):869–878
53. Ances BM, Christensen JJ, Teshome M, Taylor J, Xiong C, Aldea P, Fagan AM, Holtzman DM, Morris JC, Mintun MA, Clifford DB (2010) Cognitively unimpaired HIV-positive subjects do not have increased 11C-PiB: a case-control study. Neurology 75(2):111–115
54. Chang L, Feger U, Ernst TM (2011) Bioimaging. In: The neurology of AIDS, 3rd edn. Oxford University Press, Oxford
55. Birken DL, Oldendorf WH (1989) N-Acetyl-L-aspartic acid: a literature review of a compound prominent in 1H-NMR spectroscopic studies of brain. Neurosci Biobehav Rev 13(1):23–31
56. Simmons ML, Frondoza CG, Coyle JT (1991) Immunocytochemical localization of N-acetyl-aspartate with monoclonal antibodies. Neuroscience 45(1):37–45
57. Moffett JR, Namboodiri MA, Cangro CB, Neale JH (1991) Immunohistochemical localization of N-acetylaspartate in rat brain. Neuroreport 2(3):131–134
58. Urenjak J, Williams SR, Gadian DG, Noble M (1993) Proton nuclear magnetic resonance spectroscopy unambiguously identifies different neural cell types. J Neurosci 13(3):981–989
59. Menon DK, Ainsworth JG, Cox IJ, Coker RC, Sargentoni J, Coutts GA, Baudouin CJ, Kocsis AE, Harris JR (1992) Proton MR spectroscopy of the brain in AIDS dementia complex. J Comput Assist Tomogr 16(4):538–542
60. Meyerhoff DJ, MacKay S, Bachman L, Poole N, Dillon WP, Weiner MW, Fein G (1993) Reduced brain N-acetylaspartate suggests neuronal loss in cognitively impaired human immunodeficiency virus-seropositive individuals: in vivo 1H magnetic resonance spectroscopic imaging. Neurology 43(3 Pt 1):509–515
61. Chong WK, Sweeney B, Wilkinson ID, Paley M, Hall-Craggs MA, Kendall BE, Shepard JK, Beecham M, Miller RF, Weller IV et al (1993) Proton spectroscopy of the brain in HIV infection: correlation with clinical, immunologic, and MR imaging findings. Radiology 188(1):119–124
62. Barker PB, Lee RR, McArthur JC (1995) AIDS dementia complex: evaluation with proton MR spectroscopic imaging. Radiology 195(1):58–64
63. Meyerhoff DJ, Bloomer C, Cardenas V, Norman D, Weiner MW, Fein G (1999) Elevated subcortical choline metabolites in cognitively and clinically asymptomatic HIV+ patients. Neurology 52(5):995–1003
64. Brand A, Richter-Landsberg C, Leibfritz D (1993) Multinuclear NMR studies on the energy metabolism of glial and neuronal cells. Dev Neurosci 15(3–5):289–298
65. Laubenberger J, Haussinger D, Bayer S, Thielemann S, Schneider B, Mundinger A, Hennig J, Langer M (1996) HIV-related metabolic abnormalities in the brain: depiction with proton MR spectroscopy with short echo times. Radiology 199(3):805–810

66. Lopez-Villegas D, Lenkinski RE, Frank I (1997) Biochemical changes in the frontal lobe of HIV-infected individuals detected by magnetic resonance spectroscopy. Proc Natl Acad Sci U S A 94(18):9854–9859
67. Chang L, Ernst T, Leonido-Yee M, Walot I, Singer E (1999) Cerebral metabolite abnormalities correlate with clinical severity of HIV-1 cognitive motor complex. Neurology 52 (1):100–108
68. Saunders DE, Howe FA, van den Boogaart A, Griffiths JR, Brown MM (1999) Aging of the adult human brain: in vivo quantitation of metabolite content with proton magnetic resonance spectroscopy. J Magn Reson Imaging 9(5):711–716
69. Chang L, Ernst T, Witt MD, Ames N, Gaiefsky M, Miller E (2002) Relationships among brain metabolites, cognitive function, and viral loads in antiretroviral-naive HIV patients. Neuroimage 17(3):1638–1648
70. Ratai EM, Pilkenton SJ, Greco JB, Lentz MR, Bombardier JP, Turk KW, He J, Joo CG, Lee V, Westmoreland S, Halpern E, Lackner AA, Gonzalez RG (2009) In vivo proton magnetic resonance spectroscopy reveals region specific metabolic responses to SIV infection in the macaque brain. BMC Neurosci 10:63
71. Ratai EM, Annamalai L, Burdo T, Joo CG, Bombardier JP, Fell R, Hakimelahi R, He J, Lentz MR, Campbell J, Curran E, Halpern EF, Masliah E, Westmoreland SV, Williams KC, Gonzalez RG (2011) Brain creatine elevation and N-acetylaspartate reduction indicates neuronal dysfunction in the setting of enhanced glial energy metabolism in a macaque model of NeuroAIDS. Magn Reson Med 66(3):625–634
72. Sailasuta N, Shriner K, Ross B (2009) Evidence of reduced glutamate in the frontal lobe of HIV-seropositive patients. NMR Biomed 22(3):326–331
73. Harezlak J, Buchthal S, Taylor M, Schifitto G, Zhong J, Daar E, Alger J, Singer E, Campbell T, Yiannoutsos C, Cohen R, Navia B (2011) Persistence of HIV-associated cognitive impairment, inflammation, and neuronal injury in era of highly active antiretroviral treatment. AIDS 25(5):625–633
74. Ernst T, Jiang CS, Nakama H, Buchthal S, Chang L (2010) Lower brain glutamate is associated with cognitive deficits in HIV patients: a new mechanism for HIV-associated neurocognitive disorder. J Magn Reson Imaging 32(5):1045–1053
75. Mohamed MA, Barker PB, Skolasky RL, Selnes OA, Moxley RT, Pomper MG, Sacktor NC (2010) Brain metabolism and cognitive impairment in HIV infection: a 3-T magnetic resonance spectroscopy study. Magn Reson Imaging 28(9):1251–1257
76. Lentz MR, Kim WK, Lee V, Bazner S, Halpern EF, Venna N, Williams K, Rosenberg ES, Gonzalez RG (2009) Changes in MRS neuronal markers and T cell phenotypes observed during early HIV infection. Neurology 72(17):1465–1472
77. Roc AC, Ances BM, Chawla S, Korczykowski M, Wolf RL, Kolson DL, Detre JA, Poptani H (2007) Detection of human immunodeficiency virus induced inflammation and oxidative stress in lenticular nuclei with magnetic resonance spectroscopy despite antiretroviral therapy. Arch Neurol 64(9):1249–1257
78. Lee PL, Yiannoutsos CT, Ernst T, Chang L, Marra CM, Jarvik JG, Richards TL, Kwok EW, Kolson DL, Simpson D, Tang CY, Schifitto G, Ketonen LM, Meyerhoff DJ, Lenkinski RE, Gonzalez RG, Navia BA (2003) A multi-center 1H MRS study of the AIDS dementia complex: validation and preliminary analysis. J Magn Reson Imaging 17(6):625–633
79. Chang L, Lee PL, Yiannoutsos CT, Ernst T, Marra CM, Richards T, Kolson D, Schifitto G, Jarvik JG, Miller EN, Lenkinski R, Gonzalez G, Navia BA (2004) A multicenter in vivo proton-MRS study of HIV-associated dementia and its relationship to age. Neuroimage 23 (4):1336–1347
80. Sacktor N, Skolasky RL, Ernst T, Mao X, Selnes O, Pomper MG, Chang L, Zhong K, Shungu DC, Marder K, Shibata D, Schifitto G, Bobo L, Barker PB (2005) A multicenter study of two magnetic resonance spectroscopy techniques in individuals with HIV dementia. J Magn Reson Imaging 21(4):325–333
81. Greco JB, Sakaie KE, Aminipour S, Lee PL, Chang LL, He J, Westmoreland S, Lackner AA, Gonzalez RG (2002) Magnetic resonance spectroscopy: an in vivo tool for monitoring cerebral injury in SIV-infected macaques. J Med Primatol 31(4–5):228–236

82. Jacobs MA, Horska A, van Zijl PC, Barker PB (2001) Quantitative proton MR spectroscopic imaging of normal human cerebellum and brain stem. Magn Reson Med 46(4):699–705
83. Soher BJ, van Zijl PC, Duyn JH, Barker PB (1996) Quantitative proton MR spectroscopic imaging of the human brain. Magn Reson Med 35(3):356–363
84. Michaelis T, Merboldt KD, Bruhn H, Hanicke W, Frahm J (1993) Absolute concentrations of metabolites in the adult human brain in vivo: quantification of localized proton MR spectra. Radiology 187(1):219–227
85. Barker PB, Soher BJ, Blackband SJ, Chatham JC, Mathews VP, Bryan RN (1993) Quantitation of proton NMR spectra of the human brain using tissue water as an internal concentration reference. NMR Biomed 6(1):89–94
86. Christiansen P, Henriksen O, Stubgaard M, Gideon P, Larsson HB (1993) In vivo quantification of brain metabolites by 1H-MRS using water as an internal standard. Magn Reson Imaging 11(1):107–118
87. Ernst T, Kreis R, Ross BD (1993) Absolute quantitation of water and metabolites in the human brain; Part I: Compartments and water. J Magn Res B102:1–8
88. Tofts PS, Wray S (1988) A critical assessment of methods of measuring metabolite concentrations by NMR spectroscopy. NMR Biomed 1(1):1–10
89. Hennig J, Pfister H, Ernst T, Ott D (1992) Direct absolute quantification of metabolites in the human brain with in vivo localized proton spectroscopy. NMR Biomed 5(4):193–199
90. Toggas SM, Masliah E, Rockenstein EM, Rall GF, Abraham CR, Mucke L (1994) Central nervous system damage produced by expression of the HIV-1 coat protein gp120 in transgenic mice. Nature 367(6459):188–193
91. Persidsky Y, Limoges J, McComb R, Bock P, Baldwin T, Tyor W, Patil A, Nottet HS, Epstein L, Gelbard H, Flanagan E, Reinhard J, Pirruccello SJ, Gendelman HE (1996) Human immunodeficiency virus encephalitis in SCID mice. Am J Pathol 149(3):1027–1053
92. Persidsky Y, Buttini M, Limoges J, Bock P, Gendelman HE (1997) An analysis of HIV-1-associated inflammatory products in brain tissue of humans and SCID mice with HIV-1 encephalitis. J Neurovirol 3(6):401–416
93. Persidsky Y, Gendelman HE (2002) Murine models for human immunodeficiency virus type 1-associated dementia: the development of new treatment testing paradigms. J Neurovirol 8 (Suppl 2):49–52
94. Zink WE, Anderson E, Boyle J, Hock L, Rodriguez-Sierra J, Xiong H, Gendelman HE, Persidsky Y (2002) Impaired spatial cognition and synaptic potentiation in a murine model of human immunodeficiency virus type 1 encephalitis. J Neurosci 22(6):2096–2105
95. Jacobson S, Henriksen SJ, Prospero-Garcia O, Phillips TR, Elder JH, Young WG, Bloom FE, Fox HS (1997) Cortical neuronal cytoskeletal changes associated with FIV infection. J Neurovirol 3(4):283–289
96. Podell M, Oglesbee M, Mathes L, Krakowka S, Olmstead R, Lafrado L (1993) AIDS-associated encephalopathy with experimental feline immunodeficiency virus infection. J Acquir Immune Defic Syndr 6(7):758–771
97. Phillips TR, Prospero-Garcia O, Wheeler DW, Wagaman PC, Lerner DL, Fox HS, Whalen LR, Bloom FE, Elder JH, Henriksen SJ (1996) Neurologic dysfunctions caused by a molecular clone of feline immunodeficiency virus, FIV-PPR. J Neurovirol 2(6):388–396
98. Silvotti L, Corradi A, Brandi G, Cabassi A, Bendinelli M, Magnan M, Piedimonte G (1997) FIV induced encephalopathy: early brain lesions in the absence of viral replication in monocyte/macrophages. A pathogenetic model. Vet Immunol Immunopathol 55(4):263–271
99. Prospero-Garcia O, Herold N, Waters AK, Phillips TR, Elder JH, Henriksen SJ (1994) Intraventricular administration of a FIV-envelope protein induces sleep architecture changes in rats. Brain Res 659(1–2):254–258
100. Poli A, Abramo F, Di Iorio C, Cantile C, Carli MA, Pollera C, Vago L, Tosoni A, Costanzi G (1997) Neuropathology in cats experimentally infected with feline immunodeficiency virus: a morphological, immunocytochemical and morphometric study. J Neurovirol 3(5):361–368
101. Podell M, March PA, Buck WR, Mathes LE (2000) The feline model of neuroAIDS: understanding the progression towards AIDS dementia. J Psychopharmacol 14(3):205–213

102. Podell M, Maruyama K, Smith M, Hayes KA, Buck WR, Ruehlmann DS, Mathes LE (1999) Frontal lobe neuronal injury correlates to altered function in FIV-infected cats. J Acquir Immune Defic Syndr 22(1):10–18
103. Desrosiers RC (1990) The simian immunodeficiency viruses. Annu Rev Immunol 8:557–578
104. Zink MC, Amedee AM, Mankowski JL, Craig L, Didier P, Carter DL, Munoz A, Murphey-Corb M, Clements JE (1997) Pathogenesis of SIV encephalitis. Selection and replication of neurovirulent SIV. Am J Pathol 151(3):793–803
105. Zink MC, Spelman JP, Robinson RB, Clements JE (1998) SIV infection of macaques—modeling the progression to AIDS dementia. J Neurovirol 4(3):249–259
106. Murray EA, Rausch DM, Lendvay J, Sharer LR, Eiden LE (1992) Cognitive and motor impairments associated with SIV infection in rhesus monkeys. Science 255 (5049):1246–1249
107. Burudi EME, Fox HS (2001) Simian immunodeficiency virus model of HIV-induced central nervous system dysfunction. Adv Virus Res 56:435–468
108. Bruce-Keller AJ, Chauhan A, Dimayuga FO, Gee J, Keller JN, Nath A (2003) Synaptic transport of human immunodeficiency virus-Tat protein causes neurotoxicity and gliosis in rat brain. J Neurosci 23(23):8417–8422
109. Okamoto S, Kang YJ, Brechtel CW, Siviglia E, Russo R, Clemente A, Harrop A, McKercher S, Kaul M, Lipton SA (2007) HIV/gp120 decreases adult neural progenitor cell proliferation via checkpoint kinase-mediated cell-cycle withdrawal and G1 arrest. Cell Stem Cell 1(2):230–236
110. Zelivyanskaya ML, Nelson JA, Poluektova L, Uberti M, Mellon M, Gendelman HE, Boska MD (2003) Tracking superparamagnetic iron oxide labeled monocytes in brain by high-field magnetic resonance imaging. J Neurosci Res 73(3):284–295
111. Nelson JA, Dou H, Ellison B, Uberti M, Xiong H, Anderson E, Mellon M, Gelbard HA, Boska M, Gendelman HE (2005) Coregistration of quantitative proton magnetic resonance spectroscopic imaging with neuropathological and neurophysiological analyses defines the extent of neuronal impairments in murine human immunodeficiency virus type-1 encephalitis. J Neurosci Res 80(4):562–575
112. Phipps AJ, Hayes KA, Buck WR, Podell M, Mathes LE (2000) Neurophysiologic and immunologic abnormalities associated with feline immunodeficiency virus molecular clone FIV-PPR DNA inoculation. J Acquir Immune Defic Syndr 23(1):8–16
113. Podell M, Hayes K, Oglesbee M, Mathes L (1997) Progressive encephalopathy associated with CD4/CD8 inversion in adult FIV-infected cats. J Acquir Immune Defic Syndr Hum Retrovirol 15(5):332–340
114. Belman AL (1997) Pediatric neuro-AIDS. Update. Neuroimaging Clin N Am 7(3):593–613
115. Johnston JB, Silva C, Hiebert T, Buist R, Dawood MR, Peeling J, Power C (2002) Neurovirulence depends on virus input titer in brain in feline immunodeficiency virus infection: evidence for activation of innate immunity and neuronal injury. J Neurovirol 8 (5):420–431
116. Reimann KA, Tenner-Racz K, Racz P, Montefiori DC, Yasutomi Y, Lin W, Ransil BJ, Letvin NL (1994) Immunopathogenic events in acute infection of rhesus monkeys with simian immunodeficiency virus of macaques. J Virol 68(4):2362–2370
117. Staprans SI, Dailey PJ, Rosenthal A, Horton C, Grant RM, Lerche N, Feinberg MB (1999) Simian immunodeficiency virus disease course is predicted by the extent of virus replication during primary infection. J Virol 73(6):4829–4839
118. Greco JB, Westmoreland SV, Ratai EM, Lentz MR, Sakaie K, He J, Sehgal PK, Masliah E, Lackner AA, Gonzalez RG (2004) In vivo ^{1}H MRS of brain injury and repair during acute SIV infection in the macaque model of neuroAIDS. Magn Reson Med 51(6):1108–1114
119. Chakrabarti L, Hurtrel M, Maire MA, Vazeux R, Dormont D, Montagnier L, Hurtrel B (1991) Early viral replication in the brain of SIV-infected rhesus monkeys. Am J Pathol 139 (6):1273–1280

120. Sharer LR, Baskin GB, Cho ES, Murphey-Corb M, Blumberg BM, Epstein LG (1988) Comparison of simian immunodeficiency virus and human immunodeficiency virus encephalitides in the immature host. Ann Neurol 23(Suppl):S108–S112
121. Simon MA, Chalifoux LV, Ringler DJ (1992) Pathologic features of SIV-induced disease and the association of macrophage infection with disease evolution. AIDS Res Hum Retroviruses 8(3):327–337
122. Fox HS, Weed MR, Huitron-Resendiz S, Baig J, Horn TFW, Dailey PJ, Bischofberger N, Henriksen SJ (2000) Antiretroviral treatment normalizes neurophysiological but not movement abnormalities in simian immunodeficiency virus-infected monkeys. J Clin Invest 106(1):37–45
123. Williams K, Westmoreland S, Greco J, Ratai E, Lentz M, Kim WK, Fuller RA, Kim JP, Autissier P, Sehgal PK, Schinazi RF, Bischofberger N, Piatak M, Lifson JD, Masliah E, Gonzalez RG (2005) Magnetic resonance spectroscopy reveals that activated monocytes contribute to neuronal injury in SIV neuroAIDS. J Clin Invest 115(9):2534–2545
124. Altfeld M, Rosenberg ES, Shankarappa R, Mukherjee JS, Hecht FM, Eldridge RL, Addo MM, Poon SH, Phillips MN, Robbins GK, Sax PE, Boswell S, Kahn JO, Brander C, Goulder PJ, Levy JA, Mullins JI, Walker BD (2001) Cellular immune responses and viral diversity in individuals treated during acute and early HIV-1 infection. J Exp Med 193(2):169–180
125. Rosenberg E, Cotton D (1997) Primary HIV infection and the acute retroviral syndrome. AIDS Clin Care 9(3):19, 23–25
126. McArthur JC, Hoover DR, Bacellar H, Miller EN, Cohen BA, Becker JT, Graham NM, McArthur JH, Selnes OA, Jacobson LP et al (1993) Dementia in AIDS patients: incidence and risk factors. Multicenter AIDS Cohort Study. Neurology 43(11):2245–2252
127. Lentz MR, Kim JP, Westmoreland SV, Greco JB, Fuller RA, Ratai EM, He J, Sehgal PK, Halpern EF, Lackner AA, Masliah E, Gonzalez RG (2005) Quantitative neuropathologic correlates of changes in ratio of N-acetylaspartate to creatine in macaque brain. Radiology 235(2):461–468
128. Blusztajn JK, Wurtman RJ (1983) Choline and cholinergic neurons. Science 221 (4611):614–620
129. Kim JP, Lentz MR, Westmoreland SV, Greco JB, Ratai EM, Halpern E, Lackner AA, Masliah E, Gonzalez RG (2005) Relationships between astrogliosis and ^{1}H MR spectroscopic measures of brain choline/creatine and myo-inositol/creatine in a primate model. AJNR Am J Neuroradiol 26(4):752–759
130. Fuller RA, Westmoreland SV, Ratai E, Greco JB, Kim JP, Lentz MR, He J, Sehgal PK, Masliah E, Halpern E, Lackner AA, Gonzalez RG (2004) A prospective longitudinal in vivo ^{1}H MR spectroscopy study of the SIV/macaque model of neuroAIDS. BMC Neurosci 5(1):10
131. Westmoreland SV, Halpern E, Lackner AA (1998) Simian immunodeficiency virus encephalitis in rhesus macaques is associated with rapid disease progression. J Neurovirol 4 (3):260–268
132. Zink MC, Suryanarayana K, Mankowski JL, Shen A, Piatak M Jr, Spelman JP, Carter DL, Adams RJ, Lifson JD, Clements JE (1999) High viral load in the cerebrospinal fluid and brain correlates with severity of simian immunodeficiency virus encephalitis. J Virol 73 (12):10480–10488
133. Zink M, Clements J (2000) A rapid, reproducible model of AIDS and encephelitis in SIV infected macaques demonstrates the role of viral load in CNS disease. NeuroAIDS 3(5): online—October
134. Zink MC, Clements JE (2002) A novel simian immunodeficiency virus model that provides insight into mechanisms of human immunodeficiency virus central nervous system disease. J Neurovirol 8(Suppl 2):42–48
135. Weed MR, Hienz RD, Brady JV, Adams RJ, Mankowski JL, Clements JE, Zink MC (2003) Central nervous system correlates of behavioral deficits following simian immunodeficiency virus infection. J Neurovirol 9(4):452–464

136. Schmitz JE, Kuroda MJ, Santra S, Sasseville VG, Simon MA, Lifton MA, Racz P, Tenner-Racz K, Dalesandro M, Scallon BJ, Ghrayeb J, Forman MA, Montefiori DC, Rieber EP, Letvin NL, Reimann KA (1999) Control of viremia in simian immunodeficiency virus infection by CD8+ lymphocytes. Science 283(5403):857–860
137. Schmitz JE, Simon MA, Kuroda MJ, Lifton MA, Ollert MW, Vogel CW, Racz P, Tenner-Racz K, Scallon BJ, Dalesandro M, Ghrayeb J, Rieber EP, Sasseville VG, Reimann KA (1999) A nonhuman primate model for the selective elimination of CD8+ lymphocytes using a mouse-human chimeric monoclonal antibody. Am J Pathol 154(6):1923–1932
138. Williams K, Alvarez X, Lackner AA (2001) Central nervous system perivascular cells are immunoregulatory cells that connect the CNS with the peripheral immune system. Glia 36 (2):156–164
139. Domercq M, Matute C (2004) Neuroprotection by tetracyclines. Trends Pharmacol Sci 25 (12):609–612
140. Baptiste DC, Fehlings MG (2006) Pharmacological approaches to repair the injured spinal cord. J Neurotrauma 23(3–4):318–334
141. Stirling DP, Koochesfahani KM, Steeves JD, Tetzlaff W (2005) Minocycline as a neuroprotective agent. Neuroscientist 11(4):308–322
142. Tikka T, Fiebich BL, Goldsteins G, Keinanen R, Koistinaho J (2001) Minocycline, a tetracycline derivative, is neuroprotective against excitotoxicity by inhibiting activation and proliferation of microglia. J Neurosci 21(8):2580–2588
143. Lin S, Zhang Y, Dodel R, Farlow MR, Paul SM, Du Y (2001) Minocycline blocks nitric oxide-induced neurotoxicity by inhibition p38 MAP kinase in rat cerebellar granule neurons. Neurosci Lett 315(1–2):61–64
144. Si Q, Cosenza M, Kim MO, Zhao ML, Brownlee M, Goldstein H, Lee S (2004) A novel action of minocycline: inhibition of human immunodeficiency virus type 1 infection in microglia. J Neurovirol 10(5):284–292
145. Dutta K, Mishra MK, Nazmi A, Kumawat KL, Basu A (2010) Minocycline differentially modulates macrophage mediated peripheral immune response following Japanese encephalitis virus infection. Immunobiology 215(11):884–893
146. Zink MC, Uhrlaub J, DeWitt J, Voelker T, Bullock B, Mankowski J, Tarwater P, Clements J, Barber S (2005) Neuroprotective and anti-human immunodeficiency virus activity of minocycline. JAMA 293(16):2003–2011
147. Ratai EM, Bombardier JP, Joo CG, Annamalai L, Burdo TH, Campbell J, Fell R, Hakimelahi R, He J, Autissier P, Lentz MR, Halpern EF, Masliah E, Williams KC, Westmoreland SV, Gonzalez RG (2010) Proton magnetic resonance spectroscopy reveals neuroprotection by oral minocycline in a nonhuman primate model of accelerated NeuroAIDS. PLoS One 5(5):e10523
148. Ratai EM, Fell R, He J, Piatak M, Lifson JD, Burdo TH, Campbell J, Autissier P, Annamalai L, Masliah E, Westmoreland SV, Williams KC, Gilberto Gonzalez R (2012) Reduction of brain virus by minocycline and combination anti-retroviral therapy produces neuronal protection in a primate model of AIDS. In: ISMRM 20th Annual Meeting & Exhibition, Melbourne
149. Campbell JH, Burdo TH, Autissier P, Bombardier JP, Westmoreland SV, Soulas C, Gonzalez RG, Ratai EM, Williams KC (2011) Minocycline inhibition of monocyte activation correlates with neuronal protection in SIV NeuroAIDS. PLoS One 6(4):e18688
150. Lentz MR, Kim WK, Kim H, Soulas C, Lee V, Venna N, Halpern EF, Rosenberg ES, Williams K, Gonzalez RG (2011) Alterations in brain metabolism during the first year of HIV infection. J Neurovirol 17(3):220–229
151. Sailasuta N, Ross W, Ananworanich J, Chalermchai T, DeGruttola V, Lerdlum S, Pothisri M, Busovaca E, Ratto-Kim S, Jagodzinski L, Spudich S, Michael N, Kim JH, Valcour V (2012) Change in brain magnetic resonance spectroscopy after treatment during acute HIV infection. PLoS One 7(11):e49272

152. Valcour V, Chalermchai T, Sailasuta N, Marovich M, Lerdlum S, Suttichom D, Suwanwela NC, Jagodzinski L, Michael N, Spudich S, van Griensven F, de Souza M, Kim J, Ananworanich J (2012) Central nervous system viral invasion and inflammation during acute HIV infection. J Infect Dis 206(2):275–282
153. Peluso MJ, Meyerhoff DJ, Price RW, Peterson J, Lee E, Young AC, Walter R, Fuchs D, Brew BJ, Cinque P, Robertson K, Hagberg L, Zetterberg H, Gisslen M, Spudich S (2013) Cerebrospinal fluid and neuroimaging biomarker abnormalities suggest early neurological injury in a subset of individuals during primary HIV infection. J Infect Dis 207(11):1703–1712
154. Masters MC, Ances BM (2014) Role of neuroimaging in HIV-associated neurocognitive disorders. Semin Neurol 34(1):89–102
155. Menon DK, Baudouin CJ, Tomlinson D, Hoyle C (1990) Proton MR spectroscopy and imaging of the brain in AIDS: evidence of neuronal loss in regions that appear normal with imaging. J Comput Assist Tomogr 14(6):882–885
156. Salvan AM, Vion-Dury J, Confort-Gouny S, Nicoli F, Lamoureux S, Cozzone PJ (1997) Brain proton magnetic resonance spectroscopy in HIV-related encephalopathy: identification of evolving metabolic patterns in relation to dementia and therapy. AIDS Res Hum Retroviruses 13(12):1055–1066
157. Tracey I, Carr CA, Guimaraes AR, Worth JL, Navia BA, Gonzalez RG (1996) Brain choline-containing compounds are elevated in HIV-positive patients before the onset of AIDS dementia complex: a proton magnetic resonance spectroscopic study. Neurology 46 (3):783–788
158. Marcus CD, Taylor-Robinson SD, Sargentoni J, Ainsworth JG, Frize G, Easterbrook PJ, Shaunak S, Bryant DJ (1998) 1H MR spectroscopy of the brain in HIV-1-seropositive subjects: evidence for diffuse metabolic abnormalities. Metab Brain Dis 13(2):123–136
159. Moller HE, Vermathen P, Lentschig MG, Schuierer G, Schwarz S, Wiedermann D, Evers S, Husstedt IW (1999) Metabolic characterization of AIDS dementia complex by spectroscopic imaging. J Magn Reson Imaging 9(1):10–18
160. von Giesen HJ, Wittsack HJ, Wenserski F, Koller H, Hefter H, Arendt G (2001) Basal ganglia metabolite abnormalities in minor motor disorders associated with human immunodeficiency virus type 1. Arch Neurol 58(8):1281–1286
161. Anthony IC, Ramage SN, Carnie FW, Simmonds P, Bell JE (2005) Influence of HAART on HIV-related CNS disease and neuroinflammation. J Neuropathol Exp Neurol 64(6):529–536
162. Anthony IC, Bell JE (2008) The neuropathology of HIV/AIDS. Int Rev Psychiatry 20 (1):15–24
163. Chang L, Ernst T, Leonido-Yee M, Witt M, Speck O, Walot I, Miller EN (1999) Highly active antiretroviral therapy reverses brain metabolite abnormalities in mild HIV dementia. Neurology 53(4):782–789
164. Schifitto G, Navia BA, Yiannoutsos CT, Marra CM, Chang L, Ernst T, Jarvik JG, Miller EN, Singer EJ, Ellis RJ, Kolson DL, Simpson D, Nath A, Berger J, Shriver SL, Millar LL, Colquhoun D, Lenkinski R, Gonzalez RG, Lipton SA (2007) Memantine and HIV-associated cognitive impairment: a neuropsychological and proton magnetic resonance spectroscopy study. AIDS 21(14):1877–1886
165. Schifitto G, Yiannoutsos CT, Ernst T, Navia BA, Nath A, Sacktor N, Anderson C, Marra CM, Clifford DB (2009) Selegiline and oxidative stress in HIV-associated cognitive impairment. Neurology 73(23):1975–1981
166. Pfefferbaum A, Adalsteinsson E, Sullivan EV (2005) Cortical NAA deficits in HIV infection without dementia: influence of alcoholism comorbidity. Neuropsychopharmacology 30 (7):1392–1399
167. Chang L, Ernst T, Speck O, Grob CS (2005) Additive effects of HIV and chronic methamphetamine use on brain metabolite abnormalities. Am J Psychiatry 162(2):361–369
168. Marra CM, Zhao Y, Clifford DB, Letendre S, Evans S, Henry K, Ellis RJ, Rodriguez B, Coombs RW, Schifitto G, McArthur JC, Robertson K (2009) Impact of combination

antiretroviral therapy on cerebrospinal fluid HIV RNA and neurocognitive performance. AIDS 23(11):1359–1366
169. Schweinsburg BC, Taylor MJ, Alhassoon OM, Gonzalez R, Brown GG, Ellis RJ, Letendre S, Videen JS, McCutchan JA, Patterson TL, Grant I (2005) Brain mitochondrial injury in human immunodeficiency virus-seropositive (HIV+) individuals taking nucleoside reverse transcriptase inhibitors. J Neurovirol 11(4):356–364
170. Keller MA, Venkatraman TN, Thomas A, Deveikis A, LoPresti C, Hayes J, Berman N, Walot I, Padilla S, Johnston-Jones J, Ernst T, Chang L (2004) Altered neurometabolite development in HIV-infected children: correlation with neuropsychological tests. Neurology 62(10):1810–1817
171. Mbugua K, Holmes MJ, Hess AT, Little F, Cotton MF, Dobbels E, van der Kouwe AJ, Laughton B, Meintjes EM (2014) Effects of ART timing and HIV progression on neurometabolite levels in basal ganglia at age 5 years. In: 20th Annual Meeting of the Organization for Human Brain Mapping, Hamburg, Germany
172. Holmes MJ, Mbugua K, Little F, Cotton MF, van der Kouwe AJ, Laughton B, Meintjes EM (2014) A longitudinal study of the effects of HIV exposure on metabolite levels in the Midfrontal Gray Matter in children: at 5 and 7 years. In: 20th Annual AIDS conference, Melbourne, Australia
173. Banakar S, Thomas MA, Deveikis A, Watzl JQ, Hayes J, Keller MA (2008) Two-dimensional ^{1}H MR spectroscopy of the brain in human immunodeficiency virus (HIV)-infected children. J Magn Reson Imaging 27(4):710–717
174. Thomas MA, Yue K, Binesh N, Davanzo P, Kumar A, Siegel B, Frye M, Curran J, Lufkin R, Martin P, Guze B (2001) Localized two-dimensional shift correlated MR spectroscopy of human brain. Magn Reson Med 46(1):58–67
175. Trabesinger AH, Boesiger P (2001) Improved selectivity of double quantum coherence filtering for the detection of glutathione in the human brain in vivo. Magn Reson Med 45(4):708–710
176. Trabesinger AH, Weber OM, Duc CO, Boesiger P (1999) Detection of glutathione in the human brain in vivo by means of double quantum coherence filtering. Magn Reson Med 42(2):283–289
177. Terpstra M, Henry PG, Gruetter R (2003) Measurement of reduced glutathione (GSH) in human brain using LCModel analysis of difference-edited spectra. Magn Reson Med 50(1):19–23
178. Terpstra M, Marjanska M, Henry PG, Tkac I, Gruetter R (2006) Detection of an antioxidant profile in the human brain in vivo via double editing with MEGA-PRESS. Magn Reson Med 56(6):1192–1199
179. Srinivas A, Dias BF (2008) Antioxidants in HIV positive children. Indian J Pediatr 75(4):347–350
180. Markovic I, Clouse KA (2004) Recent advances in understanding the molecular mechanisms of HIV-1 entry and fusion: revisiting current targets and considering new options for therapeutic intervention. Curr HIV Res 2(3):223–234
181. Provencher SW (1993) Estimation of metabolite concentrations from localized in vivo proton NMR spectra. Magn Reson Med 30(6):672–679
182. Tkac I, Oz G, Adriany G, Ugurbil K, Gruetter R (2009) In vivo ^{1}H NMR spectroscopy of the human brain at high magnetic fields: metabolite quantification at 4T vs. 7T. Magn Reson Med 62(4):868–879
183. Ratai EM, Pilkenton S, He J, Fell R, Bombardier JP, Joo CG, Lentz MR, Kim WK, Burdo TH, Autissier P, Annamalai L, Curran E, O'Neil SP, Westmoreland SV, Williams KC, Masliah E, Gilberto Gonzalez R (2011) CD8+ lymphocyte depletion without SIV infection does not produce metabolic changes or pathological abnormalities in the rhesus macaque brain. J Med Primatol 40(5):300–309
184. Gonen O, Liu S, Goelman G, Ratai EM, Pilkenton S, Lentz MR, Gonzalez RG (2008) Proton MR spectroscopic imaging of rhesus macaque brain in vivo at 7T. Magn Reson Med 59 (4):692–699

185. Caruso PA, Johnson J, Thibert R, Rapalino O, Rincon S, Ratai EM (2013) The use of magnetic resonance spectroscopy in the evaluation of epilepsy. Neuroimaging Clin N Am 23(3):407–424
186. Hess AT, Tisdall MD, Andronesi OC, Meintjes EM, van der Kouwe AJ (2011) Real-time motion and B_0 corrected single voxel spectroscopy using volumetric navigators. Magn Reson Med 66(2):314–323
187. Tisdall MD, Hess AT, Reuter M, Meintjes EM, Fischl B, van der Kouwe AJ (2012) Volumetric navigators for prospective motion correction and selective reacquisition in neuroanatomical MRI. Magn Reson Med 68(2):389–399
188. Bogner W, Hess AT, Gagoski B, Tisdall MD, van der Kouwe AJ, Trattnig S, Rosen B, Andronesi OC (2013) Real-time motion- and B-correction for LASER-localized spiral-accelerated 3D-MRSI of the brain at 3T. Neuroimage 88C:22–31
189. Kuperman JM, Brown TT, Ahmadi ME, Erhart MJ, White NS, Roddey JC, Shankaranarayanan A, Han ET, Rettmann D, Dale AM (2011) Prospective motion correction improves diagnostic utility of pediatric MRI scans. Pediatr Radiol 41(12):1578–1582
190. Brown TT, Kuperman JM, Erhart M, White NS, Roddey JC, Shankaranarayanan A, Han ET, Rettmann D, Dale AM (2010) Prospective motion correction of high-resolution magnetic resonance imaging data in children. Neuroimage 53(1):139–145
191. Andrews-Shigaki BC, Armstrong BS, Zaitsev M, Ernst T (2011) Prospective motion correction for magnetic resonance spectroscopy using single camera Retro-Grate reflector optical tracking. J Magn Reson Imaging 33(2):498–504
192. Keating B, Ernst T (2012) Real-time dynamic frequency and shim correction for single-voxel magnetic resonance spectroscopy. Magn Reson Med 68(5):1339–1345

Chapter 12
Magnetic Resonance Spectroscopy in Epilepsy

Jullie W. Pan

Abstract This review discusses the utilization of MR spectroscopy and spectroscopic imaging for epilepsy from a clinical localization and research perspective. As a relatively common neurological problem that affects the entire age range, the understanding and management of epilepsy has benefited substantially from the recent past improvements in anatomical MRI quality and resolution. With multiple facets of epilepsy dysfunction identified metabolically and neurophysiologically, the sensitivity of metabolic and functional MR imaging to such processes suggest that continued MR development can be important as well. Metabolically and spectroscopically, much of the challenge for the most common type of clinical epilepsy (localization related) is the sizable interpatient variability for both location of abnormality and severity of injury as well as the need to adequately evaluate the neocortical ribbon. These factors combine to place emphasis on developments at high field for SNR and voxel size, acceleration, and adequate lipid suppression. From a basic science perspective, substantial work has shown that metabolic and cellular changes are well detected by MRS early and late in the process of epileptogenesis, consistent with major shifts in neuronal and astrocytic processes. Thus, the role of MR spectroscopy has much room to progress for clinical and research applications in epilepsy.

Keywords Localization-related epilepsy • Primary generalized epilepsy • Medial temporal lobe • Bioenergetics • Glial/neuronal unit • GABA • Glutamate

Epilepsy: The Role for MR Spectroscopy in the Clinic and the Bench

Epilepsy is a chronic neurological disorder characterized by spontaneous recurrent seizures. It has a prevalence of about 6–7/1000 people, varying between 0.5 and 0.8 % in developed countries to less affluent countries, respectively [1, 2]

J.W. Pan (✉)
Department of Neurology, University of Pittsburgh School of Medicine,
3471 5th Ave, Ste 811, Pittsburgh, PA 15213, USA
e-mail: JWP44@pitt.edu

G. Öz (ed.), *Magnetic Resonance Spectroscopy of Degenerative Brain Diseases*,
Contemporary Clinical Neuroscience, DOI 10.1007/978-3-319-33555-1_12

Jallon P, 1997. Multiple types of epilepsy exist, and in its more common form of "localization related" or focal epilepsy, seizures are typically believed to arise from a discrete cerebral location. This nomenclature is used to distinguish it from "primary generalized epilepsy," which is substantially less common than focal epilepsy (15–20 % of all the epilepsies), [2]. In primary generalized epilepsy, the seizures are not thought to arise from any specific brain region and can instead appear to involve the entire brain simultaneously. However, it should be noted that many focal onset seizures can evolve into what appears to be a generalized seizure. In focal epilepsy, the transient behavior that is seen in a seizure is variable between patients, depending on the location of seizure onset, e.g., a momentary lapse of attention, staring to whole body convulsive events. Importantly, the region of seizure onset is believed to be at, adjacent or linked to a direct site of brain dysfunction or injury. Through this region, the brain's electrical activity is abnormal, and the overt rapid appearance of the seizure results from propagation of abnormal electrical activity through the brain.

While the majority of patients with epilepsy can be managed with medications, ~35 % of patients find that medications are insufficient for seizure control [2, 3]. In these medically intractable cases, many studies have shown that if the region of seizure onset can be identified, surgical resection of that region can be highly effective to treat and possibly cure the epilepsy. One of the most common types of focal epilepsy that is readily identified as such is medial temporal lobe epilepsy, MTLE. MTLE has a characteristic seizure semiology and with improvements in high field (3 T) structural imaging, is generally well identified with T1-weighted and FLAIR imaging, especially as it is also anatomically restricted to the regions of the hippocampi and amygdala. Success of temporal lobe epilepsy surgery is excellent, at approximately ~66 % with a range of 50–85 % seizure freedom at 2 years follow-up [6, 7].

Thus, as MTLE is relatively straightforward to identify and has an outstanding response to surgery, the belief is that other localization-related epilepsies may or should be similarly responsive. Data have supported this view (e.g., Hauptman and Mathern [8] with focal cortical dysplasias); however, in non-MTLE or extratemporal epilepsy, structural imaging can be commonly negative or ambiguous for identification of seizure onset. With careful clinical description of seizure behavior, the lobe or region of seizure onset can generally be hypothesized; however, extensive testing is still required for adequate surgical planning. Multiple imaging methods are used to attempt to identify the region of greatest cerebral "irritability," now commonly including FDG-PET, magnetoencephalography (MEG), SPECT, and MR spectroscopy. For this goal, our group and others have suggested that MRSI can provide pertinent data. While this role is clearest on the major question of clinical localization, it should be noted that other spectroscopic measurements such as GABA and glutamate are also of obvious interest for human epilepsy research.

From a basic science perspective, metabolic dysfunction has been continually considered in the evaluation of epilepsy, seizures, and epileptogenesis. A seizure, defined as that acute process of aberrant cerebral electrical propagation, has long been known to be an energetically demanding event (e.g., [9] and earlier). The fact that the ketogenic diet, which shifts the brain from its normally preferred fuel of glucose to ketone bodies, can strongly affect seizure frequency and severity, has been a very compelling basis to better understand the interaction between metabolic function and seizures. It is therefore not surprising that MR spectroscopy is of particular interest for basic science studies of epilepsy. This is exemplified by early animal model studies, e.g., [10–12], finding that seizures induce specific changes in lactate, NAA, myo-inositol, and also possibly in high energy phosphates.

In studying the development of epilepsy, it is important to recall that from both human and animal model experience, the pathophysiology of most focal epilepsy is initiated by a cerebral insult (e.g., the fairly common event of fever-induced or febrile seizures, head trauma or infection), followed by a latent period that precedes the onset of overt spontaneous recurrent seizures, i.e., epilepsy. Given several clinical studies (e.g., [13]) that have shown that treating the immediate seizure or status epilepticus (i.e., a prolonged seizure, occasionally less responsive to medications) does not necessarily prevent the subsequent development of epilepsy, it is clear that epileptogenesis is not identical to ictogenesis, i.e., the process that leads to spontaneous recurrent seizures is not the same as the acute process that leads to the occurrence of any given seizure. Thus in animal models, much interest is focused on epileptogenesis, hypothesized to occur during the latent period between insult to recurrent seizures. Several physiologic and pathophysiologic processes have been identified, including development of novel aberrant circuits (e.g., mossy fiber sprouting), abnormal neuroregeneration, GABA receptor shifts, inflammation, glial dysfunction and mitochondrial injury (for review, see [14, 15]), or a combination therein. Given the brain's obvious complexity, its endogenous regulation and the variety of insults that can all result in epilepsy, the response to a major injury of this kind may very well be multifactorial, exhibiting several of these pathophysiologic processes. From an imaging and spectroscopic perspective, there is interest in studying the process from a metabolic dysfunction viewpoint, although it is clear that identification of an imaging biomarker that characterizes this process is also of obvious importance.

Thus overall, the role of MR spectroscopy in epilepsy has at least two major aspects. First, several groups [16–20] have suggested that MR spectroscopic imaging can contribute toward the challenging problem for localization of regions of brain dysfunction, i.e., identify possible regions of seizure onset. It is also clear that in human research, measurements of GABA, glutamate may provide similar localization information and/or clarify the nature of seizure control or drug effect. Second, from a basic science view, evaluating the role and consistency of MR detectable dysfunction in epileptogenesis may also be of major importance [21, 22]. Not only might it better define the metabolic distortions in epileptogenesis, it and other MRI methods of early detection of epileptogenesis [23] may be especially useful given the recent developments of novel neuroprotective therapies that target epileptogenesis. For both of these avenues, an interdisciplinary approach is needed, given that the scientific, clinical, and imaging aspects of the problem are still evolving.

Epilepsy as a Neurodegenerative Disorder

Before further discussion on MR spectroscopic imaging in epilepsy, it should be stated that the classification of epilepsy as a neurodegenerative disorder, characterized by progressive neuronal death commonly linked with mitochondrial dysfunction, can be debated. To address this, two factors may be considered. First, as stated earlier, mitochondrial dysfunction with its attendant abnormalities in calcium handling and free radical oxygen species generation (for review, [24]) is hypothesized as a contributing factor in epileptogenesis which with its key function for cellular bioenergetics has been strongly implicated in a self-sustaining cycle that can propagate into progressive injury. While a single brief seizure itself is not strongly associated with neuronal death (for review, [25]), recurrent and severe seizures (e.g., status epilepticus) are. Thus, even during the latent period before onset of spontaneous recurrent seizures, clinical experience and many animal models of epilepsy require a sustained period or multiple episodes of seizures or status to initiate the process of epileptogenesis. Second, it is clear that under conditions of mitochondrial disease such as Leigh syndrome, MELAS, MERRF, epilepsy is commonly a key phenotype, implying that epilepsy and recurrent seizures can be considered a symptom of mitochondrial dysfunction.

Thus, in either the development of epilepsy or the recurrent seizure condition, neuronal injury is present. Given that many clinical research and animal studies have shown there to be a strong correlation between seizure frequency and several pertinent indices, e.g., subsequent seizure control [26] and cognitive performance [27, 28], these data suggest that the injury from continuing seizures has broad impact, for both the epilepsy and brain function itself. As a result, approaches that can evaluate and quantify such degenerative injury remain of significant interest.

What Can Be Evaluated

In this chapter, we review the state of the art for MR spectroscopy in epilepsy. We stay with ^{1}H MR spectroscopy, as the availability of X-nucleus channels is generally limited with many human spectrometers. It should already be apparent that with the detection sensitivity of MR spectroscopy being in the range of 1–10 mM of small molecules, this means that the compounds of interest are most frequently components of metabolic pathways. Given the hypothesized dysfunction of neuronal, astrocytic, and mitochondrial associated injury in epilepsy [21, 29, 30], the interest in the MR spectroscopic evaluation of epilepsy is logical.

The most commonly studied ^{1}H spectroscopic compounds in MRSI are *N*-acetyl aspartate (NAA), creatine, myo-inositol, glutamate, glutamine, and GABA. As stated in earlier chapters, NAA is synthesized only in neuronal mitochondria [31, 32] and is strongly correlated with oxidative metabolism [33–35]. As a result, many studies of a variety of brain disorders have found NAA to be an informative measure of neuronal function. Creatine, as a key component of phosphocreatine, is highly useful as a normalization factor for many bioenergetic parameters, being consistent over a cross section of species and within a given tissue type, as discussed by Connett [36]. While found in both neurons and astrocytes, the highest concentration of creatine is in astrocytes and thus when considering an integrated unit of neuronal function, many groups have used the ratio of NAA/Cr as a normalized parameter. We characterize NAA/Cr as reflecting the bioenergetics of the glial/neuronal unit ("bGNU"), finding it to be very informative for identifying regions of energetic and neuronal dysfunction [37–39].

The comparison between MRSI and 18-fluoro-deoxyglucose-PET (FDG-PET) is inevitable, and while these are clearly complementary measurements, it is worthwhile considering some of the pertinent differences and similarities. FDG-PET evaluates total glucose consumption and has been long used to measure differences in glucose consumption in varying tissue types (FDG uptake in gray matter is approximately 3× that of white matter [40]) and with various states of activation. The amount of uptake reflects the amount of tissue present in the voxel of interest, and thus decreases in uptake can result from both decreased cerebral consumption and/or tissue atrophy (with potential for distortion of white and gray matter contributions). There is commonly variation in voxel size or point spread function with location in PET, but given improvements with human high resolution research tomograph (HRRT) cameras, the voxel resolution has improved to better than 2–3 mm in-plane resolution [41]. For application to epilepsy, the success rate of FDG-PET in identifying the region of seizure onset very much depends on the population studied with the best localization rates of ~80–90 % in temporal lobe epilepsy, although it is probably less successful in the nonlesional neocortical epilepsy patients (for review, [6, 42, 43]). This lesser success rate reflects the challenge in neocortical epilepsy, with the region of seizure onset commonly much less well defined, with variations in propagation paths and volume of injury.

In MRSI, the use of the NAA/Cr ratio largely eliminates the sensitivity of the parameter to tissue volume due to minimal NAA and Cr in the CSF; however, there is sensitivity to tissue type (gray, white matter), with the majority of workers finding that NAA/Cr is smaller in gray than in white matter (reflecting primarily a higher creatine concentration in gray [44–46]). With most sampling strategies, the sampling volume of MRSI is regionally constant, and for practicable time limits of study, are typically 0.64–2 cc. Thus in epilepsy, where tissue atrophy can be variably gross or subtle, the ability of MRSI to detect dysfunction using the bGNU with tissue type correction is potentially excellent as it does not require use of asymmetry indices and is relatively independent of tissue atrophy.

MRSI of Medial Temporal Lobe Epilepsy

Studies initially performed in the 90s established the ability of ^{1}H-MRS to lateralize the seizure focus in temporal lobes and to test the validity of MRSI to already established methods of localization, using video-EEG as gold standard as well as MRI-volumetry (MRIV), FDG-PET, and SPECT. Hugg et al. [47] initially demonstrated a significant asymmetry of NAA left/right metabolite ratios and further studies have demonstrated comparable results, e.g., Cendes et al. [48] and Kuzniecky et al. [49] in larger patient groups. In addition, ^{1}H-MRS was also sensitive in detecting bilateral dysfunction. The degree of asymmetry in NAA/Cr ratios correlated with the degree of one sidedness of EEG abnormalities. These results support the fact that ^{1}H-MRS is a valid method even in bilateral cases presenting a high concordance to the degree of bilateral EEG findings in patients with TLE. Our recent data [50–52] in temporal lobe epilepsy has demonstrated that the decrements in NAA are not just localized to the ipsilateral hippocampus, but consistent with existing PET studies, are also found in a network of involved limbic and subcortical nuclei.

Such studies, pertinent for seizure localization, have also been integrated with physiological research studies of epilepsy. For example from animal studies of mitochondrial dysfunction, it has been known that mild to moderate oxidative stress can also cause major abnormalities in GABA (and glutamate). Saransaari and Oja [53] studied the mouse hippocampus with variable levels of peroxide stress to show increases in basal GABA release, ranging from 30 to more than 550 %. These observations are of particular interest for epilepsy where GABA is thought to be a key component underlying the abnormal hyperexcitability. Given this and the known high energetic cost of neurotransmission and synaptic activity [54], we anticipated that GABA neurotransmission and metabolic function might be correlated, certainly in the seizure onset zone. This may be especially important given the recent in vitro work that has suggested GABA function may be either anti- or proconvulsant [55, 56]. This was evaluated in human epilepsy patients undergoing intracranial EEG and microdialysis analysis. Their microdialysis measurements of extracellular GABA (ecGABA) were compared to preoperative measures of the bGNU or NAA/Cr. All data from this study were acquired from the hippocampus, including patients with either hippocampal epilepsy or nonhippocampal (neocortical) epilepsy. Figure 12.1 shows the correlations from this study, finding very strong correlations between ecGABA and NAA/Cr. In the MTLE patients, ecGABA strongly negatively correlated with decreasing NAA/Cr, $R = -0.94$, $p < 0.001$ (Fig. 12.1) and implies that ecGABA and mitochondrial function are largely representing parallel processes, and appears consistent with the conclusions from Saransaari and Oja [53], and Woo et al. [56]. In contrast, with the neocortical epilepsy patients (and outside of the seizure onset zone), the relationship between ecGABA and NAA/Cr is also significant but positive with $R = +0.70$, $p < 0.015$. In these neocortical patients, the hippocampus being studied was ipsilateral to the cortical seizure onset region and thus may be a site of proximal propagation.

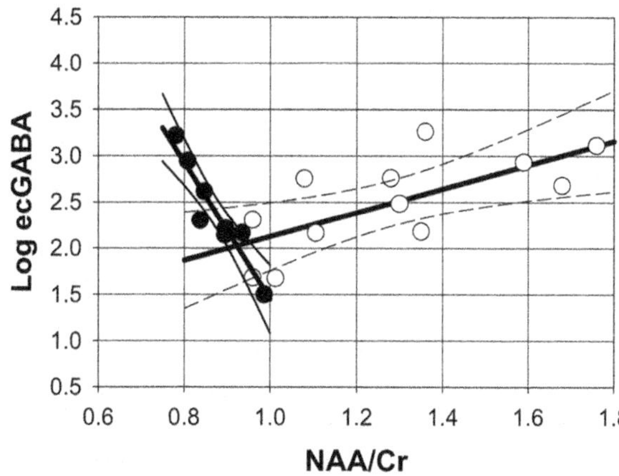

Fig. 12.1 Interictal extracellular GABA (ecGABA) levels are measured by quantitative zero flow microdialysis, which estimates the true basal concentrations of extracellular fluid neurochemicals such as glutamate, glutamine, and GABA [57]. *Closed circles* from hippocampal epilepsy patients; *open circles* from nonhippocampal (neocortical) epilepsy patients

Nonetheless, as it is not the seizure focus, these data remain conceptually similar to the results of Petroff et al. [58], who found that well outside the seizure focus (studying the occipital lobe), patients with better seizure control have higher tissue GABA levels.

While these data cannot specify which or several of these injurious processes may be ongoing in the seizure onset zone, it is evident that the relationship of ecGABA with NAA/Cr within the ipsilateral hippocampus is distinctive in comparison to ecGABA correlations in the non-MTLE group. It is notable that these two groups are largely separated by the NAA/Cr measurement, given that the dynamic range of ecGABA is similar. This suggests in the healthier (non-MTLE) vs. the diseased (MTLE) hippocampus, there is a real change in the function of ecGABA, e.g., potentially proconvulsant as suggested by Woo et al. [56].

Neocortical Epilepsy

The localization of extra-temporal and neocortical epilepsies is commonly much more challenging than solely temporal and medial temporal lobe epilepsy, much in part due to given the large volume of neocortex that has potential for seizure onset, which can be MRI negative or MRI ambiguous. Furthermore, there is also the potential for variable propagation paths and variation between different subjects for volume of seizure onset. Nonetheless, given that the success of surgery is high with accurate identification of seizure onset zone [6, 43, 59, 60], additional information that can help guide the localization process is desirable. As now routinely performed between neurosurgery and neurology, intracranial EEG monitoring is used to more accurately localize seizure onset; however, it is clear that positioning of electrodes is a critical step. If electrodes do not adequately sample the seizure onset zone, the likelihood for success for seizure localization is low.

Thus, it is obvious that for such challenging cases, there is a potential strong role for spectroscopic imaging to assist in the identification of candidate regions of seizure onset. As discussed earlier, the high sensitivity of the bGNU measure is a robust parameter to develop for this target. However starting from the relative success in studies of medial temporal lobe epilepsy, the required jump to neocortical epilepsy poses a major technologic challenge. The challenges include a commonly large target region that varies between different patients, thus requiring extended volume coverage. For example, temporal lobe (with its functions for memory and learning) seizure onset is a very common site for epilepsy (>50 % of all cases), but even so, is frequently not well distinguished from frontal lobe onset. The size of the seizure onset zone is variable and unknown as to the requisite voxel size of MRSI measurement, implying that to maintain adequate SNR, studies may use larger voxel sizes that can then dilute the metabolic abnormality. The requirement to visualize the cortical ribbon imposes a need for outstanding extracerebral lipid suppression which can be difficult for large volume spectroscopic imaging. While overall, epilepsy patients span the entire age range and are a highly cooperative patient group, very lengthy acquisition times commonly considered for spectroscopic imaging studies are intrinsically difficult. In spite of these many significant challenges, there have been a few groups that have targeted work in large volume spectroscopic imaging for epilepsy [17–19, 61, 62]. Maudsley et al. [17] developed whole brain echo-planar spectroscopic imaging in epilepsy at 3 T, achieving study durations of 25 min, FOV $280 \times 280 \times 180$ with a final sampling matrix of $50 \times 50 \times 18$, giving 70 ± 6 % coverage of the whole brain. The approach uses a moderate echo (TE 70 ms), global inversion recovery for lipid suppression (TI 198 ms) supplemented with a k-space extrapolation based on a scalp mask to reduce lipid ringing artifact. However after filtering the data based on spectral quality (<13 Hz linewidth), much of the entire temporal and inferior frontal lobes is excluded, explained as resulting from the well-known problems in field homogeneity with consequent difficulties for linewidth and spectral quality. This report on $n = 14$ patients concluded that the MRSI was helpful although varied substantially between patients, with detected dysfunction found in 7/14 subjects, the other 7 subjects with inconclusive findings. A figure from Maudsley et al. [17] is reshown here (Fig. 12.2) and shows the spectral variations in the temporal and parietal lobes.

Whether the 50 % detection result is due to the variability in patient severity of bGNU abnormality or sensitivity of the study is unknown. However, it is known that more than 50 % of surgical epilepsy patients have their seizure onset in the temporal lobe, with either initiation from or propagation into the medial temporal region and thus methods need to be able to consistently evaluate this region. This challenge is still present even at 1.5 T where the residual inhomogeneity is less; however, as reported by Mueller et al. [61], there can be problematic spectra in up to 50 % of all pixels in the temporal region.

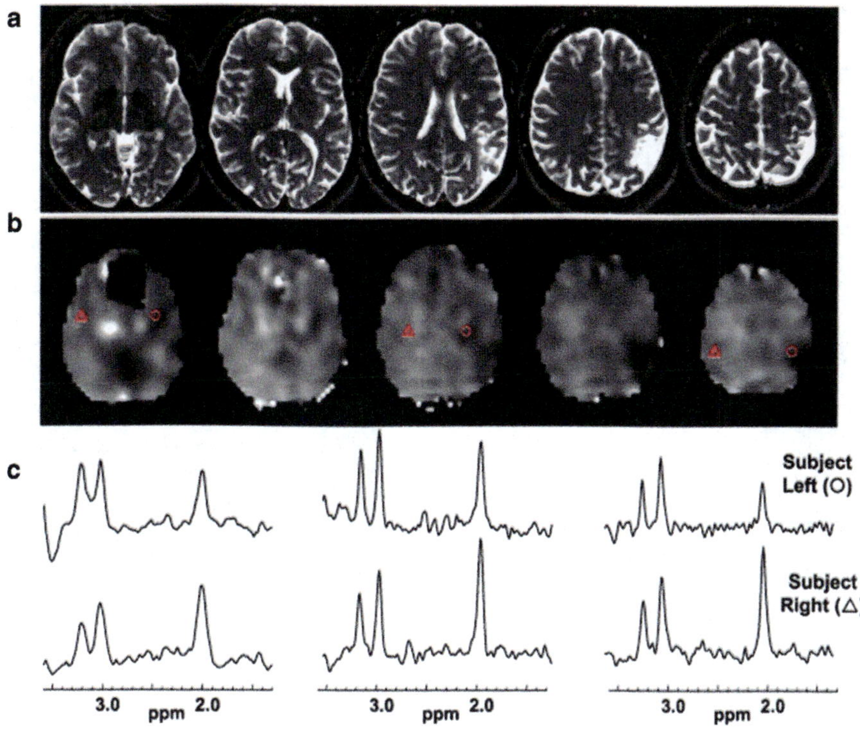

Fig. 12.2 From Maudsley et al. [17]. Results for subject 17 that exhibits a clearly defined left parietal lesion seen on the T2-weighted MRI (**a**) and by decreased NAA/Cr (**b**). Spectra selected from contralateral anatomical locations indicated by the *triangle* (subject right) and *circle* (subject left) symbols are shown in (**c**) for the *left*, *middle*, and *rightmost slices* shown

With the demonstrated improvement of SNR available at 7 T and the uncertainty on the severity of metabolic dysfunction in epilepsy, our group has developed 7 T MRSI for neocortical epilepsy. This initial work, performed to establish the methodology and to assess the range of severity of dysfunction, performed targeted whole slice MRSI studies from regions determined with available clinical information. The extent of overlap between the NAA/Cr abnormality with surgical resection (none, partial, or complete) was compared with patient outcome (International League Against Epilepsy (ILAE) classification dichotomized to I–III and IV–VI). These studies used TE/TR 40 ms/1.5 s and single slice acquisitions (acquisition time 14 min/slice), with high degree and order (3rd and 4th) B_0 shimming to accommodate large volumes of study as well as B_1 RF shimming to overcome known problems with B_1 inhomogeneity and amplitude present at 7 T [63, 64]. Figure 12.3a shows data from a patient with neocortical epilepsy with a history of meningitis who had unilateral (left) intracranial EEG coverage based on semiology. The resulting resection surgery included the L precuneus, which did overlap with the bGNU abnormality. Postoperatively this patient did well initially for 3 months

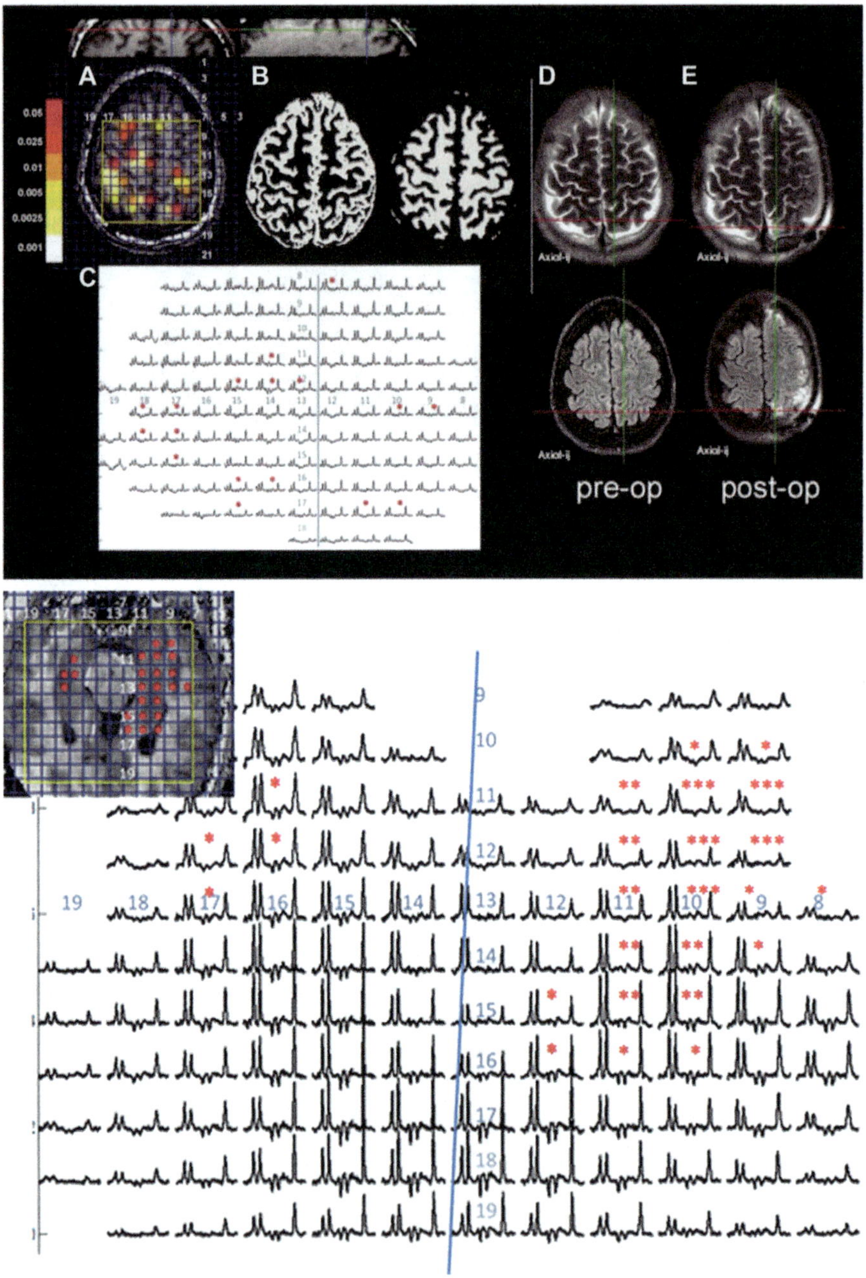

Fig. 12.3 (**a**) With permissions from Pan et al. [18, 19]. Data from a neocortical epilepsy patient: (*A*) scout with statistical overlay, (*B*) segmentation data showing gray and white matter masks, (*C*) spectra and (*D* and *E*) pre-op and post-op clinical imaging showing the area of resection in the precuneus. There are multiple areas of NAA/Cr abnormality seen in both hemispheres (shown with *stars*). (**b**) An MTLE patient who was MRI negative with a large extent of metabolic dysfunction over the left MTL greater than right MTL

but had an ILAE class IV outcome. Figure 12.3b shows temporal lobe data from a MTLE patient with a negative MRI, having had intractable epilepsy for a relatively short 3 years. These data were consistent with intracranial monitoring results, and as a group, all of the MTLE patients undergoing anterior temporal lobectomy resection had ILAE class I or II outcomes. In this initial group of 25 patients all of whom eventually underwent resective surgical treatment, we have assessed the coherence of the bGNU with resection region (i.e., did the region of resection overlap with metabolic abnormality or not) and eventual outcome. While it is clear that this is a small patient group, a Fisher's exact 3×2 contingency statistical test found that the concordance between MRSI and surgical resection was significantly related to good outcome, $p < 0.001$ [18, 19].

Thus, the question on whether the spectroscopic imaging of NAA/Cr has the adequate sensitivity to detect pertinent dysfunction in neocortical epilepsy seems promising. Nonetheless, it seems to be highly variable; in this short literature review, recent studies seem to range from ~50 % detection of abnormality [17] to higher rates of dysfunction detection [18, 19, 62]. In fact, as suggested by Mueller et al. [62], the widespread distribution of injury seen by MRSI in fact raises the possibility that the bGNU parameter could be too sensitive for seizure localization. As stated, at least part of this wide variability may be due to acquisition methodology; however, it is also clear that epilepsy patients can be highly variable. Depending on the nature of the epilepsy type (e.g., malformations, meningitis, traumatic, mesial temporal lobe, etc.) the extent and nature of metabolic aberrancy may be expected to differ. For example, as a group, malformations can be highly variable from both a structural imaging as well as etiologic view, some are subtly MRI detectable vs. those with multiple and large lesions, some are highly familial while others are most likely due to in utero insult. Large systematic MRSI studies of various epilepsy etiologies have not been done but an eight patient study [65] of several types of malformations (including polymicrogyria, dysplasias, heterotopia) found variable NAA/Cr abnormalities within and surrounding the lesion. Another cause for variability is the reasonably well-established view of epilepsy as a network disorder [50, 66] which could be expected to manifest in a distribution of injury that includes the propagation path(s) for the seizure. The practical relevance of this network is pertinent; e.g., propagated injury through the seizure network (which can be hypothesized as a deviant form of normal brain connectivity) is possibly a key basis for the relatively common presence of dual pathology, that condition in which medial temporal lobe epilepsy is identified in concert with a separate (typically ipsilateral) neocortical lesion. Thus, while not all abnormalities of the bGNU are necessarily epileptic, it is not surprising that a widespread distribution of abnormalities is seen and needs to be assessed individually in the context of the patient and their epilepsy. The data thus far have suggested that the identified abnormal NAA/Cr regions are informative and may be more viewed as candidate regions of seizure onset with additional study needed for their classification (e.g., seizure onset, multiple seizure onset vs. propagation vs. unrelated injury).

Measurements of GABA and Glutamate

Other important resonances identified for epilepsy include glutamate, GABA, myo-inositol, lactate, and glutamine. Glutamate and GABA, in particular, are important targets given their neurotransmission roles. However as coupled resonances, these compounds generally require more care in detection and analysis, either via short echo spectroscopy or through editing sequences. While short echo spectroscopy is a relatively successful approach, there are potential significant problems due to the variable macromolecule baseline that is present [67, 68]. Our group has targeted the detection of coupled resonances using a J-refocused coherence transfer (double echo) sequence that enables longer echo times, which minimizes the macromolecule baseline due to T2 decay and yet retains excellent sensitivity to coupled resonances. Figure 12.4a shows the performance of the coherence transfer sequence in comparison to a short echo acquisition and shows the elevated glutamine resonance in an epilepsy patient being treated with valproic

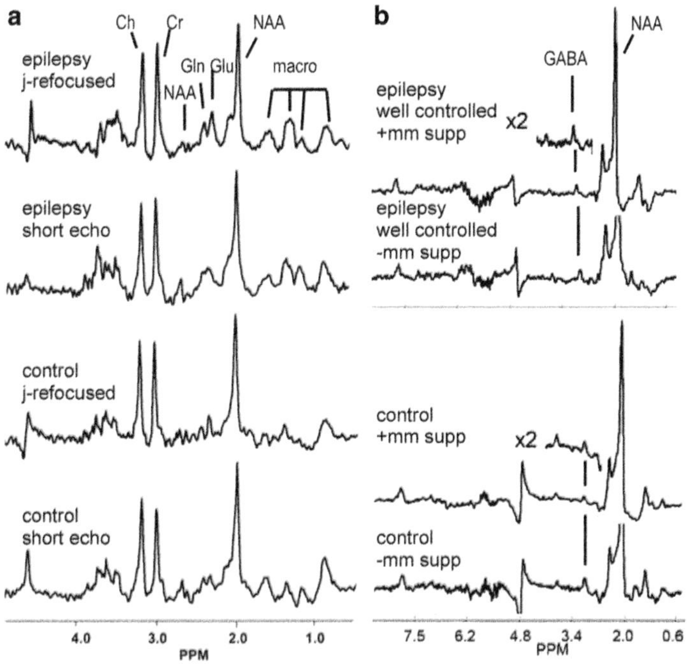

Fig. 12.4 (**a**) (*left*) J-refocused spectra from a control (*bottom*) and epilepsy patient on valproic acid (*top*). For both the control and patient, short echo and j-refocused spectra are shown. Key resonances are identified. The epilepsy patient shows a substantially increased glutamine resonance in comparison to control. (**b**)(*right*) GABA spectra from a control (*bottom*) and epilepsy patient whose seizures are well controlled (*top*). For both control and patient spectra, macromolecule (mm) suppressed and nonsuppressed spectra are shown. The epilepsy patient shows a substantially increased GABA resonance

acid. This glutamine change is consistent with the known hyperammonemia and hepatic changes that occur with use of valproic acid.

As the major inhibitory neurotransmitter, GABA detection is also of obvious interest for epilepsy. However, with its resonances obscured by many other metabolites (creatine 3.0 ppm, NAA 1.9 ppm, and amino acids 2.28 ppm) and relatively low concentrations of typically 0.8–2 mM, for optimal detection, a selection process such as spectral editing or a multiple quantum selection is preferred. Spectral editing has been used by several groups in epilepsy, such as Petroff et al. [58], finding that tissue GABA is increased in patients whose seizures are well controlled in comparison to poorly controlled patients. We have adapted the J-refocused approach to select for the C4 3.0 ppm resonance of GABA. Based on an initial inversion recovery suppression of the 3.0 ppm region, the coherence transfer sequence induces magnetization transfer from the C3 1.9 ppm to its coupled partner at C4 3.0 ppm. The GABA C4 3.0 ppm resonance is therefore detected without the overlapping creatine resonance. As a single shot acquisition (no scan-to-scan differencing needed) GABA is detected with an efficiency of ~50 %. As studied in the thalamus in a small epilepsy group (example spectra Fig. 12.4b), this study found that the thalamic GABA/Cr in well controlled, poorly controlled epilepsy patients vs. controls was 0.119 ± 0.013, 0.054 ± 0.013, and 0.071 ± 0.015, respectively [51, 52].

Animal Models of Epileptogenesis

MR spectroscopy has long been proposed to be useful to noninvasively evaluate the process of epileptogenesis [11, 12, 69]. The more recent work of Filibian et al. [21] used a pilocarpine rat model to study the process of epileptogenesis, using TE 10 ms PRESS and 16 μl voxel sizes in the hippocampus. This report found progressive increases in myo-inositol, glutathione with decreases in NAA in the immediate days after status epilepticus, most likely characterizing glial activation, edema, and neuronal injury (Fig. 12.5). In the chronic epileptic rat, similar findings were reported. It is clear that in this time course, many of the changes are metabolic, with comparatively less change seen in GABA and glutamate. This would be reasonably expected, given the known metabolic demands of seizures. However the NAA changes, interpreted as abnormalities in neuronal mitochondrial function, when taken in comparison with the myo-inositol changes, suggest a difference in the temporal response of neuronal vs. glial processes. This interesting study raises the question on what these changes may mean for epileptogenesis as a better understanding these early processes will suggest which cellular pools are most dynamic and may suggest avenues of approach and prediction for epileptogenesis. Given these well-established methods in animals, it will become more compelling to consider evaluating these metabolites in patients who are at risk for development of epilepsy.

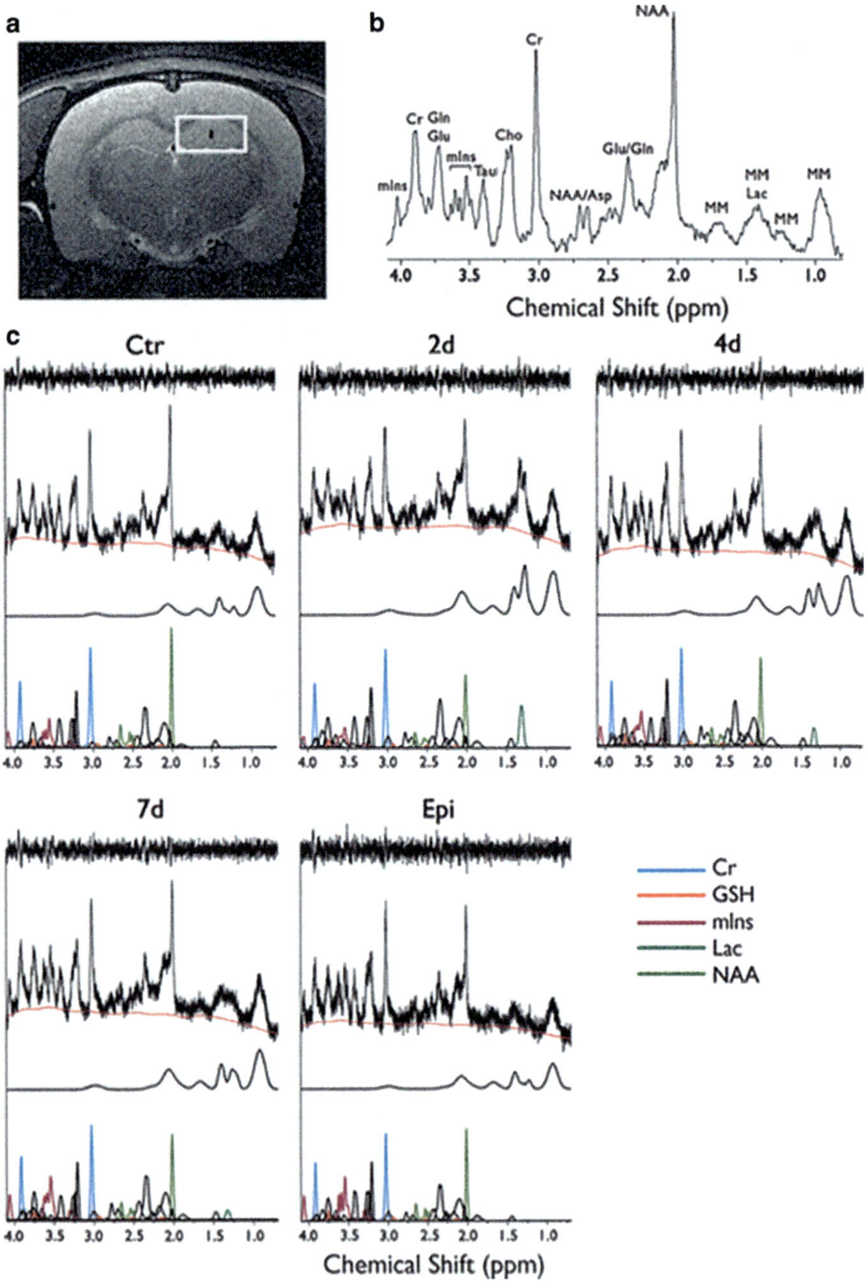

Fig. 12.5 From Filibian et al. [21]: ^{1}H spectra (7 T 16 μl single voxel TE10ms PRESS) from a pilocarpine rat model of epilepsy, showing progression of abnormalities. Fitted data from Panel (**c**): spectrum, macromolecule, fitted resonances (*top noise line* represents difference residual)

Conclusions: Imaging Challenges in Humans and Animals and Promises

Altogether, there are many aspects to the metabolic dysfunction seen in the brain injury of seizures and epilepsy that result from the close physiological relationship between brain function and metabolism. From a basic science view, it is clear that there are sizable metabolic shifts occurring early in the process of seizures and epileptogenesis. The changes seen by MR spectroscopy are consistent with dynamic glial and neuronal responses to seizure injury [11, 21]. This type of work opens an avenue toward defining and understanding the pathologic and pathophysiologic responses (e.g., the roles of NAA, myo-inositol, glutamine, and/or glutathione) to epileptogenic injury which may lead to a better understanding of the in vivo target for therapeutic intervention and to establish potential biomarkers for predicting the development of epilepsy.

From a human imaging perspective, the use of the MRSI measures for purposes of seizure localization in surgical planning remains an important question with both technological and clinical epilepsy aspects. While measurements of GABA, glutamate, and myo-inositol remain of high clinical interest for their roles in epilepsy toward neurotransmission and astrocytic function, consistent measurements of these require substantially more care because of their j-modulating signal and spectral overlap. With its robustness of acquisition and sensitivity to injury, the utility of NAA/Cr remains strong for its use for seizure localization. Given the variable nature of epilepsy, it is clear that spectroscopic imaging, rather than single voxel spectroscopy is necessary; that high SNR needs to be maintained; and finally, given that seizures are not thought to arise from white matter and subcortical nuclei, studies need to achieve excellent coverage of the cortical ribbon. These basic requirements make consistent human studies challenging.

Technologically, there has been variability on how such studies are being and will be performed, specifically with regards to field strength, hardware, pulse sequences, and analysis methods. The SNR at 7 T is clearly at least linearly better than 3 T; however whether the increased SNR is requisite for the singlet NAA/Cr measurements in comparison to the much more commonly available 3 T platform will depend on the severity and volume extent of metabolic dysfunction seen in epilepsy (e.g., it is possible that in some malformations, the volume extent of NAA/Cr decline may be very small). This needs to be balanced against the push to accelerate the acquisitions (e.g., <15 min) which if incurred at the sacrifice of SNR will make detection difficult and may ultimately come back and make the case for needing greater field strength. As discussed earlier [17, 70] and well known by MRS practitioners, at 7 T and even at 3 T, a major challenge facing consistent epilepsy MRSI is field homogeneity, with problems in the temporal lobe, inferior frontal, and lower brain regions. It should be noted however that the high degree and order shim inserts that have been developed for 7 T field homogeneity have immediate impact for 3 T, and at the present writing, developments are under way to implement additional shim hardware for 3 T. Nonetheless whether for 3 or 7 T, the field distortions for the temporal region are substantially different from the frontal–parietal regions and thus ability to obtain true simultaneous and equivalent performance in temporal and frontal–parietal regions from whole brain acquisitions will be difficult. Finally, pulse sequence design and analysis approaches substantially depend on the above factors and local equipment, with many groups finding lipid suppression provided most robustly through inversion recovery and moderate echo times to reduce j-modulating coherences while maintaining sensitivity. Accelerated methods (e.g., EPSI, spiral, multiband SI) will depend on gradient, RF coil equipment, and experience; however, it is noted that acceleration performed at the cost of SNR is likely to reduce sensitivity of detection.

So the dilemma for epilepsy is therefore reasonably clear. As a disorder that can vary substantially between different patients, commonly requires clinical and imaging scrutiny over large portions of the brain, requires high performance imaging equipment and expertise for acquisition and analysis, MRSI for epilepsy is challenging. In comparison with other difficult neurological problems, many of these other conditions are more forgiving, e.g., the disorder is generally found in the same locus in all patients, does not require visualization in the cortical ribbon and/or caudal brain regions, and is not surgically quickly verifiable. Thus, at this writing MRSI is performed for epilepsy localization only in a handful of interested academic imaging centers. For the difficult problem of neocortical epilepsy, we have taken a positive position, that MRSI can be highly informative. However, we also recognize that typical for any complex multicomponent undertaking, MRSI is demanding in that each component needs to have a high rate of success. In all reality, whether or not MRSI can be implemented for broader use in epilepsy will depend on not just the above clinical and imaging requirements, but also multiple economic factors from within neurology, neurosurgery, and radiology.

References

1. Wiebe S, Camfield P, Jetté N, Burneo JG (2009) Epidemiology of epilepsy: prevalence, impact, comorbidity and disparities. Can J Neurol Sci 36(Suppl 2):S7–S16
2. Jallon P (1997) Epilepsy in developing countries. Epilepsia 38:1143–1151
3. Jallon P, Latour P (2005) Epidemiology of idiopathic generalized epilepsies. Epilepsia 46 (Suppl 9):10–4
4. Cockerell OC, Johnson AL, Sander JW, Shorvon SD (1997) Prognosis of epilepsy: a review and further analysis of the first nine years of the British National General Practice Study of Epilepsy, a prospective population-based study. Epilepsia 38(1):31–46
5. Schuele SU, Lüders HO (2008) Intractable epilepsy: management and therapeutic alternatives. Lancet Neurol 7(6):514–524
6. Téllez-Zenteno JF, Dhar R, Wiebe S (2005) Long-term seizure outcomes following epilepsy surgery: a systematic review and meta-analysis. Brain 128(Pt 5):1188–1198
7. Spencer SS, Berg AT, Vickrey BG, Sperling MR, Bazil CW, Shinnar S, Langfitt JT, Walczak TS, Pacia SV, Multicenter Study of Epilepsy Surgery (2005) Predicting long-term seizure outcome after resective epilepsy surgery: the multicenter study. Neurology 65:912–8
8. Hauptman JS, Mathern GW (2012) Surgical treatment of epilepsy associated with cortical dysplasia: 2012 update. Epilepsia 53(Suppl 4):98–104

9. Kato M, Malamut BL, Hosokawa S, O'Neill RR, Wakisaka S, Caveness WF (1978) Local glucose utilization in cortical and subcortical structures during focal motor seizures. Trans Am Neurol Assoc 103:39–42
10. Behar KL, Fitzpatrick SM, Hetherington HP, Shulman RG (1993) Cerebral metabolic studies in vivo by combined 1H/31P and 1H/13C NMR spectroscopic methods. Acta Neurochir Suppl (Wien) 57:9–20
11. Najm IM, Wang Y, Hong SC, Lüders HO, Ng TC, Comair YG (1997) Temporal changes in proton MRS metabolites after kainic acid-induced seizures in rat brain. Epilepsia 38(1):87–94
12. van Eijsden P, Notenboom RG, Wu O, de Graan PN, van Nieuwenhuizen O, Nicolay K, Braun KP (2004) In vivo 1H magnetic resonance spectroscopy, T2-weighted and diffusion-weighted MRI during lithium-pilocarpine-induced status epilepticus in the rat. Brain Res 1030(1):11–18
13. Temkin NR (2001) Antiepileptogenesis and seizure prevention trials with antiepileptic drugs: meta-analysis of controlled trials. Epilepsia 42(4):515–524
14. Dudek FE, Sutula TP (2007) Epileptogenesis in the dentate gyrus: a critical perspective. Prog Brain Res 163:755–773
15. Pitkänen A, Lukasiuk K (2009) Molecular and cellular basis of epileptogenesis in symptomatic epilepsy. Epilepsy Behav 14(Suppl 1):16–25
16. Hammen T, Kuzniecky R (2012) Magnetic resonance spectroscopy in epilepsy. Handb Clin Neurol 107:399–408
17. Maudsley AA, Domenig C, Ramsay RE, Bowen BC (2010) Application of volumetric MR spectroscopic imaging for localization of neocortical epilepsy. Epilepsy Res 88(2–3):127–138
18. Pan JW, Duckrow RB, Gerrard J, Ong C, Hirsch LJ, Resor SR Jr, Zhang Y, Petroff O, Spencer S, Hetherington HP, Spencer D (2013) 7T spectroscopic imaging in surgically treated epilepsy. Epilepsia 54(9):1668–1678
19. Pan JW, Duckrow RB, Spencer D, Avdievich N, Hetherington HP (2013) Spectroscopic imaging of GABA in human brain at 7T. Magn Reson Med 69(2):310–316
20. Pittau F, Grouiller F, Spinelli L, Seeck M, Michel CM, Vulliemoz S (2014) The role of functional neuroimaging in pre-surgical epilepsy evaluation. Front Neurol 5:31
21. Filibian M, Frasca A, Maggioni D, Micotti E, Vezzani A, Ravizza T (2012) In vivo imaging of glia activation using 1H-magnetic resonance spectroscopy to detect putative biomarkers of tissue epileptogenicity. Epilepsia 53(11):1907–1916
22. Pearce PS, deLanerolle N, Rapuano A, Hitchens K, Pan JW (2014) Spectroscopic imaging in early post-status epilepticus in a rodent model. American Epilepsy Society Annual Meeting Seattle WA, 3.072, December 2014
23. Choy M, Dubé CM, Patterson K, Barnes SR, Maras P, Blood AB, Hasso AN, Obenaus A, Baram TZ (2014) A novel, noninvasive, predictive epilepsy biomarker with clinical potential. J Neurosci 34(26):8672–8684
24. Baron M, Kudin AP, Kunz WS (2007) Mitochondrial dysfunction in neurodegenerative disorders. Biochem Soc Trans 35(Pt 5):1228–1231
25. Dingledine R, Varvel NH, Dudek FE (2014) When and how do seizures kill neurons, and is cell death relevant to epileptogenesis? Adv Exp Med Biol 813:109–122
26. Hauser WA, Lee JR (2002) Do seizures beget seizures? Prog Brain Res 135:215–219
27. Black LC, Schefft BK, Howe SR, Szaflarski JP, Yeh HS, Privitera MD (2010) The effect of seizures on working memory and executive functioning performance. Epilepsy Behav 17 (3):412–419
28. Voltzenlogel V, Vignal JP, Hirsch E, Manning L (2014) The influence of seizure frequency on anterograde and remote memory in mesial temporal lobe epilepsy. Seizure 23(9):792–798
29. de Lanerolle NC, Lee TS, Spencer DD (2010) Astrocytes and epilepsy. Neurotherapeutics 7 (4):424–438
30. Folbergrová J, Kunz WS (2012) Mitochondrial dysfunction in epilepsy. Mitochondrion 12 (1):35–40
31. Patel TB, Clark JB (1979) Synthesis of N-acetyl-L-aspartate by rat brain mitochondria and its involvement in mitochondrial/cytosolic carbon transport. Biochem J 184:539–546

32. Urenjack J, Williams SR, Gadian DG, Noble M (1992) Specific expression of N-acetyl aspartate in neurons, oligodendrocyte type 2 astroyte progenitors and immature oligodendrocytes in vitro. J Neurochem 59:55–61
33. Bates TE, Strangward M, Keelan J, Davey GP, Munro PM, Clark JB (1996) Inhibition of N-acetylaspartate production: implications for ^{1}H MRS studies in vivo. Neuroreport 7:1397–1400
34. Goldstein FB (1969) The enzymatic synthesis of N-acetyl-aspartatic acid by sub-cellular preparation of rat brain. J Biol Chem 244:4257–4260
35. Heales SJR, Davies SEC, Bates TE, Clark JB (1995) Depletion of brain glutathione is accompanied by impaired mitochondrial function and decreased N-acetylaspartate concentration. Neurochem Res 20:31–38
36. Connett RJ (1988) Analysis of metabolic control: new insights using scaled creatine kinase model. Am J Physiol 254(6 Pt 2):R949–R959
37. Guevara CA, Blain CR, Stahl D, Lythgoe DJ, Leigh PN, Barker GJ (2010) Quantitative magnetic resonance spectroscopic imaging in Parkinson's disease, progressive supranuclear palsy and multiple system atrophy. Eur J Neurol 17(9):1193–1202
38. Muñoz Maniega S, Cvoro V, Chappell FM, Armitage PA, Marshall I, Bastin ME, Wardlaw JM (2008) Changes in NAA and lactate following ischemic stroke: a serial MR spectroscopic imaging study. Neurology 71(24):1993–1999
39. Suhy J, Rooney WD, Goodkin DE, Capizzano AA, Soher BJ, Maudsley AA, Waubant E, Andersson PB, Weiner MW (2000) ^{1}H MRSI comparison of white matter and lesions in primary progressive and relapsing-remitting MS. Mult Scler 6(3):148–155
40. Huisman MC, van Golen LW, Hoetjes NJ, Greuter HN, Schober P, Ijzerman RG, Diamant M, Lammertsma AA (2012) Cerebral blood flow and glucose metabolism in healthy volunteers measured using a high-resolution PET scanner. EJNMMI Res 2(1):63
41. Eggers C, Hilker R, Burghaus L, Schumacher B, Heiss WD (2009) High resolution positron emission tomography demonstrates basal ganglia dysfunction in early Parkinson's disease. J Neurol Sci 276(1–2):27–30. doi:10.1016/j.jns.2008.08.029
42. Spencer S, Huh L (2008) Outcomes of epilepsy surgery in adults and children. Lancet Neurol 7 (6):525–537
43. Spencer SS (1996) Long-term outcome after epilepsy surgery. Epilepsia 37(9):807–813
44. Hetherington HP, Pan JW, Mason GF, Adams D, Vaughn MJ, Twieg DB, Pohost GM (1996) Quantitative ^{1}H spectroscopic imaging of human brain at 4.1 T using image segmentation. Magn Reson Med 36(1):21–29
45. Pouwels PJ, Frahm J (1998) Regional metabolite concentrations in human brain as determined by quantitative localized proton MRS. Magn Reson Med 39(1):53–60
46. Schuff N, Ezekiel F, Gamst AC, Amend DL, Capizzano AA, Maudsley AA, Weiner MW (2001) Region and tissue differences of metabolites in normally aged brain using multislice ^{1}H magnetic resonance spectroscopic imaging. Magn Reson Med 45(5):899–907
47. Hugg JW, Laxer KD, Matson GB, Maudsley AA, Weiner MW (1993) Neuron loss localizes human temporal lobe epilepsy by in vivo proton magnetic resonance spectroscopic imaging. Ann Neurol 34(6):788–794
48. Cendes F, Caramanos Z, Andermann F, Dubeau F, Arnold DL (1997) Proton magnetic resonance spectroscopic imaging and magnetic resonance imaging volumetry in the lateralization of temporal lobe epilepsy: a series of 100 patients. Ann Neurol 42(5):737–746
49. Kuzniecky R, Hugg JW, Hetherington H, Butterworth E, Bilir E, Faught E, Gilliam F (1998) Relative utility of ^{1}H spectroscopic imaging and hippocampal volumetry in the lateralization of mesial temporal lobe epilepsy. Neurology 51(1):66–71
50. Hetherington HP, Kuzniecky RI, Vives K, Devinsky O, Pacia S, Luciano D, Vasquez B, Haut S, Spencer DD, Pan JW (2007) A subcortical network of dysfunction in TLE measured by MR spectroscopy. Neurology 69:2256–2265
51. Pan JW, Lo KM, Hetherington HP (2012) Role of high degree and order B_0 shimming for spectroscopic imaging at 7T. Magn Reson Med 68(4):1007–1017

52. Pan JW, Spencer DD, Kuzniecky R, Duckrow RB, Hetherington H, Spencer SS (2012) Metabolic networks in epilepsy by MR spectroscopic imaging. Acta Neurol Scand 126 (6):411–420
53. Saransaari P, Oja SS (1997) Enhanced GABA release in cell-damaging conditions in the adult and developing mouse hippocampus. Int J Dev Neurosci 15(2):163–174
54. Atwell D, Laughlin S (2001) An energy budget for signaling in the gray matter of the brain. J Cereb Blood Flow Metab 21(10):1133–1145
55. Palma E, Amici M, Sobrero F, Spinelli G, Di Angelantonio S, Ragozzino D, Mascia A, Scoppetta C, Esposito V, Miledi R, Eusebi F (2006) Anomalous levels of Cl- transporters in the hippocampal subiculum from temporal lobe epilepsy patients make GABA excitatory. Proc Natl Acad Sci U S A 103(22):8465–8468
56. Woo N, Lu J, England R, McClellan R, Dufour S, Mount D, Deutch A, Lovinger D, Delpire E (2002) Hyperexcitability and epilepsy associated with disruption of the mouse neuronal specific K-Cl cotransporter gene. Hippocampus 12(2):258–268
57. Cavus I, Kasoff WS, Cassaday MP, Jacob R, Gueorguieva R, Sherwin RS, Krystal JH, Spencer DD, Abi-Saab WM (2005) Extracellular metabolites in the cortex and hippocampus of epileptic patients. Ann Neurol 57(2):226–235
58. Petroff OA, Hyder F, Rothman DL, Mattson RH (2001) Homocarnosine and seizure control in juvenile myoclonic epilepsy and complex partial seizures. Neurology 56(6):709–715
59. Jayakar P, Dunoyer C, Dean P, Ragheb J, Resnick T, Morrison G, Bhatia S, Duchowny M (2008) Epilepsy surgery in patients with normal or nonfocal MRI scans: integrative strategies offer long-term seizure relief. Epilepsia 49(5):758–764
60. Siegel AM, Jobst BC, Thadani VM, Rhodes CH, Lewis PJ, Roberts DW, Williamson PD (2001) Medically intractable, localization-related epilepsy with normal MRI: presurgical evaluation and surgical outcome in 43 patients. Epilepsia 42(7):883–888
61. Mueller SG, Laxer KD, Suhy J, Lopez RC, Flenniken DL, Weiner MW (2003) Spectroscopic metabolic abnormalities in mTLE with and without MRI evidence for mesial temporal sclerosis using hippocampal short-TE MRSI. Epilepsia 44:977–980
62. Mueller SG, Ebel A, Barakos J, Scanlon C, Cheong I, Finlay D, Garcia P, Weiner MW, Laxer KD (2011) Widespread extrahippocampal NAA/(Cr + Cho) abnormalities in TLE with and without mesial temporal sclerosis. J Neurol 258(4):603–612
63. Avdievich NI, Pan JW, Baehring JM, Spencer DD, Hetherington HP (2009) Short echo spectroscopic imaging of the human brain at 7T using transceiver arrays. Magn Reson Med 62:17–25
64. Hetherington HP, Avdievich NI, Kuznetsov AM, Pan JW (2010) RF shimming for spectroscopic localization in the human brain at 7T. Magn Reson Med 63(1):9–19
65. Mueller SG, Laxer K, Barakos J, Cashdollar N, Flenniken D, Vermathen P, Matson G, Weiner M (2005) Metabolic characteristics of cortical malformations causing epilepsy. J Neurol 252 (9):1082–1092
66. Dansereau CL, Bellec P, Lee K, Pittau F, Gotman J, Grova C (2014) Detection of abnormal resting-state networks in individual patients suffering from focal epilepsy: an initial step toward individual connectivity assessment. Front Neurosci 8:419. doi:10.3389/fnins.2014.00419
67. Hofmann L, Slotboom J, Boesch C, Kreis R (2001) Characterization of the macromolecule baseline in localized ^{1}H-MR spectra of human brain. Magn Reson Med 46(5):855–863
68. Hwang JH, Graham GD, Behar KL, Alger JR, Prichard JW, Rothman DL (1996) Short echo time proton magnetic resonance spectroscopic imaging of macromolecule and metabolite signal intensities in the human brain. Magn Reson Med 35(5):633–639
69. Gomes WA, Lado FA, de Lanerolle NC, Takahashi K, Pan C, Hetherington HP (2007) Spectroscopic imaging of the pilocarpine model of human epilepsy suggests that early NAA reduction predicts epilepsy. Magn Reson Med 58(2):230–235
70. Maudsley AA, Domenig C, Sheriff S (2010) Reproducibility of serial whole-brain MR spectroscopic imaging. NMR Biomed 23:251–256

Index

G. Öz (ed.), *Magnetic Resonance Spectroscopy of Degenerative Brain Diseases*, Contemporary Clinical Neuroscience, DOI 10.1007/978-3-319-33555-1

I

K

L

M

R

S

Printed by Printforce, the Netherlands